T5-BAM-463

Advances in Physical Geochemistry

Volume 3

Kinetics and Equilibrium in Mineral Reactions

Edited by
Surendra K. Saxena

With Contributions by
L. Ya. Aranovich A. L. Boettcher S. R. Bohlen
H. P. Eugster E. S. Ilton A. C. Lasaga
I. V. Lavrent'eva T. P. Loomis
L. L. Perchuk K. K. Podlesskii
R. Powell S. K. Saxena V. J. Wall

With 99 Illustrations

Springer-Verlag
New York Berlin Heidelberg Tokyo

Series Editor
Surendra K. Saxena
Department of Geology
Brooklyn College
City University of New York
Brooklyn, NY 11210
U.S.A.

Production: Richard Ruzycka

Library of Congress Cataloging in Publication Data
Main entry under title:
Kinetics and equilibrium in mineral reactions.
(Advances in physical geochemistry ; v. 3)
Bibliography: p.
Includes index.
1. Mineralogical chemistry. 2. Chemical reaction, Rate of. 3. Chemical equilibrium. I. Saxena, Surendra Kumar, 1936– . II. Aranovich, L. Ya. III. Series.
QE371.K56 1983 551.9 83-14612

Typeset by University Graphics, Inc., Atlantic Highlands, New Jersey.
Printed and bound by R. R. Donnelley & Sons, Harrisonburg, Virginia.
Printed in the United States of America.

9 8 7 6 5 4 3 2 1

ISBN 0-387-90865-X Springer-Verlag New York Berlin Heidelberg Tokyo
ISBN 3-540-90865-X Springer-Verlag Berlin Heidelberg New York Tokyo

Preface

The third volume in this series consists of eight chapters. The first three deal with kinetic aspects of compositional variations both within individual phases and across crystal boundaries. Basically, the authors use the kinetic theory and the sparsely available rate data to explain the formation of various types of zoning and the exsolution processes in silicates. Loomis rightly argues that "the kinetic inhibitions to reequilibration that preserve primary igneous crystals and high-grade metamorphic assemblages also affect the crystallization and prograde metamorphism of these rocks." These "kinetic inhibitions" appear in the form of zoned crystals, reaction rims and disequilibrium assemblages. Their proper recognition and quantitative characterization leads to an understanding of the physico-chemical history of the rock.

On a similar theme, I examine possible relationships between the exsolution processes in Ca–Fe–Mg pyroxenes and the cation order–disorder on nonequivalent crystallographic sites. A multi-technique study of exsolutions in crystals employing electron microscopy and X-ray structural refinements should contribute greatly in understanding the thermal history of the rock.

Many geothermometric studies result in discordant temperatures when the estimates are done using serveral coexisting pairs of minerals in a single specimen. Lasaga uses the kinetic rates of diffusion of various chemical species and explains the discordance through his "geospeedometric" approach.

Before the geospeedometer can be applied quantitatively, it is important that the criteria of chemical equilibria in a mineral assemblage are definitely established. In the next four articles the authors accomplish this by presenting their latest experimental results on compositions of coexisting phases. Eugster and Ilton examine the Fe–Mg distribution between a fluid phase and a coexisting solid solution and clearly document the need for much more additional data on metamorphic fluids. In the fifth article, Bohlen, Wall and Boettcher review their recent experimental data and present the synthesis of their results as geobarometers applicable to granulites. The solution model for garnet is still being actively researched and this chapter provides useful reference material and data to aid in further development of the garnet model and the garnet geobarometer. Besides the author's own valuable experimental data on the olivine–orthopyroxene–quartz, garnet–fayalite–anorthite and garnet–rutile–ilmenite–sillimanite–quartz systems, this chapter contains many applications to natural assemblages.

The next two chapters contain experimental data on the cordierite–garnet–sillimanite–quartz and the cordierite–garnet–biotite systems. Although the style of presentation is more appropriate to speciality journals, the chapters have been included here because the reviewers considered that the authors have contributed importantly to the solution of a long-term experimental problem. While the experimental details should be of great interest to experimenters, other geochemists will find the results useful in geothermometry and geobarometry.

In the final chapter, Powell advocates the development of the quasichemical model for expressing activity–composition relationships in silicates. For solutions showing substantial non-ideality with strong short-range ordering the regular/subregular models are inadequate. Powell suggests that the use of the quasichemical model may pave the way toward a close approximation to physical reality.

I am confident that the geochemists will find this volume highly engrossing and informative. Thanks are due to Ralph Kretz, R. C. Newton, J. Ganguly and A. B. Thompson for their reviews of several chapters and to the staff of Springer-Verlag New York for their cooperation in the production of this volume.

S. K. SAXENA

Contents

Contributors

Aranovich, L. Ya.	Institute of Experimental Mineralogy, USSR Academy of Sciences, 142432 Chernogolovka, U.S.S.R.
Boettcher, A. L.	Institute of Geophysics and Planetary Physics and Department of Earth and Space Sciences, University of California, Los Angeles, California 90024 U.S.A.
Bohlen, S. R.	Department of Earth and Space Sciences, State University of New York at Stony Brook, Stony Brook, New York 11794 U.S.A.
Eugster, H. P.	Department of Earth and Planetary Sciences, The Johns Hopkins University, Baltimore, Maryland 21218 U.S.A.
Ilton, E. S.	Department of Earth and Planetary Sciences, The Johns Hopkins University, Baltimore, Maryland 21218 U.S.A.
Lasaga, A. C.	Department of Geosciences, College of Earth and Mineral Sciences, The Pennsylvania State University, University Park, Pennsylvania 16802 U.S.A.
Lavrent'eva, I. V.	Institute of Experimental Mineralogy, USSR Academy of Sciences, 142432 Chernogolovka, U.S.S.R.

Loomis, T. P.	Department of Geosciences, University of Arizona, Tucson, Arizona 85721 U.S.A.
Perchuk, L. L.	Institute of Experimental Mineralogy, USSR Academy of Sciences, 142432 Chernogolovka, U.S.S.R.
Podlesskii, K. K.	Institute of Experimental Mineralogy, USSR Academy of Sciences, 142432 Chernogolovka, U.S.S.R.
Powell, R.	Department of Earth Sciences, University of Leeds, Leeds LS2 9JT England
Saxena, S. K.	Department of Geology, Brooklyn College, Brooklyn, New York 11210 U.S.A.
Wall, V. J.	Department of Earth Sciences, Monash University Clayton, Victoria 3168 Australia

Chapter 1
Compositional Zoning of Crystals: A Record of Growth and Reaction History

Timothy P. Loomis

Introduction

Under some geological conditions of formation, most common minerals capable of solid solution have produced compositionally zoned crystals. In fact, the formation of zoned crystals during growth and dissolution may be much more common than generally believed because growth zoning can be readily purged by diffusion in plutonic and metamorphic rocks during their long residence time at high temperature. Thus, many of the reaction and growth kinetic processes studied using only a few minerals may be of more general significance to the crystallization of rocks than is obvious from the annealed end-products that we usually examine.

The emphasis in this work is on compositional zoning of garnet in metamorphic rocks and on plagioclase zoning in igneous ones for several reasons. These two minerals form the most commonly zoned crystals in nature and have been studied in greatest detail by other workers and by myself. Both minerals are found in a wide range of geological bulk compositions and environments, making them useful tracers of geological history rather than curiosities. Finally, we are developing enough understanding of the kinetics of growth and reaction to interpret compositional zoning of these minerals in terms of important geological variables such as cooling and heating rate, pressure variations, changes of fluid content, and convection in magnas.

My objectives in this chapter are to (1) present an overview of the types of geologically useful information that are available from zoning, (2) illustrate the fundamental physicochemical processes that must be understood to interpret zoning, (3) present my views of how zoning in plagioclase and garnet forms, and (4) suggest directions of future research. My intent here is to present these ideas in a readable manner and to leave the tedium of mathematical and numerical analysis necessary to calculate simulations to the referenced works and to future publications.

Requirements for Compositional Zoning

Partitioning

There are two basic requirements for the formation and preservation of compositional zoning. The first requirement is that the "effective partitioning" of an element between the zoned phase and the rest of the assemblage must be unequal if zoning is produced during growth or consumption, or that "effective partitioning" must change with time if zoning is produced by simple exchange of elements between phases. Here, we are concerned with elements that form solid solutions in the zoned phase, such as Mn in garnet or Ca in plagioclase. "Effective partitioning" refers to the actual distribution of a component, which may be governed by disequilibrium processes as well as equilibrium ones; effective partitioning does not necessarily correspond to "equilibrium partitioning." Also I will often use the term "reaction" to include growth of the zoned mineral as well as consumption by a reaction that produces other products.

Since the experimental work of Bowen (1913), fractionation paths that can produce zoned plagioclase crystals have been traced on equilibrium phase diagrams. The marked, unequal distribution of $NaAlSi_3O_8$ and $CaAl_2Si_2O_8$ (henceforth Ab and An components) between the crystal and the melt is obvious in Fig. 1. It causes the melt to develop a higher Ab/An ratio as crystallization progresses if previously formed crystal does not react with the melt. The changing melt composition causes more Ab-rich layers of plagioclase to be deposited on plagioclase crystals, generating normal zoning. Despite the experimental basis of this explanation for zoning and its popularity with teachers, several basic problems with this model appear if it be applied quantitatively. The reverse and oscillatory zoning often found in natural crystals, as shown in Fig. 2, cannot be explained by the model. Moreover, fractionation of An and Ab between the crystal and the melt found experimentally in the binary system is much greater than observed in natural melts, as illustrated in Fig. 1. Evidently, natural zoning can tell us a lot more about partitioning during natural crystallization of plagioclase than we can predict from equilibrium experiments in simple systems. However, the basic idea that plagioclase crystals are often zoned because Ab and An components are unequally partitioned between melt and crystal is fundamentally correct.

A second example of growth zoning caused by unequal partitioning is the distribution of Fe and Mg between garnet and the lower-grade minerals chlorite and biotite. It is usually observed that garnet has a higher Fe/Mg ratio than coexisting chlorite or biotite. Consequently, a prograde reaction, such as described in the Fe–Mg system by

$$\text{biotite} + \text{chlorite} + \text{quartz} \rightarrow \text{garnet} + \text{muscovite} + \text{water}$$

must produce zones in garnet with successively higher Mg/Fe ratios as the reaction proceeds. This divariant reaction may take place close to equilibrium at each step ("divariant shift"), or it may occur under significant disequilibrium conditions if T, P, or P_{H_2O} change faster than equilibrium can be maintained in the

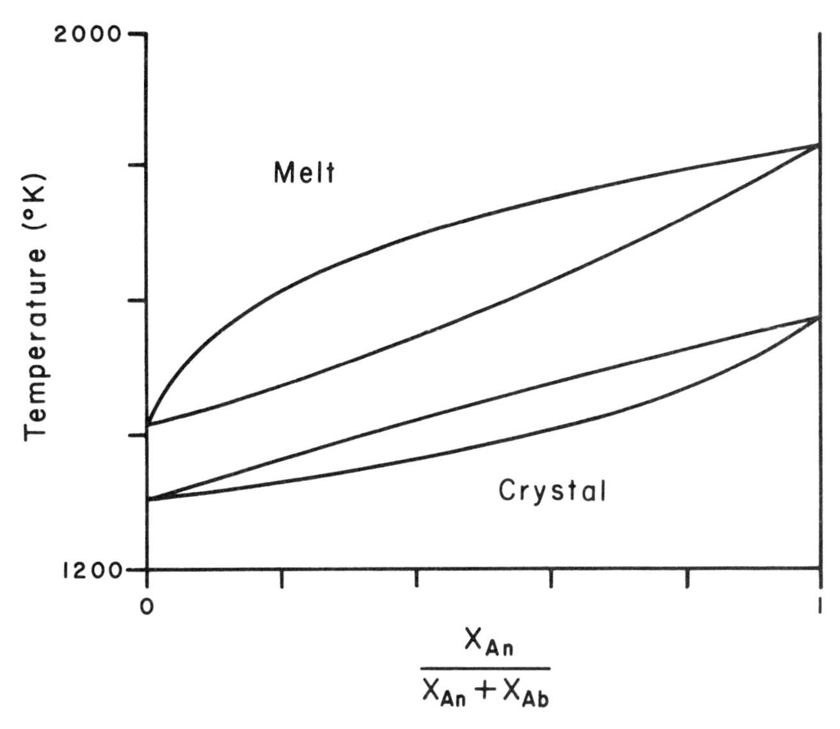

Fig. 1. Plagioclase phase diagrams (mole fraction) calculated by the methods described by Loomis (1979b). *Top:* Loop consistent with Bowen's data for a plagioclase melt. *Bottom:* Loop for system containing a residual component and water equivalent to a granodiorite with 2 wt % water.

matrix of the rock. Probably a more important element to consider for zoning of garnet at low grade is Mn. The common bell-shaped profile of spessartine in garnet (Harte and Henley, 1966; Hollister, 1966), as shown in Fig. 3, is pronounced mainly because Mn is concentrated in garnet relative to chlorite or biotite by a factor of around 40 (Hollister, 1969; Loomis, 1982).

As in the case of plagioclase, there are a number of characteristics of natural growth zoning in garnet that are difficult to explain quantitatively by using a simple equilibrium partitioning model. For example, it is observed that the core compositions of garnets in different-grade rocks of similar bulk composition are significantly different, even though the cores should have all formed under approximately the same conditions during heating. Other difficulties include representing end-member components in minerals such as chlorite and biotite, and calculating the equilibrium (and disequilibrium) partitioning of several components among phases simultaneously. Thus, the simulation of garnet zoning under natural conditions severely tests our knowledge of equilibrium and disequilibrium partitioning in metamorphic rocks.

Diffusion zoning is also caused by unequal partitioning or changes of partitioning with time. A homogeneous crystal can become zoned by diffusion only if the

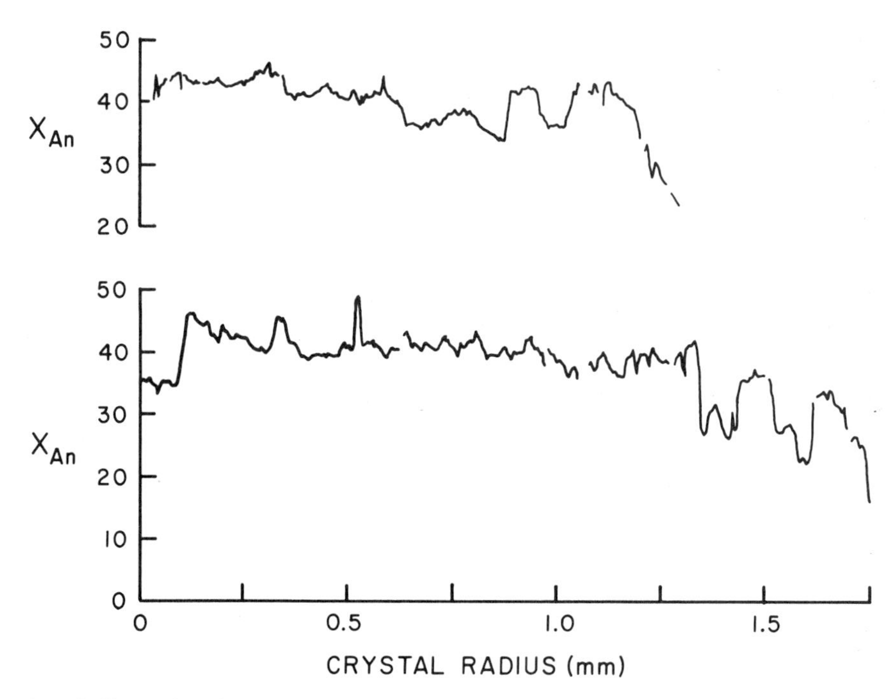

Fig. 2. Examples of zoning profiles in two plagioclase phenocrysts in the same sample from a small trondhjemite pluton (Caribou Mountain, California). X_{An} is mol % An in plagioclase.

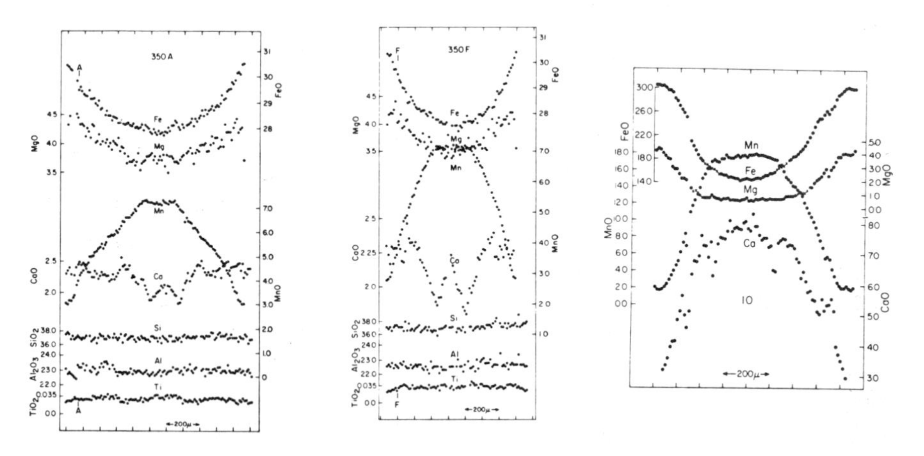

Fig. 3. Examples of growth zoning in garnet (reproduced with permission of The Geological Society of America from Hollister, 1969, Figs. 7 and 8). The vertical scales are wt % for the indicated oxides.

composition imposed on the edge of the crystal by equilibration with other phases (including a fluid) differs from the existing composition within the crystal. This change of composition can occur during reaction if partitioning is unequal, or during exchange of elements if partitioning changes with time.

Many other examples of the influence of partitioning on zoning could be cited. The general conclusion is that the compositional zoning of crystals cannot be understood without a proper understanding of the equilibrium and disequilibrium partitioning of components among phases throughout the progress of reaction. Consequently, substantial space is devoted to models of equilibrium and disequilibrium partitioning below.

Diffusion

Diffusion controls compositional zoning in two ways. First, it tends to eliminate growth zoning that occurred by natural fractionation processes. As noted in the introduction, many common minerals, such as micas, chlorite, amphiboles, and alkali feldspars, are usually unzoned because diffusion homogenized the crystals during growth or afterward, rather than because fractionation could not occur. The second control of diffusion is in the creation of zoning in a homogeneous crystal during consumption by reaction. This type of "diffusion zoning" relies for its existence on enough diffusion to propagate inward compositional changes produced on the edge of a crystal, but not enough to homogenize the crystal. Diffusion zoning has been studied most extensively in garnet because the rate of diffusion is appropriate for the rate of change of geological conditions in some high-grade metamorphic rocks. Moreover, garnet has the properties of diffusion isotropy, physical strength, abundance, and a composition accurately measurable with the microprobe that make it ideal for the study of diffusion. Dissolution of an old crystal and precipitation of a crystal with a new composition seems to be the alternative mechanism of reaction preferred by plagioclase (Maaloe, 1976; Loomis, 1977).

The amount of diffusive transfer of material within a crystal, measured by the flux J, depends on both the diffusion rate, measured by D, and the compositional gradient ∇C, induced by the change of partitioning with time:

$$J = -D\nabla C. \qquad (1)$$

The distance inside a crystal that the compositional disturbance on the edge propagates can be solved for by simple or complicated mathematical means, depending on the details of the problem. However, a common method used for comparison purposes to estimate this distance is to assume the following: the crystal surface is an infinite plane, the crystal is initially homogeneous and has infinite depth, and a new composition is instantaneously imposed at the surface and held constant over time. The "penetration depth" is the distance from the surface at which the composition intermediate between the initial and new surface compositions is found. It is calculated from the equation

$$d = (Dt)^{1/2}, \tag{2}$$

where D is the diffusivity and t is time. A penetration distance of about 1 μm (log d = 0) can be considered to be the threshold value at which induced diffusion profiles can be measured with the microprobe and at which diffusion starts to affect growth zoning. In Fig. 4, I assume that the diffusion rate for exchange is given by the standard kinetic formula, after Arrhenius, for a thermally activated process:

$$D = Do\ \exp(-A/RT), \tag{3}$$

where Do is a constant, R is the gas constant, T is absolute temperature, and A is the activation enthalpy. The values of Do and A in Fig. 4 for Mg–Fe and Mn–Fe exchange in common aluminous garnets are preliminary estimates for crustal pressures from the experimental data of Elphick *et al.* (1981, 1982), and probably slightly overestimate the effectiveness of diffusion due to the error of measuring diffusion profiles generated in these experiments. Figure 4 shows that the exchange diffusion of Mg and Fe should become effective at granulite temperatures if these temperatures are encountered for millions of years. Contact metamorphic rocks should also show diffusion zoning of garnet if they reached very high temperatures. Mn zoning should be significantly destroyed by diffusion at

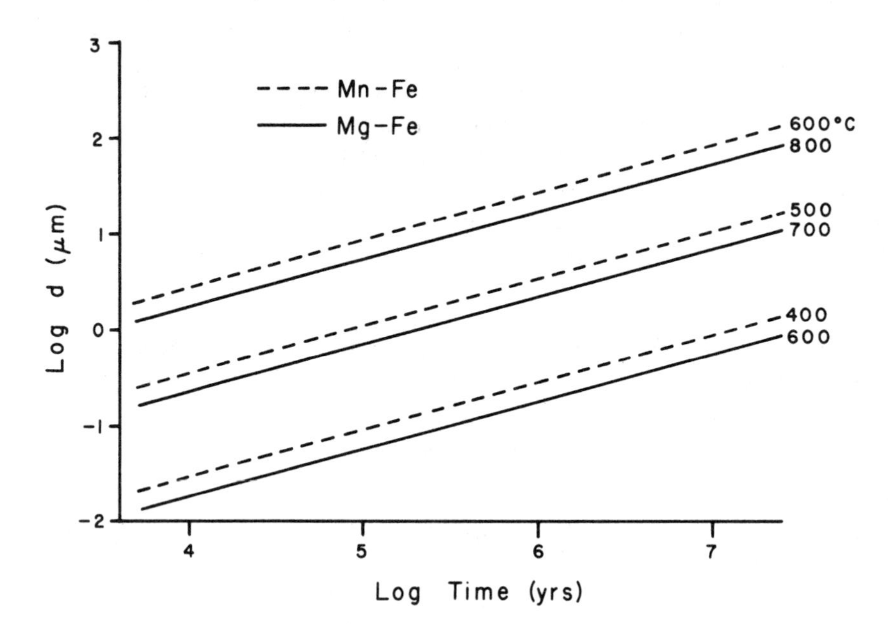

Fig. 4. Diffusion penetration depth (d) as a function of time and temperature for Mn–Fe and Mg–Fe exchange in almandine-rich compositions. The activation enthalpies and Do's are based on *preliminary* estimates from the data of Elphick *et al.* (1981, 1982), and are 85 Kcal, 0.0227 cm^2/sec for Mg–Fe exchange and 55 Kcal, 0.000015 cm^2/sec for Mn–Fe exchange.

these conditions. Diffusion zoning of Mg and Fe has been studied in these regional and very high grade contact rocks. Figure 4 also suggests that rocks of staurolite grade (550 °C) or lower that remained at that temperature a short time (as in contact metamorphism) would suffer little diffusion alteration of any growth zoning profiles; these are the rocks used to study growth zoning.

Information from Zoning

Equilibrium

In an apparent contradiction, zoned crystals both preserve past equilibration history and simultaneously are one of the best indicators of the recent state of equilibrium. The reason for this dual role is that limited diffusivity allows a single zoned crystal to behave like a series of tiny homogeneous crystals appearing through time. At a given time, only the composition of the edge of the crystal need be changed to reach equilibrium.

The transfer of material that is necessary to maintain the edge of the garnet at the equilibrium composition is expressed mathematically as follows:

$$Cr\,v = Ce\,v + D\nabla C. \tag{4}$$

Ce is the composition of the edge of the crystal (usually equilibrium), *Cr* is the composition of the crystal that the matrix or magma "sees" as it reacts with the crystal, and v is the velocity with which the crystal edge is growing or being dissolved. The product of the diffusion rate, *D*, times the steepness of zoning, ∇C, is the flux of material flowing to the edge of the crystal from the interior due to diffusion. Equation (4) shows that if the diffusion flux is small, the crystal appears to the rest of the rock to have a composition similar to the equilibrium edge composition. The slower diffusion is, the less reaction of composition *Cr* is required to maintain the edge composition at *Ce*. At slow diffusion rates, only a small amount of intergranular exchange is sufficient to maintain the edge composition of a zoned mineral at equilibrium. At the same time, information on earlier stages of equilibration is preserved in the zoning profile in the interior of the crystal.

An example of the type of equilibration information on the complete history of a sample that can be preserved in diffusion zoning of garnet is discussed in detail in Loomis (1976). Figure 5 shows the chemographic relations of a rock in which the assemblage garnet–biotite–kyanite (and sillimanite)–K-feldspar–quartz was preserved in the process of reacting to produce cordierite and hercynite. Both garnet and biotite changed compositions as they reacted by disequilibrium processes with kyanite and sillimanite to form cordierite and hercynite. Moreover, the extent of reaction varied from place to place (called reaction domains) in a single thin section, and it is possible to measure how the compositions of phases changed with the extent of reaction.

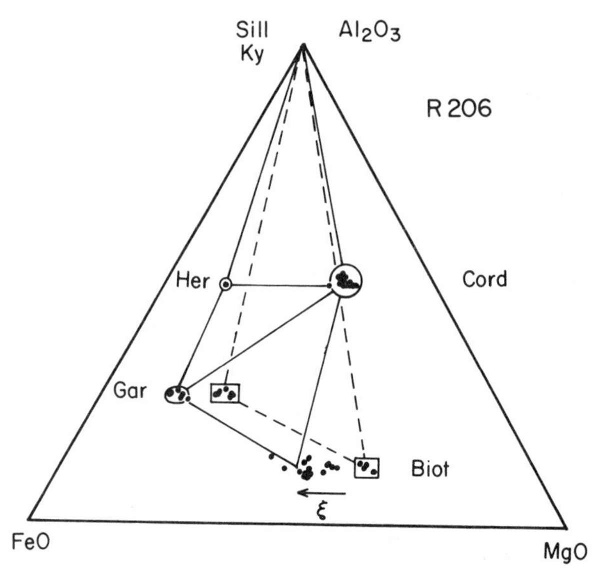

Fig. 5. AFM projection from quartz and K–feldspar of phase composition data from a granulite sample preserved in the process of reaction. The initial composition of garnet, preserved in crystal cores and unreacted grains, and of biotite, preserved as inclusions within garnet, are enclosed by boxes. Other data are from matrix phases and garnet rims. The dashed lines indicate the initial assemblage, and the solid lines potential resulting assemblages, depending on the bulk composition. The migration of the matrix biotite composition with extent of reaction (χ) is shown. The figure is modified after Loomis (1976, Fig. 10).

Garnet and biotite responded to reaction in different ways. Rapid intracrystalline diffusion caused biotite crystals to remain homogeneous. Consequently, the composition of the biotite in a domain could change only slowly as it reacted incongruently because a great deal of mass transfer was required to change the composition of all the biotite. The composition of biotite was observed to shift slowly to more Fe-rich compositions as the extent of reaction increased. In contrast, slow diffusion within garnet crystals isolated the edge of the crystals from the interior and allowed the edge of crystals to equilibrate rapidly, even though the bulk of the crystal retained the prereaction composition. Figure 5 shows that the edges of all garnet crystals that have undergone any reaction have the same composition. Thus, the composition of the garnet edge is a better indicator of the state of equilibration in the intergranular medium than is biotite, even though biotite equilibrates easily. Furthermore, the composition of the cores of garnet crystals, together with biotite and other phases trapped inside, preserve the state of the system before reaction, whereas the biotite in the matrix of the rock does not preserve this information.

If we accept the premise that the edge of a zoned crystal equilibrates readily during reaction and then preserves this composition as the crystal grows, what information of value can be sought? I will mention a few examples in the following sections of the usefulness of zoned crystals, without intending to present a complete review of published work.

Assemblage and Volatile History

The most obvious use of zoning, together with inclusions of other phases, is to identify assemblage history. One example of the use of garnet zoning and inclusions to identify a previously stable metamorphic assemblage was just described. In the original work on these rocks (Loomis, 1972a), the preservation of inclusions of biotite and kyanite in the pyropic cores of garnets, while cordierite and hercynite were found only in the matrix, was cited as evidence of unloading during contact metamorphism. Combined with structural, gravity, and plate-tectonic analysis, these data were used to construct a model of mass flow and crustal evolution during the emplacement of the Ronda high-temperature peridotite (Loomis, 1972b; 1974). Basic rocks in the same metamorphic aureole contain garnet in granulite assemblages. Some evidence of the past amphibolite history of these samples is preserved as inclusions of sodic plagioclase and amphibole within the garnet, while calcic plagioclase and no amphibole are found in the matrix (Loomis, 1977).

Another example of the use of zoning and inclusions in garnet to trace assemblage history was described by Thompson *et al.* (1977). They found inclusions of chloritoid, margarite, rutile, magnetite, and ilmenite that are absent in the matrix of the rock and were able to deduce the sequence of prograde reactions that occurred as the garnet grew. Many authors have suggested correlations between zoning features or bands of inclusions in garnet and reactions that occurred while the garnet was growing.

Zoning and inclusions in igneous plagioclase have been used occasionally to document the appearance of phases during crystallization. Delaney *et al.* (1978) found that glass-vapor inclusions in plagioclase crystals in the glassy rims of pillow basalts had volatile-element compositions significantly different than measured in the glass outside the plagioclase. This was interpreted as evidence of the alteration of the exterior melt composition after the plagioclase grew. Anderson (1982, personal communication) has correlated vapor and glass inclusions in zoned plagioclase crystals with the occurrence of irregular growth zones and proposed that both were caused by the same physical disruption of the crystal. Vance (1962, 1965), Wiebe (1968), Pringle *et al.* (1974), Maaloe (1976), McDowell (1978), and Kuo and Kirkpatrick (1982) have explained common zoning patterns in plagioclase in terms of intrusion, magma mixing, and water saturation. There is an extensive literature on fluid inclusions in igneous, metamorphic, and hydrothermal rocks that I will not presume to review here.

There are a number of problems to be considered in the use of inclusions as indicators of previously stable assemblages. Among these are the propensity of growing crystals to include some minerals and not others, the possibility that trapped phases could react to produce other assemblages, and the uncertainty of knowing the shape of the crystal as it grew to be able to interpret the stratigraphy of inclusions.

Geothermometry

The compositions of zoned crystals and their inclusions can be used for geothermometry and geobarometry in at least two ways. First, the composition of a solid-solution phase in a divariant assemblage is directly determined by the value of the intensive variables controlling the system and is not controlled by the bulk composition of the system. For example, if we consider the assemblage garnet–biotite–sillimanite–quartz–K–feldspar–ilmenite–graphite (shown by the dashed lines in Fig. 5) to have been the divariant assemblage in which garnet grew, then the composition of the garnet core along with that of biotite places a constraint on the values of temperature, pressure, and P_{H_2O} imposed on the system. In this particular case, the Mg/Fe ratio in both garnet and biotite would be expected to become larger as the temperature increased.

The second approach to geothermometry is to apply geothermometers based on ion exchange equilibria to inclusions and the adjacent composition of zoned host crystal. Inclusions of biotite and other minerals in metamorphic garnets have been used for this purpose. For example, Tracy *et al.* (1976) and Ghent *et al.* (1979) compared calculated garnet–biotite equilibration temperatures for included biotite–interior garnet pairs and matrix biotite–garnet rim pairs. A variation of this approach is the use of pressure-induced strain halos around quartz in garnet by Rosenfeld (1969) as a geobaromenter. In igneous rocks, Watson (1976) and Delaney *et al.* (1978) are two studies in which the reaction of glass inclusions within phenocrysts after entrapment has been considered in an attempt to determine the composition of the original melt from which the crystal grew; the unreacted compositions of phenocrysts and the glass inclusions could then be used for olivine-melt and plagioclase-melt geothermometry. In another igneous example, pairs of mineral inclusions in diamonds in kimberlite pipes have been used by Gurney *et al.* (1979) and Harte *et al.* (1980) to constrain the temperature and pressure of crystallization of diamond.

There are numerous problems and uncertainties that complicate geothermometry based on the composition of crystals in divariant assemblages and on the partitioning of ions. A common problem is to demonstrate that assemblages are really divariant, given the complicating effects of O, S, Mn, Zn, and other minor elements. If the assemblage were not truly divariant, then variations of the equilibrating bulk composition would affect the composition of the mineral and would have to be taken into account in the interpretation of zoning. For all geothermometers it is difficult to be sure that an inclusion has not communicated with the matrix through a small crack that may have been subsequently closed or that is not visible in the thin section examined. An obvious example of this problem is visible in zoned plagioclase because An-rich regions of crystals commonly alter more readily than Ab-rich regions. Zoned phenocrysts of plagioclase in plutonic rocks often display selective alteration of their An-rich cores and even of thin growth layers surrounded by unaltered crystal even though there are no visible avenues for exchange with the rest of the rock.

A problem with using partitioning geothermometers on inclusions has been documented in one of the examples mentioned above, in which the distribution of Mg and Fe between garnet–biotite pairs found inside and outside of garnet crystals was used to calculate temperature. Tracy *et al.* (1976) found that interior pairs gave higher temperatures than matrix pairs, but Ghent (1975) and Ghent *et al.* (1979) found just the opposite. Presumably, interior pairs should record a lower temperature of equilibration if the garnet crystal grew near equilibrium during prograde metamorphism and if the equilibration composition during growth were preserved unaltered by the subsequent thermal history. Tracy *et al.* (1976) suggested that garnet rim–matrix biotite pairs suffered retrograde reequilibration. If the exchange of Mg^{2+} and Fe^{2+} between garnet and biotite is simply controlled by the rate of thermally activated diffusion within the crystals, then garnet–biotite pairs both inside and outside garnet should have reequilibrated to approximately the same temperature (there may be some pressure differences). For example, I found that the compositions of biotite inclusions located at different distances from the edge of a garnet that had developed an induced diffusion profile by reaction with the matrix varied sympathetically with the composition of the surrounding garnet. The diffusion process that had affected the garnet composition had also modified the included biotite compositions. The different calculated temperatures found for included and external mineral pairs may indicate that the rate of exchange of ions between mineral pairs is sensitive to conditions other than simply temperature and pressure. For example, Yund and Tullis (1980) found that trace amounts of water and strain affect the rate of order–disorder reactions in feldspar. Then the difference between the temperatures calculated from included and external mineral pairs may be a better measure of differences of P_{H_2O} history than the temperature difference during formation of the crystals. Another possibility is that inclusions have reequilibrated with the surrounding crystal but that the zoning in the host crystal around the inclusion can not be detected. In that case, the calculated temperature is meaningless. As mentioned above, Watson (1976) has shown how the original compositions of glass inclusions in host phenocrysts can be determined after subsolidus reaction has significantly modified the inclusions. Studies of inclusions provide one method of investigating the cation exchange process under natural conditions to determine what is really being measured by the calculated geothermometer temperature.

Other possible problems with the use of the compositions of inclusions in zoned crystals for quantitative calculations are related to disequilibrium growth processes. If a crystal grows under disequilibrium conditions, as seems likely for most igneous crystals and probably some metamorphic ones (as discussed below), then the composition of matrix minerals or melt samples that become included may have been neither initially in equilibrium with the host crystal nor representative of the composition of the bulk matrix or melt. The effect of disequilibrium partitioning during growth on the calculated temperatures of the garnet–biotite geothermometer is considered below.

Growth Mechanisms

Crystal zoning is one of the primary sources of information on disequilibrium growth processes for several reasons. First, a major source of disequilibrium may be the difficulty of nucleating new crystals. Then the record of disequilibrium is contained in the composition of a crystal at the time of nucleation because partitioning of elements during disequilibrium growth should be different than during equilibrium growth. Zoned crystals retain this nucleation composition in their core, preserved from later reequilibration if we are lucky. A related opportunity to detect disequilibrium partitioning is presented by the composition of inclusions trapped near the core during growth, as discussed above. Second, the zoning pattern of crystals, as a function of the metamorphic grade of a rock, position in a pluton, bulk composition, etc., can be compared with equilibrium crystallization models to detect the action of disequilibrium processes. Third, zoning may preserve some information on the shape of a crystal during growth. The shape of a crystal is an indicator of the effect of interface processes on growth, and sector zoning may be directly related to the differences of growth processes on different crystal faces. Finally, the directional variation of zoning profiles in a crystal may be related to the presence of neighboring crystals or convective motions during growth, indicating the influence of these factors on the composition of a growing crystal. A number of examples of the possible influence of disequilibrium processes on the growth of garnet and plagioclase will be considered in later sections of this work.

One topic not reviewed in any detail below is the origin of sector zoning in staurolite and pyroxenes (for example: Hollister, 1970; Hollister and Gancarz, 1971; Nakumura, 1973; Downes, 1974; Dowty, 1976; Harkins and Hollister, 1977; Carpenter, 1980b; Larsen, 1981). Sector zoning is also useful for investigating the process of growth of crystal faces. A discussion of the various models of crystallographically controlled growth and geochemistry is given by Dowty (1977).

Reaction or Growth Rate

Diffusion zoning provides one means of determining the rate of a reaction that causes zoning in a crystal if the rate of diffusion can be estimated. Equation (4) expresses the relationship between the velocity of growth or consumption of the crystal surface (v) and the steepness of the zoning profile (∇C); Ce and Cr can be measured with the microprobe and calculated from the stoichiometry of reaction. Then the zoning profile becomes a measure of v if D is known. It is important to point out that even though we are using zoning in a crystal to measure the rate of reaction, the rate limiting step in the overall reaction may not be the diffusion process in the crystal. The velocity can be limited by other processes, such as the dissolution or growth of another participant in the heterogeneous reaction.

I have used zoning in garnet to attempt to estimate the rate of some disequilibrium reactions in garnet–granulite-grade rocks (Loomis, 1976, 1979a). It was concluded that the rate-limiting step in these reactions was probably the dissolution of the aluminum silicate phases. Compositional zoning in igneous plagioclase attributable to growth under disequilibrium conditions may be useful in predicting the rate of crystal growth (Loomis, 1981). A related topic, beyond the scope of this work, is the use of exsolution microstructures to study the diffusion processes in feldspars and pyroxenes, and to trace the environmental history of rocks (for example, some recent papers are: Yund and Ackermand, 1979; Carpenter, 1980a, 1981; Day and Brown, 1980; Yund and Tullis, 1980; McCallister and Nord, 1981; and Grove, 1982).

Kinetic Control of Geophysical and Geochemical Properties

Some examples of important geophysical changes that are controlled by the rate of reactions are the rise and fall of continental crust in response to changes of heat flow, the increasing density of oceanic crust as it undergoes the gabbro to eclogite transition during subduction, the increasing density of crustal rocks as they undergo blueschist–facies reactions near subduction zones, and the exsolution of volatiles from melts as they cool. Examples of geochemical properties in high-grade rocks that are controlled by reactions include the gain and loss of volatile elements during reaction and, of course, crystal fractionation of melts. Many of these processes have been modelled assuming that a state of equilibrium exists throughout the rock or melt. While reasonable as a first approach, the presence of zoned crystals and disequilibrium textures in many of these rock types suggests that kinetics could have a significant effect on fractionation.

As an example, Ahrens and Schubert (1975) argue that the importance of the gabbro–eclogite transition as a driving force for plate-tectonic motions depends heavily on the rate of the reaction; they suggest that the rate of reaction is controlled, in turn, by the presence in minor quantities of hydrous minerals. Based on my studies of disequilibrium reactions involving zoned garnets in garnet granulite rocks, I have suggested that the rate of reaction to form higher pressure assemblages is greater than the reverse reactions at the same temperature because zoning in garnet impedes reaction (Loomis, 1977). The result could be a tectonic "hysteresis" effect whereby the density of lower crustal rocks can be increased by an increase of pressure or decrease of temperature at a faster rate than the density can be decreased as the temperature rises or pressure falls. The importance of disequilibrium partitioning during growth of phenocrysts in melts to the evolution of trace elements in a magma was emphasized by Albarede and Bottinga (1972) and has been considered by a number of others since. The significant effect of high cooling rates on the phase assemblage to crystallize from a basalt melt was demonstrated by Walker *et al.* (1976). Delayed nucleation of crystals upon slow cooling could be responsible for the sudden release of volatiles and even heating of a magma when rapid crystal growth begins. The kinetic path taken by a metamorphic reaction can cause an intermediate gain or loss of volatiles contrary to the overall result of reaction (Loomis, 1979a).

I am sure that there are many examples of the use of crystal zoning data to understand kinetic processes and to solve geological problems that I have overlooked in the literature or neglected. As our understanding of igneous and metamorphic petrology advances, and our interest turns increasingly to details, it is possible that the importance of disequilibrium, kinetic processes will be recognized increasingly.

Equilibration

Equilibrium

Before we attempt to treat a variety of equilibrium and disequilibrium reaction models, it will be useful to review the various types of equilibrium states that can be envisioned to occur in heterogeneous assemblages (see also Mueller, 1967; Ganguly, 1977). First, a state of total or "complete equilibrium" can be defined for the entire rock. It requires that the chemical potential of each component in all phases be equal. In terms of measurable quantities, complete equilibrium requires that all crystals be homogeneous, that the number of phases satisfy the phase rule, and that all crystals of a given phase have the same composition. Obviously, total equilibrium can be approximated only in a system small enough that elements can be transported among all crystals, but large enough to accommodate distinct crystals of all the necessary phases. The equilibrating bulk composition is simply the bulk composition of the system.

The concept of "local equilibrium" (Korzhinskii, 1959; Thompson, 1959) is well established in the literature and is a fundamental assumption in models of diffusion-controlled reaction. Local equilibrium requires that each small domain in a rock have an equilibrium assemblage appropriate for the bulk composition at that domain, but that the assemblage can vary with the local bulk composition on the scale of interest. Local equilibrium is easily defined in simple systems, but the concept is more difficult to apply to heterogeneous assemblages in which the bulk composition is represented by several different crystals of several phases scattered over a finite volume of rock.

Another state of equilibration, in which only some of the phases in a rock are in equilibrium with each other, can be called "partial equilibrium." This state requires that all possible reactions among the matrix phases are reversible, but that some reactions with certain "disequilibrium" phases are kinetically impeded. Partial equilibrium could arise if the reaction of a phase is inhibited by surface kinetics. In this case, the material included within the phase is simply removed from the equilibrating bulk composition. Similarly, partial equilibrium can be applied to systems containing armoring coronas around crystals or containing zoned crystals. In these cases, a portion of the system is restricted from the equilibrating bulk composition by the limitations of intracrystalline diffusion. The interior of zoned crystals can be considered to comprise an infinite number of phases (different compositions) that are not in equilibrium with the rest of the system.

Still another concept of equilibrium is "partitioning equilibrium." If different components have different mobilities within the matrix or melt, then two crystals may be able to exchange rapidly some components but not others. Moreover, it is also possible that the strong bonding that characterizes the aluminum–silicate framework of some common mineral structures restricts growth and dissolution of these minerals even though they may exchange other cations. In either case, two crystals may be in equilibrium with respect to the exchange of some cations, but in disequilibrium with respect to their abundances for the bulk composition. A probable example of this state was cited above (Fig. 5). The partitioning of Mg and Fe among garnet rims, cordierite, and hercynite is constant in all domains in this sample even though the extent of reaction that consumed garnet and produced cordierite and hercynite varied markedly among domains. A reasonable interpretation is that Mg and Fe equilibrated among product minerals in the matrix while the extent of reaction was restrained by other kinetic processes. Because many studies rely on partitioning of cations among solid-solution phases to trace the path of disequilibrium reactions, the existence of partitioning equilibrium may falsely lead to the conclusion that local equilibrium of all components obtained in the system.

Measurement of Disequilibrium

Obviously, disequilibrium can be defined as any state that is not equilibrium. However, it is convenient to measure disequilibrium somehow, and several means have been used. The common term used in igneous systems is undercooling, defined as the difference between the equilibrium temperature of crystallization of a phase from a melt of given composition (T_1) minus the actual temperature; it is usually signified by ΔT. Undercooling is useful for visualizing disequilibrium conditions on phase diagrams. In Fig. 6, the undercooling of the melt shown by the box is $T_1 - T$. The metamorphic equivalent of undercooling is "overstepping" of a solid-state reaction. An alternative view to undercooling that is more convenient for illustrating disequilibrium on plots of compositional gradients is "constitutional supercooling," after Rutter and Chalmers (1953). The amount of disequilibrium can be measured in terms of the difference between the equilibrium and actual composition in the melt. Constitutional supercooling is less convenient for multicomponent melts for which there are an infinite number of possible equilibrium compositions.

The last measure of disequilibrium I will mention is affinity. Affinity is equivalent to the negative change of the thermodynamic free energy with the extent of reaction [see Prigogine and Defay (1954) for the history and rigorous definition of the term]. Affinity can be rigorously calculated for any compositional or thermal variations of a system if appropriate thermodynamic data are available. The formulation of irreversible thermodynamics includes the assumption that the rate of a reaction should be proportional to the affinity of reaction. Fisher (1978) has used this assumption and estimated kinetic constants to predict the relative importance of various reaction processes. However, proponents of irreversible thermo-

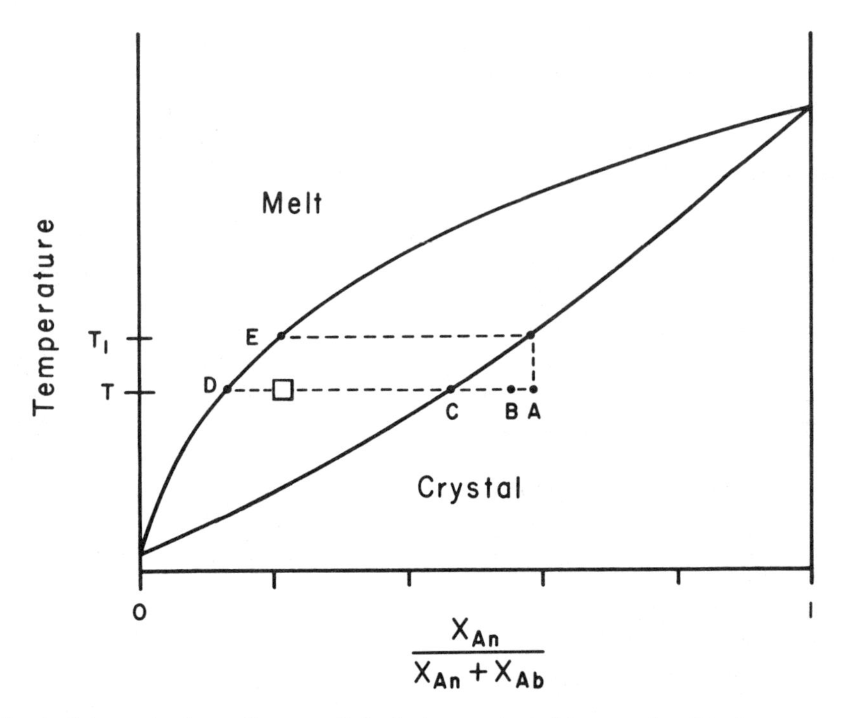

Fig. 6. Schematic phase diagram of plagioclase projected from melts of various compositions. See text for discussion of labels.

dynamics caution that the linear rate law assumptions of the theory may not apply to the large energy differences of chemical reactions (Fitts, 1962; Katchalsky and Curran, 1965; deGroot and Mazur, 1969), and I have shown that the predictions of the theory do not explain the progress of a metamorphic reaction (Loomis, 1976). For these reasons, and the fact that affinity cannot be shown on the diagrams we usually use to represent phase equilibrium, I will not make use of affinity in this chapter.

Models of Partitioning

A key consideration in fractionation models is how partitioning between the zoned crystal and the rest of the assemblage changes during growth. Taking the simplest case of binary ideal solution, the distribution of a component between two phases can be described by *Ke*. For example, the distributions of An and Ab components between plagioclase and a melt at equilibrium is described by

$$Ke_{\text{An}} = f_{\text{An}}(T,P,P_{\text{H}_2\text{O}}, \text{etc.}) = X_{\text{An}}^{X1}/X_{\text{An}}^{M}, \tag{5}$$

$$Ke_{\text{Ab}} = f_{\text{Ab}}(T,P,P_{\text{H}_2\text{O}}, \text{etc.}) = X_{\text{Ab}}^{X1}/X_{\text{Ab}}^{M}, \tag{6}$$

where X_i^j is the mole fraction of component i (An or Ab) in phase j (Xl = crystal and M = melt). The arguments of the functions are the intensive variables, T being the most important usually. The values of Ke_{An} and Ke_{Ab} can be calculated at a given T and P if appropriate thermodynamic data are available. In the binary case, Eqs. (5) and (6) plus the material balance constraints define unique compositions of the melt and crystal at complete equilibrium, shown as E and A, respectively, in Fig. 6. For systems with more exchange components and more phases, the basic constraints are similar, but the solution (or solutions) is harder to calculate. The primary conclusion is that, at equilibrium at a given T and P, the partitioning equations for the distribution of each component between each pair of phases, together with the bulk composition of the system, require that there be a unique composition of each phase in the system.

The conclusion of the previous paragraph is obvious, but often overlooked in reaction models. The most common example is the popular application of Rayleigh fractionation. It is usually assumed in Rayleigh fractionation that the partitioning of a component between the zoned phase and other phases is constant at a given temperature. The distribution of An, for example, is given by Eq. (5). A convincing isothermal growth zoning profile can be calculated by holding Ke_{An} constant as plagioclase grows and fractionates the melt. This seems reasonable as long as we ignore Eq. (6) but, if both Eqs. (5) and (6) are valid, then the compositions of plagioclase and melt must be fixed and zoning is not possible. Thus, Rayleigh fractionation is possible only if the element considered is so small in abundance that mass balance can be ignored (that is, Henry's law applies). It may be appropriate for zoning of trace elements in garnet, as described by Bollingberg and Bryhni (1972). We will not consider Rayleigh fractionation further in this chapter.

If we deal with fractionation of major elements under disequilibrium conditions, an alternative to Eqs. (5) and (6) must be found. The basic problem is to find the composition of the crystal that will crystallize from a melt or matrix assemblage that does not have the equilibrium composition for the imposed temperature. Three possibilities were proposed by Hopper and Uhlmann (1974) for binary systems. I have applied the three models to the crystallization of plagioclase (Loomis, 1981) and one model to garnet (Loomis, 1982). One model that assumes that the free energy of the crystal-melt system is minimized can not be calculated for most complex systems because appropriate thermodynamic data are not available; consequently, we will investigate the two models that can be calculated in many systems.

The first model requires that the decrease of the free energy of the system be maximized for a given amount of reaction, a model compatible with the assumptions of irreversible thermodynamics. I showed that this model predicted that the effective K_D was constant and equal to the equilibrium K_D calculated from thermodynamic data for each pair of elements. For example, K_D for Ab–An exchange is calculated as follows:

$$K_D = X_{An}^{Xl} X_{Ab}^{M} / (X_{An}^{M} X_{Ab}^{Xl}) = f_{An}/f_{Ab}. \tag{7}$$

Equation (7) is a combination of Eqs. (5) and (6). Because it imposes only one constraint rather than two on partitioning, it is possible to use Eq. (7) to find the crystal composition for an arbitrary melt composition. In Fig. 6, the crystal composition calculated for the melt composition shown by the box at temperature T is B.

A second model that can be used as a disequilibrium partitioning law proposes that the crystal composition is the equilibrium one for the given temperature regardless of the actual melt composition. The predictions of this second model are easy to show in Fig. 6 because the crystal composition for *any* bulk melt composition at temperature T is C. It turns out that this crystal composition is also predicted if local equilibrium prevails in the system at temperature T because the melt composition adjacent to the crystal face is forced to assume composition D. This partitioning model is probably also appropriate for a state of partitioning equilibrium in which the exchange components that cause zoning have an equilibrium distribution among phases even though the phases are not in equilibrium with respect to all components.

Before we worry about which model is best, we must determine how the models can be applied to multicomponent systems of two phases. The first model was extended to multicomponent systems by Loomis (1982). The result is simply that K_D's for the exchange of all pairs of components between the two exchange phases are fixed by the thermodynamic data and can be used, as in the binary case, to find the crystal composition for a given matrix mineral or melt composition. The second model has also been used for multicomponent systems (Loomis, 1981, 1983a) but it is not as well defined. The problem with defining the second model is that the composition of the crystal at a given T depends on the abundance of other components in the melt. For example, if the melt in Fig. 6 contained water and other components, the position of the phase loop would move up or down as the abundance of these other components changed, thereby changing the composition C. To use the partitioning model, it is necessary to make the additional assumption that the known composition of water and other components in the melt can be used to calculate the position of the loop (even though the mole fractions of An and Ab in the melt are ignored by the partitioning model). While this works for the distribution of two components, it is not apparent how the simultaneous partitioning of three or more components can be calculated accurately for disequilibrium partitioning.

The partitioning models predict some interesting results that are observed in nature. The easiest to test is how Ke should vary during growth. If growth takes place near equilibrium and if we assume ideal solution, Ke can be calculated directly from equations like (5) and (6) and should be a function only of the T and P history of the system; Ke is not a function of bulk composition. Consequently, all crystals in rocks of any bulk composition should have experienced the same Ke at the same T of growth. The zoning in crystals can be determined directly from phase diagrams, as discussed below. The disequilibrium partitioning models differ from equilibrium models in two significant ways. First, Ke is a function of the actual temperature of growth of the crystal, which may be different than the temperature at which the crystal should have grown if it were in equilibrium with the matrix or melt. Therefore, two rocks with identical bulk compositions can give rise to different crystal compositions if they experience a different amount of disequilibrium. Secondly, *Ke* is a function of bulk composition, and *Ke* should change during growth even at a constant temperature. Two rocks experiencing the same thermal history but with different bulk compositions can produce crystals that record different values of *Ke*. Some observations that indicate the action of disequilibrium partitioning in both igneous and metamorphic rocks are discussed below when various growth models are compared.

Which partitioning model is closest to being correct? Essentially the only published experimental evidence we have of the effect of disequilibrium partitioning on crystal composition is the data of Lofgren on plagioclase growth from water-saturated, plagioclase melts at 5 kbar (Lofgren, 1972). His experimental charges were step cooled and produced discontinuously zoned plagioclase crystals. Because the residual melt fractionated during the growth process, it is necessary to match fractionation simulations to his observed profiles to establish the effect of rapid cooling on the crystal composition. I found remarkable agreement with his observations using the second partitioning model, including the appearance of reverse zoning during isothermal growth, but very poor agreement using the first model (Loomis, 1983a). My previous work on garnet (Loomis, 1982) used the first model because it was necessary to calculate the simultaneous distribution of three components between garnet and chlorite; the simulations showed the same sense of varitation of zoning with bulk composition and metamorphic grade as observed in nature, but the conditions under which the rocks formed are not well-enough known to attempt quantitative comparisons.

We are led to the general conclusion that these disequilibrium partitioning models probably give the right sense of compositional deviation from equilibrium to be expected from disequilibrium conditions, but that confidence in quantitative calculations must await the appearance of better experimental data with which to test the models. Quantitative models of partitioning are probably not required to verify the importance of disequilibrium partitioning in natural rocks, but the use of partitioning of elements as a geothermometer in disequilibrium systems will require a better understanding of this process than we now have. Other models than those reviewed here can be proposed, but I believe that the kinetics of reactions processes are not well-enough understood to prove any of them theoretically; they must be established *empirically* as more experimental data become available.

Equilibrium Growth Model

Now that the basic processes of partitioning and diffusion have been reviewed, it is possible to examine some models of crystal growth and compare simulations to natural zoning. It is convenient to treat the reaction of crystals in terms of two end-member processes. The first process is growth without any diffusion within the crystal occurring, called growth zoning, and the second is zoning induced within an initially homogeneous crystal by diffusion, called diffusion zoning. The zoning in a crystal may involve both processes acting simultaneously in high-grade rocks and magmas, but often the action of the two processes can be distinguished. In this section, the simplest growth model is discussed, and then the variations on zoning that may be caused by more complex disequilibrium processes are considered in the following section. Diffusion zoning is reviewed in the last part of this chapter.

The equilibrium growth model is the simplest model from a thermodynamic point of view, even though it is conceptually more complex than the commonly used Rayleigh fractionation model. The assumptions of the model are as follows. Nucleation can not be explicitly modelled as an equilibrium process; some kinetic factor must be responsible for a finite number of crystal nuclei to grow. Therefore, the first assumption is that a fixed number of crystals nucleate at the isograd temperature at which the phase of interest first becomes stable. The crystals are assumed to be spherical in shape. The second assumption is that growth is controlled by equilibrium in the melt or matrix of the magma or rock. Equilibrium requires that the transport of components in the matrix or melt be fast enough to keep up with any reaction required to maintain partial equilibrium in the matrix or melt, and that the edge of all crystals are in equilibrium with the matrix or melt. The rate of reaction is governed strictly by the rate of change of the intensive variables controlling equilibrium (temperature, pressure, P_{H_2O}, etc.). The third assumption is that diffusion does not occur within the growing crystal of interest. Finally, the system is assumed to be closed and the matrix or melt will be fractionated as the crystals grow. The composition of the rock in partial equilibrium outside of the growing crystals is called the equilibrating bulk composition. Trzcienski (1977) also considered garnet zoning caused by this process of continuous reaction.

The calculation of equilibrium growth simulations is most easily explained in terms of a finite-difference computer model for growing garnet (numerical simulations are usually required due to the complexity of equilibrium calculations). Starting with the system at partial equilibrium with nuclei of the growing crystal present, the temperature is raised a very small amount. The amount and composition of garnet stable in the equilibrating matrix composition is calculated and it is distributed on the nuclei as a thin shell. The amount of garnet grown is removed from the equilibrating matrix composition, and the process is repeated.

The progress of zoning in garnet during equilibrium growth in an idealized pelitic system without Mn, Ca, or staurolite can be visualized in Fig. 7. The loops shown on this phase diagram represent calculated equilibrium stability among the two ferromagnesian phases shown and an aluminum silicate phase, muscovite or K–feldspar, and quartz, at 5 kbar pressure and 2.5 kbar water pressure. As an example, start with the assemblage chlorite–kyanite–muscovite–quartz, where the ratio $X_{Mg}/(X_{Mg} + X_{Fe})$ in chlorite is 0.3. Upon heating, garnet of composition A should nucleate at about 510°C and chlorite has composition E. As heating continues and garnet is removed from the equilibrating system, the composition of the matrix projected onto this diagram will migrate up the chlorite curve until chlorite of composition F becomes unstable at about 545°C. The successive layers of garnet deposited on the nuclei will define a zoning profile from A in the center

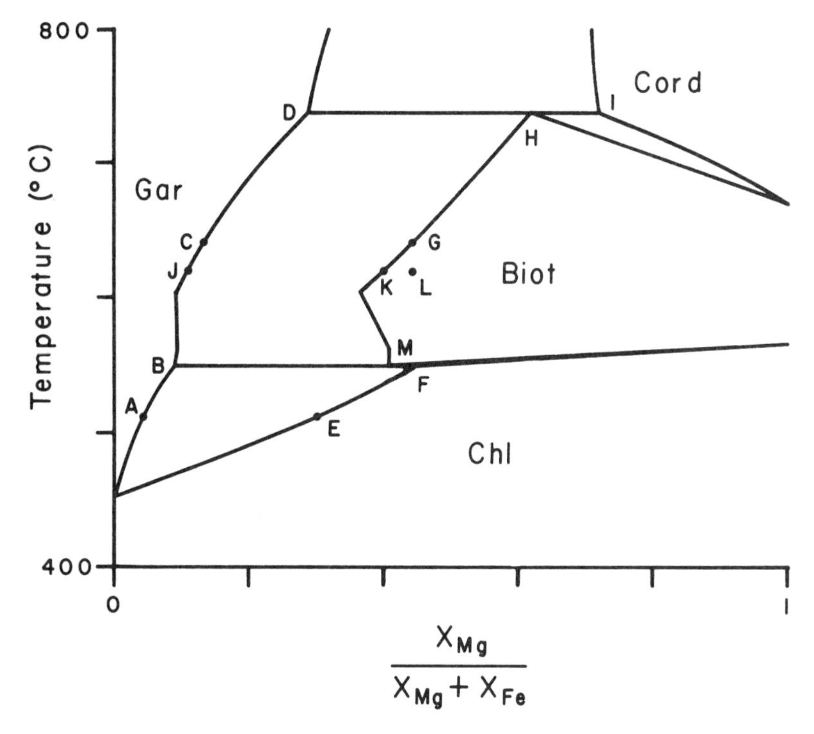

Fig. 7. Calculated phase diagram for pelitic assemblages (1) saturated in quartz, (2) containing kyanite below 561.82°C and sillimanite above, (3) containing muscovite below 603.91°C and K–feldspar above, (4) staurolite-free, (5) pressure = 5 kbar, and (6) P_{H_2O} = 2.5 kbar. The phases contain fixed mole fractions of end members other than Fe and Mg of 0.1 in garnet (Gar), 0.18 in biotite (Biot), 0.24 in chlorite (Chl), and 0 in cordierite (Cord). The thermodynamic data are unpublished, but were derived by methods similar to those used by Loomis and Nimick (1982). See text for discussion of labels.

to *B* on the edge. With further heating, the equilibrating bulk composition will enter the biotite field and the garnet will become unstable. Assuming for the moment that resorption does not occur, the next stage of garnet growth begins at about 640°C when the equilibrating bulk composition starts to enter the biotite–garnet loop at *G*. The first layer of renewed growth has composition *C*, producing a discontinuity in the zoning profile corresponding to the hiatus in growth. As the temperature increases, a zoning profile to composition *D* will be deposited on garnet and the equilibrating bulk compositions will fractionate along the biotite curve toward *H*. At approximately 735°C, biotite of composition *H* should disappear by reaction to form cordierite and garnet. This event results in the sudden growth of a layer on garnet of uniform composition *D* and the shift of the equilibrating bulk composition to the cordierite composition *I*. Further heating could cause garnet to be resorbed because the garnet edge would no longer be in equilibrium with the matrix assemblage. The discontinuous zoning profile created by this equilibrium growth process is shown in Fig. 8.

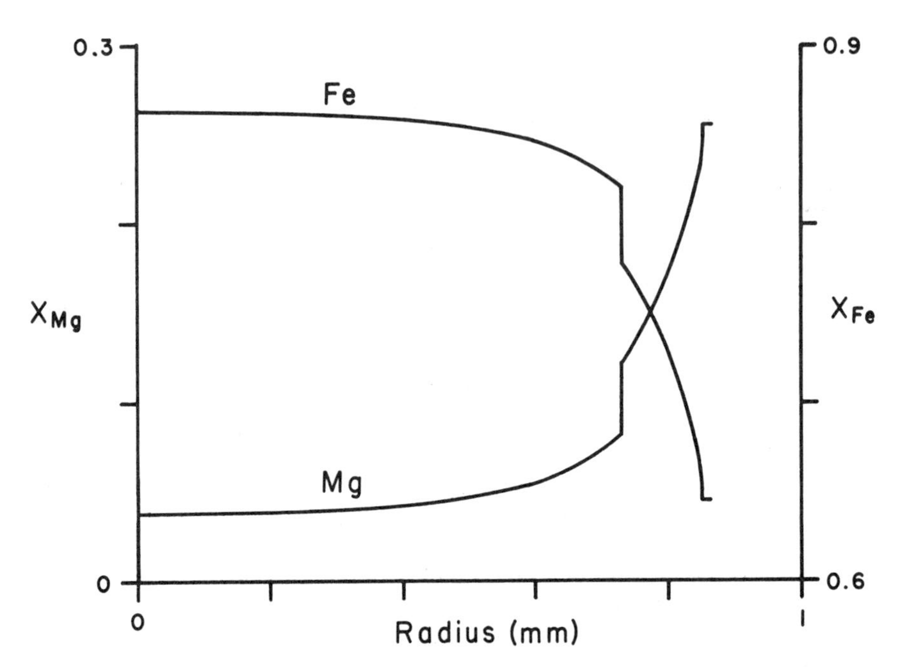

Fig. 8. Computed equilibrium growth zoning profile for garnet grown in the system shown in Fig. 7 with a ratio of Mg/(Mg + Fe) of 0.3. The nucleation density was about 56 crystals per cm^3. X_{Fe} and X_{Mg} indicate the mole fractions of almandine and pyrope, respectively, in garnet.

The growth history just described illustrates a conclusion which is founded on basic phase equilibrium. The garnet field in Fig. 7 is bounded by a continuous line. Consequently, if garnet grows continuously at near equilibrium conditions, changes of assemblage can only induce discontinuities in the slope of the zoning profile, not discontinuities in the profile. It is often implied in the literature that a discontinuous reaction that changed the rock assemblage was recorded by a discontinuity in the zoning profile of garnet (or in any other phase). This can be true only if the reaction is overstepped and becomes a disequilibrium one. A discontinuity in profile can be caused by a hiatus in growth during which no reaction of the previously formed garnet occurred. This latter situation also requires disequilibrium in the form of the metastable persistence of garnet. During the periods when the garnet edge is out of equilibrium with the matrix, the garnet could undergo dissolution reactions. If diffusion is still not possible within garnet, then the dissolution reactions would be disequilibrium ones because the garnet edge would not be in equilibrium with the matrix. It is interesting that this "equilibrium" growth process, in which partial equilibrium is maintained in the matrix, can induce disequilibrium reactions to occur in the rock. If the garnet does react and undergoes internal diffusion, then its edge composition would follow along the garnet phase boundaries in Fig. 7, and the zoning profiles created during the periods of growth would be connected by a zone of continuous profile formed by diffusion.

The equilibrium growth models must be extended to include Mn as well as Fe and Mg to describe natural zoning. An example of the zoning predicted for equilibrium growth of garnet in the assemblage garnet–chlorite–kyanite–muscovite–quartz is shown in Fig. 9. The central part of the simulated zoning profiles look very much like those observed in nature (compare Figs. 3 and 9). It was found in this system that the value of *Ke* for Mn varied only slightly over the temperature range of growth (Loomis and Nimick, 1982). Consequently, the profiles generated for Mn are very similar to those predicted by Rayleigh fractionation, which assumes that *Ke* is constant. The success of both types of simulations in matching the form of natural zoning simply indicates that we must be careful to not jump to conclusions based on morphology of zoning alone. In this case, the extreme concentration of Mn in garnet overwhelms all other factors. The proof of disequilibrium comes from the dependence of *Ke* on grade and bulk composition.

A simulation of equilibrium growth of plagioclase from a granodiorite melt with 2 wt % water is shown as curve *E* in Fig. 10. The simulation was calculated by assuming that only plagioclase grew and that water and other components accumulated in the melt as plagioclase was subtracted (water stauration was not reached). The specific composition and size of the crystal depend on the bulk composition of the melt, whether water is lost, and the assumed number of nuclei, but the general shape of this "normal" zoning profile is the same in all systems. It is characterized by a fairly flat center and increasing steepness of zoning outward.

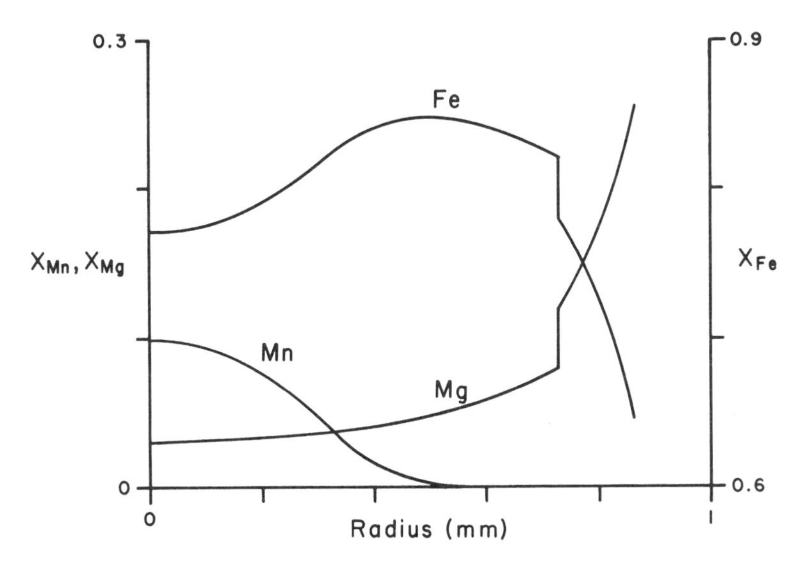

Fig. 9. Computed equilibrium growth garnet zoning profile for the same system as Fig. 7 but with the addition of 0.004 moles of MnO for each mole of MgO + FeO in the bulk composition.

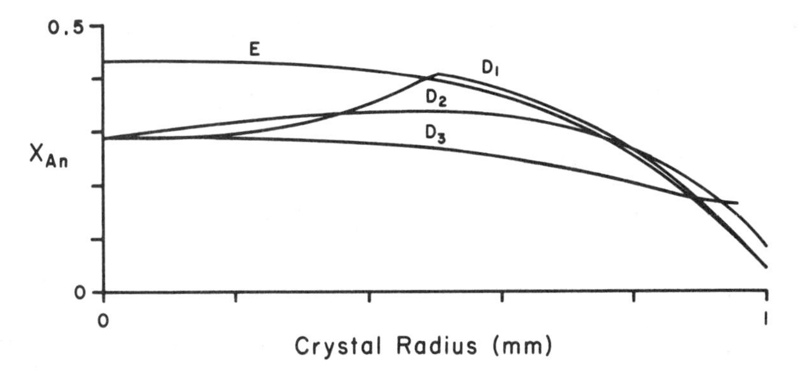

Fig. 10. Simulated zoning profiles in plagioclase grown under various conditions from a granodiorite melt with 2 wt % water at the beginning of growth, without other phases crystallizing, and with a nucleation density of about 100 crystals per cm^3. *E*: Equilibrium growth; D_1: 20 °C undercooling at the beginning of growth, then the temperature was held constant until equilibrium, then equilibrium growth; D_2: 20 °C undercooling at the beginning of growth, then the temperature was lowered so that the undercooling decreased linearly to 0 at 1 mm of growth; D_3: 20 °C undercooling maintained throughout growth. The local equilibrium partitioning model (*C* in Fig. 6) was assumed.

Disequilibrium Growth Models

Disequilibrium reactions, as I will use the term here, are defined by lack of equilibrium between the crystal surface and the major part of the matrix or melt that is not directly adjacent to the crystal, called the "bulk" matrix or melt. This definition is probably best understood by examining the two end-member models of disequilibrium processes I wish to consider, diffusion-controlled and reaction-controlled growth.

Diffusion-Controlled Growth

The diffusion-controlled growth model has long been popular in petrology because the influence of intergranular diffusion on metamorphic textures is sometimes obvious, and the mathematical formulations to describe this process are well established. This model has been applied to the development of metasomatic banding and segregation structures by many geologists, and it is usually implied that the rate of growth of individual crystals is controlled by the same mechanism. I will not attempt to review the complex models that have been proposed for metasomatic segregation and banding structures, but will try to present the basic implications of the model for crystal growth and zoning.

The main assumption of the model is illustrated in the upper part of Fig. 11. The chemical potentials of all components vary continuously in the system from the edge of the crystal out into the matrix, with the zoning probably most pronounced next to the crystal. The corresponding phase compositions may be discontinuous at phase boundaries, but the partitioning across the boundaries is governed by equilibrium partitioning (represented by E in Fig. 11). Compositional zoning of matrix crystals should reflect the continuous zoning of chemical potentials away from the crystal. In sum, a state of local equilibrium exists in each domain in the rock, even though the crystal is out of equilibrium with the distant matrix or melt.

The key assumption in the model is that local equilibrium prevails at the crystal surface. The implications of local equilibrium for growth in *binary* systems are easily visualized on phase diagrams. For example, we can predict the zoning to be generated in plagioclase growing from a plagioclase melt by reference to Fig. 6. Assume that the initial conditions in the bulk melt are indicated by the box at temperature T within the loop, representing a state of undercooling equal to $T_1 - T$. If local equilibrium is maintained at the crystal surface, then the composition of the melt and the crystal at this surface must be the unique compositions D and C, respectively, in Fig. 6. Thus, the crystal composition becomes more Ab-

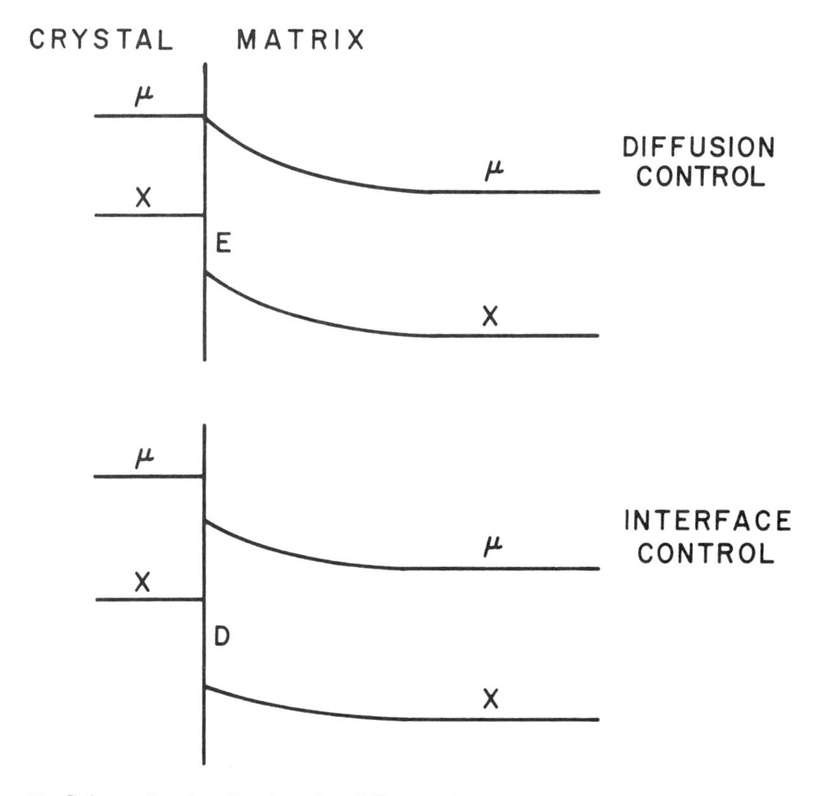

Fig. 11. Schematic plot showing the difference between growth conditions assumed in diffusion- and interface-controlled growth. X is mole fraction of a component and u is chemical potential. E indicates local equilibrium partitioning of components between the crystal and the matrix, and D indicates disequilibrium partitioning.

rich relative to the equilibrium composition (*A*) for the bulk melt as the undercooling increases. Diffusion gradients are set up in the melt between composition *D* at the surface and the composition of the bulk melt shown by the box. The rate of crystal growth will be rapid at first because the gradients are steep and the diffusion fluxes in the melt are large, but the growth rate will slow with time as the aureole of diffusion gradients spreads out into the bulk melt.

We must also consider fractionation in a complete model of crystal growth even if the rate of growth is controlled by diffusion. As growth proceeds in the system represented by Fig. 6, the diffusion halos around the crystals will intersect eventually and the bulk melt will be fractionated. Fractionation of the melt is represented in Fig. 6 by the gradual shift of the box to the left. When the box reaches composition *D*, the melt is homogeneous and growth stops because the entire melt and the crystal surface have reached partial equilibrium.

Application of the diffusion-controlled growth model to explain zoning in garnet follows the arguments above for plagioclase. Turning to Fig. 7, the garnet–chlorite loop in this Fe–Mg binary system can be treated like the plagioclase loop in Fig. 6. The composition of garnet growing from any matrix composition within the loop will be fixed by the equilibrium composition for a given temperature. As overstepping increases, the Mg/Fe ratio in the garnet, relative to the partial equilibrium composition for the equilibrating bulk composition, will increase. The important conclusion for diffusion-controlled crystal growth in *binary* systems is that the crystal composition is fixed by the temperature of growth regardless of what happens to the melt. We can think of the diffusion process as isolating the crystal from the influence of the composition of the bulk melt. Consequently, the crystal will be *unzoned* unless the temperature changes during growth. It is also apparent from Figs. 6 and 7 that the crystal composition formed will deviate from the equilibrium composition for the bulk melt or matrix composition by an amount proportional to the undercooling or overstepping.

The main difference between the predictions of the diffusion-controlled model in binary and in multicomponent systems is that there are an infinite number of melt and crystal compositional pairs that can be in equilibrium in a multicomponent system. Therefore, the composition of the crystal can change as the composition of the melt or matrix changes. The two processes that change the melt composition at the crystal surface are conveniently classified as: (1) bulk fractionation: fractionation of the bulk melt when the diffusion aureoles around crystals overlap; and (2) surface fractionation: the deviation of the melt composition at the crystal surface from the bulk composition due to concentration or depletion of components in the melt adjacent to the crystal. These two classes could also be called long and short range fractionation.

As an illustration of how these two processes can act to cause zoning during diffusion-controlled growth, let the melt in Fig. 6 contain water and a residual component in addition to An and Ab components. The residual component includes all the other anhydrous components not contained in plagioclase. First, consider the effect of bulk fractionation when the diffusion aureoles around crystals overlap. The rise of the concentration of water and the residual component in the bulk melt will cause similar changes next to the crystal, regardless of the diffusivities of components. The equilibrium loop in Fig. 6 must drop significantly as bulk fractionation occurs, and the crystal composition will migrate from the region of *C* upward along the curve toward more calcic values closer to *A*. Let me emphasize, for reference below when considering garnet zoning, that the plagioclase becomes more *calcic* even though the melt is being fractionated toward *lower* An/Ab values. The profile produced during isothermal growth by bulk fractionation will tend to be reverse zoning. Note that undercooling decreases during this isothermal process because the liquidus temperature decreases and reverse zoning can be considered to be a result of this decreasing undercooling. Growth continues until the box in Fig. 6, moving left, intersects the liquidus, moving downward, and partial equilibrium throughout the melt is attained.

The bulk fractionation zoning trend can be modified by surface fractionation as follows. As growth begins from the bulk melt signified by the box at temperature *T*, the water and residual component excluded by the crystal will build up at the crystal surface and force the equilibrium loop projected in Fig. 6 to move downward and to change shape slightly. This process of surface fractionation will cause the crystal composition produced at temperature *T* to be somewhat more An-rich than the equilibrium composition *C*. Thus, surface fractionation acts in the same way as bulk fractionation to decrease the difference between the crystal composition and the partial equilibrium composition assumed in the equilibrium growth model, and might partly offset the effect of undercooling on composition. Moreover, the composition of the crystal being deposited can vary through time as the surface composition of the melt varies in response to the interaction between the different diffusivities of components and fractionation at the interface. This complex interaction is difficult to simulate because we have little information on multicomponent diffusivities in melts, but my approximate simulations for natural melt compositions (Loomis, 1983a, Part I) suggest that (1) the deviation of the crystal composition from *C* is affected little by the water content in the melt, and (2) little compositional zoning of the crystal is caused by the surface fractionation process.

This analysis predicts that the zoning trend of plagioclase produced by diffusion-controlled growth should be determined primarily by bulk fractionation. The major zoning trend produced during isothermal, diffusion-controlled growth from complex melts will tend to be reverse zoning caused by bulk fractionation and consequent decreasing undercooling, with probably a small decrease in the magnitude of zoning, and perhaps some small variation of the zoning trend, caused by the multicomponent diffusion process. The bulk fractionation trend, of course, is also dependent on the crystallization of other phases which modify the accumulation of the residual component. Normal zoning due to decreasing temperature (increasing undercooling) during disequilibrium growth will be superimposed on the reverse zoning caused by fractionation to produce a zoning pattern indicative of the thermal and compositional history of the melt. Simulations of combined fractionation and cooling are presented below.

The analysis of the effect of fractionation on zoning in garnet is complicated by the fact that three components are partitioned between both phases, and the calculated equilibrium composition for any temperature depends on the bulk composition. The qualitative effect of overstepping can probably be estimated from the three binary phase diagrams for the Mn–Fe–Mg system. For example, the major growth zoning of pelitic garnets is represented by variations of Mn and Fe, and the pseudobinary Fe–Mn loop for the garnet–chlorite system, shown in Fig. 12, can be used to investigate the primary zoning features. As shown there, increasing overstepping should significantly decrease the Mn/Fe ratio in garnet and, for an approximately constant Mg content, raise the content of Fe and decrease that of Mn. Similar review of the other phase diagrams suggests that increasing overstepping probably causes the garnet composition to become poorer in Mn and to have a higher Mg/Fe ratio than the true partial equilibrium composition for the bulk composition of the matrix. The effect of fractionation of the bulk matrix at constant temperature on garnet zoning can be predicted also from the phase diagrams if we assume that the garnet composition crystallizing is enriched in Mn and deficient in Mg relative to the matrix. Then the Mg/Fe ratio should decrease as the Fe–Mg loop (Fig. 7) rises in response to decreasing Mn,

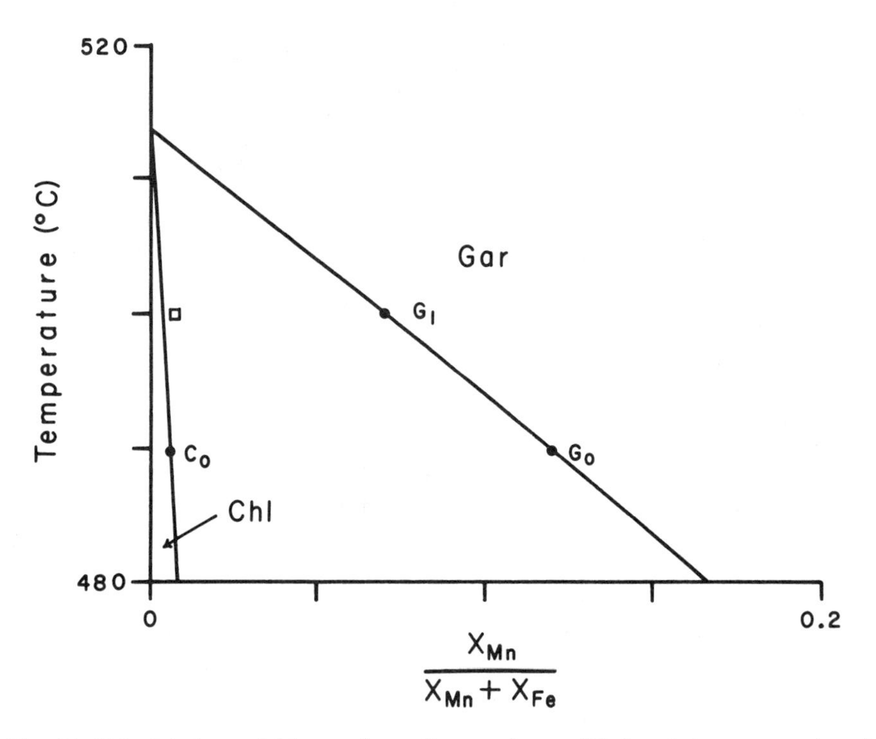

Fig. 12. Calculated pseudobinary phase diagram for equilibrium between garnet and chlorite with kyanite, muscovite, and quartz at a pressure of 5 kbar and P_{H_2O} of 2.5 kbar. Garnet and chlorite contain fixed amounts of other end members besides Fe and Mn equivalent to their compositions at 497°C in Fig. 9. For illustration, C_o and G_o indicate the equilibrium compositions of chlorite and garnet at 490°C, and the box and G_1 indicate the compositions of these phases at 500°C (10°C overstepping) under disequilibrium conditions, according to the local equilibrium partitioning model.

and the Mn content of garnet should *rise* as the matrix exchange phases are stabilized relative to garnet by increasing Mg (Fig. 12). This trend is opposite to the normal zoning predicted by the equilibrium growth models and constitutes reverse zoning (it seems opposite to the zoning we expect intuitively, at first, but is exactly the same effect predicted above for plagioclase).

Assuming that garnet has a higher content of Mn and Fe and lower content of Mg relative to matrix phases, Mn and Fe should be depleted and Mg enriched in the matrix adjacent to garnet by surface fractionation. The quantitative simulation of the process is not practical with our limited knowledge of intergranular multicomponent diffusion and the complexity of the simulation of multicomponent diffusion in spherical coordinates. However, it seems reasonable that surface fractionation will cause changes of the matrix composition next to garnet that are similar to those caused in the bulk matrix (depletion of Mn and Fe, enrichment of Mg). As in the plagioclase example, surface fractionation should reduce the effect of overstepping on the crystal composition. The magnitude of the effect of surface fractionation on crystal composition, and how zoning might be caused by transient variations of diffusion fluxes during the early stages of growth, are well beyond what I am even willing to speculate about.

Let me summarize the predictions of the diffusion-controlled growth model as follows. The effect of increasing disequilibrium (undercooling or overstepping) in the examples that we have considered is to drive the crystal composition away from the composition predicted by the equilibrium crystallization model toward compositions that would form subsequently according to the equilibrium model. Consequently, the *magnitude of zoning is reduced* relative to that predicted by the equilibrium growth model. Fractionation of the bulk melt or matrix at constant (disequilibrium) temperature probably causes zoning that is reverse to that predicted by the equilibrium growth model. The effects of surface fractionation probably reduce the compositional change due to disequilibrium and could induce unusual zoning trends as a result of multicomponent diffusion interactions; the lack of diffusion data make it very difficult to predict the effect of diffusion on crystal zoning.

It is appropriate to mention some textural implications of the diffusion-controlled growth model here. If strong diffusion gradients exist around growing crystals, metasomatic zones with fewer than the normal complement of phases may appear around the crystal. For example, quartz "halos" around garnet are cited as evidence of local depletion of ferromagnesian components. If the halos contain solid-solution phases, one would predict compositional zoning of elements toward the growing crystal. Lastly, the distribution of crystals and their relative size should reflect local compositional control.

Reaction-Controlled Growth

The simplest case of the reaction-controlled growth model occurs if reaction kinetics at the crystal surface limit the rate of growth. As shown in Fig. 11, the interface-controlled reaction model allows the chemical potentials of some components to be discontinuous at the crystal boundary. Reaction control does not require that diffusion gradients in the matrix can not exist, although we will use this assumption in the end-member model discussed in this section. I have drawn Fig. 11 with some compositional gradient in the matrix to emphasize that gradients can exist but are reduced in proportion to the kinetic importance of interface reaction processes.

The key uncertainty in the reaction-controlled growth model, if interface kinetics limits growth rate, is the process of disequilibrium partitioning at the crystal surface. We have discussed a couple of calculatable models above but, ultimately, we must confess that we know very little about partitioning under disequilibrium conditions. However, we may be able to detect disequilibrium partitioning from variations of crystal compositions even though absolute compositions cannot be calculated reliably.

Besides interface kinetics at the growing crystal interface, there are three other potential rate-limiting processes that create conditions best described by the reaction-controlled growth model. First, dissolution kinetics at the surface of reactant crystals in metamorphic rocks could limit the growth rate of product crystals. The dissolution model is widely applied in aqueous solution and hydrothermal studies of reaction. I have suggested that the rate of dissolution of kyanite and sillimanite could have limited the growth rate of garnet in some granulites (Loomis, 1976, 1979a). The coexistence of touching aluminum silicate phases in metamorphic rocks of various grades clearly demonstrates the limiting effect that dissolution and perhaps interface-controlled growth processes in these phases have on metamorphic reaction rates (Loomis, 1972c). If the dissolution model is correct, chemical potentials can remain uniform in the matrix and local equilibrium obtains at the surface of growing crystals. The partitioning of elements that cause zoning in garnet is probably best described by the second partitioning model. A second process that could limit the reaction rate is diffusion within garnet, considered in detail in the latter part of this chapter. The edge of garnet and other phases are in equilibrium, but the rate of reaction and the equilibrating bulk composition is controlled by the flux of components from the interior of garnet crystals. An example of reactions controlled by this process was examined quantitatively by Loomis (1977). Equilibrium partitioning of components among crystals is described by the second partitioning model. Third, the diffusion of components that are stoichiometric constituents of the growing crystal could be rate-limiting in igneous or metamorphic rocks. For example, if the diffusivity of Al is much slower in the matrix of metamorphic rocks than the diffusivities of Fe, Mg, and Mn, then the rate of garnet growth could be limited by the supply of Al, but the compositional zoning of garnet caused by diffusion of Fe, Mg, and Mn would be very small. Again, the distribution of components that cause zoning is probably described by the second partitioning model.

Compositional zoning of crystals in the reaction-controlled model is caused by the effect of disequilibrium partitioning on the crystal composition and by fractionation of the equilibrating bulk compostion. The isothermal growth of plagioclase in a binary system can be illustrated in Fig. 6 and follows the same analysis as for bulk fractionation during diffusion-controlled growth, except that the partitioning law may be different. In the binary case, plagioclase of composition *B* will nucleate at temperature *T* from a melt with the composition shown by the box if the first partitioning model is used, or composition *C*, if the second model is used. As the melt is fractionated and the box moves toward *D*, the crystal composition will migrate toward *C*, for the first partitioning model, or remain at *C* for the second model. Thus, whether or not zoning is produced depends on the partitioning model assumed.

In the more interesting multicomponent case, the first compositions to crystallize will also be given by *B* or *C* in Fig. 6. As water and the residual component accumulate in the melt and the loop migrates downward, the second partitioning model predicts that reverse zoning should be produced, as we discussed above for diffusion-controlled growth (but the zoning is not modified by diffusion processes). The first partitioning model predicts that the zoning will be a compromise between the normal zoning produced by movement from *B* toward *C* as the An/Ab ratio decreases in the melt, and the reverse zoning produced by the accumulation of water and the residual component.

Simulations of interface-controlled growth using the second disequilibrium partitioning model, for various thermal histories, are shown in Fig. 10. As noted in the section on partitioning, the second partitioning model provides a better fit to experimental data. It predicts that the composition of plagioclase in natural melts is very sensitive to undercooling; consequently, the reverse zoning predicted to form during isothermal crystallization by the decrease of undercooling in response to bulk fractionation is also pronounced. In contrast, the first partitioning model predicts a much smaller deviation from the equilibrium composition in response to undercooling (Loomis, 1981); consequently, the isothermal zoning profile is very close to the equilibrium growth profile but slightly reversed for granodiorite compositions.

The prediction of zoning of garnet according to the interface-controlled growth model is complicated again by partitioning of at least three components. The isothermal zoning trend has been simulated using the first disequilibrium partitioning model by Loomis (1982), shown as Fig. 13. It is predicted that garnet should become poorer in Mn, richer in Mg and Fe, and have a larger Mg/Fe ratio as overstepping increases. These simulations predict that the zoning trend is similar to that produced by equilibrium growth but that the magnitude of zoning is reduced. The second partitioning model is not well defined for three components. However, the analysis of this model presented in the preceding section (diffusion-controlled growth) suggests also that the garnet will have a decreased Mn and higher Fe and Mg content, and higher Mg/Fe ratio as overstepping increases. By analogy with the plagioclase example, it is probable that the magnitude of deviation of the composition from that predicted by the equilibrium growth model will be much larger than predicted by the first partitioning model in Fig. 13. The zoning predicted by the second partitioning model under isothermal conditions may be reversed. Both models predict also that *Ke* is a function of bulk composition and should vary with fractionation even during isothermal growth. As in the diffusion-controlled growth model, *Ke* is also dependent on the amount of overstepping or undercooling.

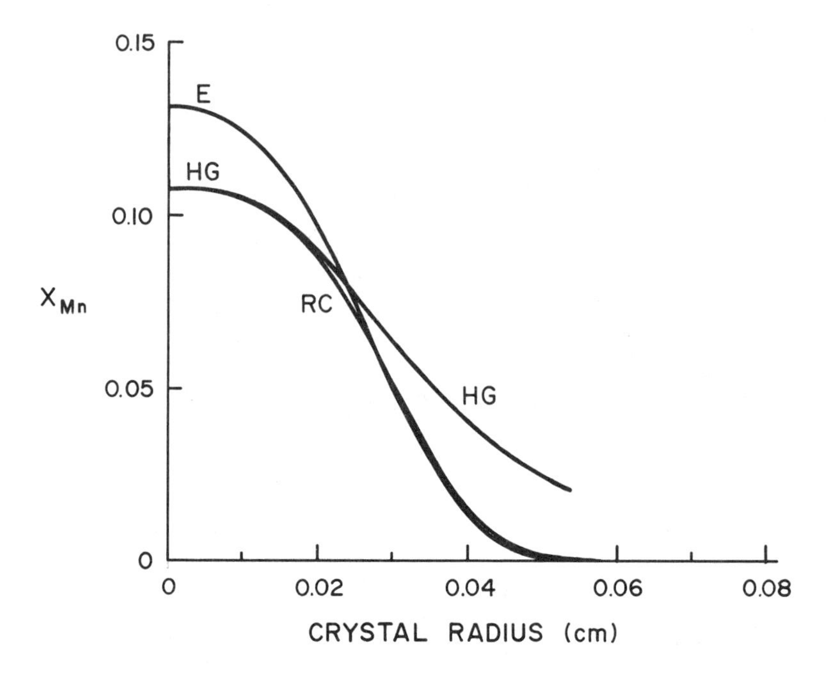

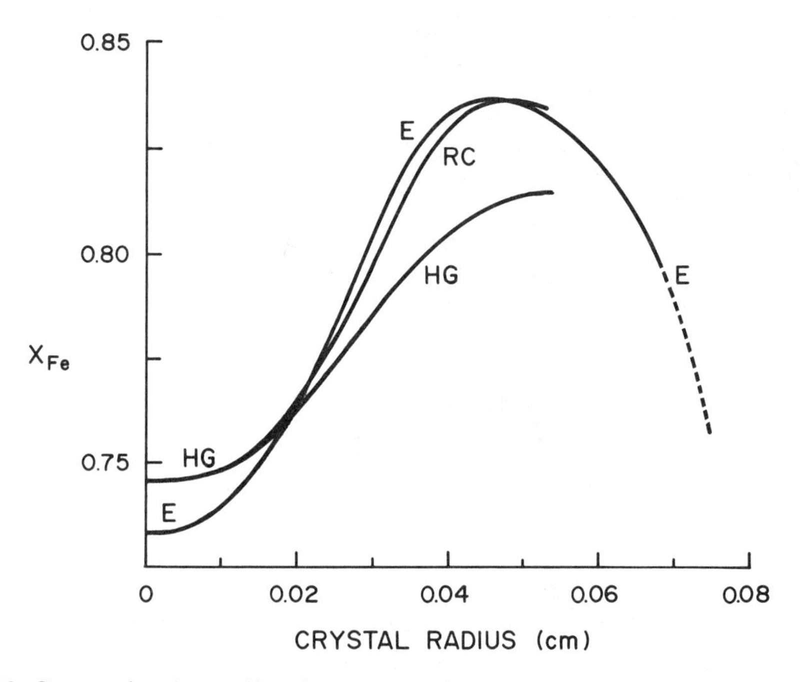

Fig. 13. Computed zoning profiles of garnet according to the equilibrium (E) and reaction-controlled (RC) growth models. The curve *HG* shows the composition of homogeneous garnet as a function of crystal radius predicted by the homogeneous growth model. The data used to calculate the simulations is given in Loomis and Nimick (1982) and Loomis (1982).

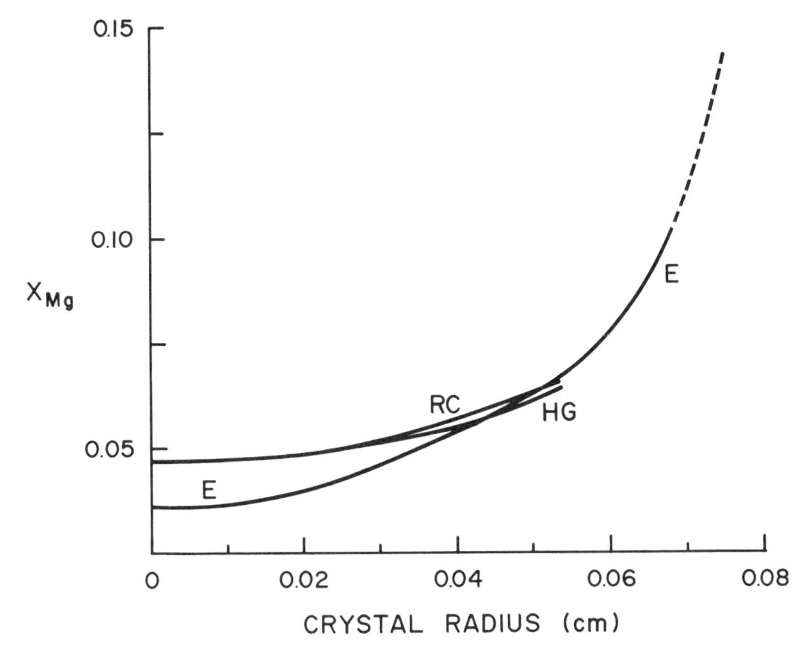

Fig. 13. (*continued*)

The suitability of the reaction-controlled model for describing crystal growth in rocks can be tested by two observations. First, the disequilibrium nature of partitioning may be evident from the variation of *Ke* with bulk composition and overstepping or undercooling not predicted by the equilibrium model. Secondly, compositional gradients of solid-solution components in the melt or matrix should be reduced from those expected in diffusion-controlled growth, and the textural evidence of diffusion gradients should not be as prominent.

Evaluation of Growth Models: The Origin of Zoning in Natural Crystals

Limited data on diffusion and disequilibrium partitioning, as well as limited thermodynamic data, hamper our ability to compare growth models. However, it is possible to suggest some semiquantitative differences in zoning, according to the three models, based on the analysis above.

The equilibrium growth model is well defined because partial equilibrium in a homogeneous melt or matrix is assumed. It predicts that zoning in plagioclase or garnet is due mainly to bulk fractionation. The predicted zoning profiles for plagioclase and garnet (Figs. 8–10) are smooth curves, representing what has come to be known as "normal zoning." Rocks with the same bulk composition should have zoning profiles with the same morphology regardless of the cooling or heating rate, or the ultimate grade of the sample. *Ke* and K_D should be nearly the same in all samples with similar bulk compositions and should not change dramatically with composition because these equilibrium parameters vary slowly with temperature and with nonideal mixing. In contrast, the disequilibrium models of growth predict that zoning profiles can vary with the amount of undercooling or overstepping, and hence with cooling and heating rate. Reverse and irregular zoning are possible. *Ke* is dependent on cooling and heating rate, and on bulk composition.

The disequilibrium growth models suggest an origin for several observed zoning features in rocks. (1) Both models indicate that the crystal composition should be depressed toward more average values (i.e., lower An/Ab in plagioclase and Mn/Fe in garnet) as the amount of disequilibrium causing growth increases. Based on this observation, it seems reasonable that the Mn core composition of garnet will be reduced in rocks of the same bulk composition as the rate of heating increases and the probability that nucleation occurs at larger overstepping increases. Similarly, plagioclase nucleated in more rapidly cooled rocks should have lower An/Ab in the core than slowly cooled rocks. (2) The compositional history of the crystal reflects the change of disequilibrium during growth. In melts, normal zoning of plagioclase will be promoted by continued cooling, by the contemporaneous crystallization of other phases, and by the transport of water out of the local system. The last two processes reduce the accumulation of water and the residual component in the melt and maintain undercooling. In contrast, stable temperatures and the absence of other phases promote reverse zoning. In metamorphic rocks, normal zoning of garnet should be promoted by a heating rate comparable to the rate of fractionation of the matrix by garnet, and reverse zoning may be possible if the heating rate is slow. (3) Because nucleation is a difficult kinetic step in a crystal's history, it is likely that the condition of strong disequilibrium exists when the crystal first starts to grow. Consequently, reverse zoning may be expected to be most common in the core of crystals or following renucleation of growth on a preexisting crystal.

The differences between the diffusion- and reaction-controlled growth models are more difficult to determine. If we assume that bulk fractionation occurs eventually during the growth of crystals, then both models produce similar predictions if the second partitioning model is used for interface-controlled growth. The first partitioning model predicts that bulk fractionation should generate more normal zoning than the second model. The effect of surface fractionation in the diffusion-controlled growth model is probably to reduce the effect of bulk fractionation on the zoning profiles because it reduces the deviation of the crystal composition away from the equilibrium composition caused by undercooling or overstepping. Consequently, the diffusion-controlled model predicts profiles that tend to be less zoned than the interface-controlled model. Other effects on the zoning profile may be caused by multicomponent diffusion because the transport rate of components to the crystal surface may vary with time. Given our inability to prove the accuracy of disequilibrium partitioning models, and the paucity of data on multicomponent diffusion in melts and intergranular space, it is probably not possible to choose between the diffusion- and reaction-controlled growth models based on zoning profiles.

Comparison of typical plagioclase zoning profiles, such as shown in Fig. 2, with the equilibrium growth simulation shown in Fig. 10(E) enables us to conclude immediately that the equilibrium growth model is not sufficient to describe the sharp changes of composition, reverse zoning, and oscillations characteristic of phenocrysts. My review of the influence of major geological variables on the *equilibrium* crystallization of plagioclase (Loomis, 1983a) indicates that variations of pressure, water content, or the crystallization of other phases probably do not have enough effect on the crystal composition to explain the observations mentioned above. Thus, a disequilibrium crystallization model is required.

The disequilibrium simulations shown in Fig. 10(D) were computed assuming that the interface-controlled growth model with partitioning model 2 applied; this model is equivalent to the diffusion-controlled growth model without the probably small effects of surface fractionation. The disequilibrium growth simulations show that a variety of reversed to normal profiles can be caused by different cooling and fractionation histories. Reverse zoning in natural profiles, such as shown in Fig. 2, may be caused by slow cooling and bulk fractionation of the melt after nucleation at tens of degrees undercooling. Reverse zoning in experimentally grown plagioclase has been successfully simulated using disequilibrium growth models (Loomis, 1982). Irregular zoning is probably produced by erratic changes of undercooling caused by local convection around growing crystals (Loomis, 1983a).

I conclude that zoning in many plagioclase phenocrysts indicates extensive disequilibrium during growth caused by the difficulty of nucleation and the limits interface processes place on growth. Turbulence probably occurs in plutons during solidification as well as in volcanic environments and during intrusion, as is indicated by the common occurrence of broken and abraded phenocrysts, by synneusis, and by the irregular zoning of plagioclase. Evidence of rapid convection during the rise and crystallization of silicic as well as basic magmas is accumulating from theoretical studies of the rise of plutons, differentiation in small plutons, and by plagioclase zoning (see Loomis and Welber, 1983, for discussion).

Simulations of the equilibrium growth of garnet in the assemblage chlorite, kyanite, muscovite, and water (Loomis and Nimick, 1982) and in other assemblages with biotite and staurolite (unpublished) indicate that the basic zoning morphologies observed in natural garnets are easily explained by equilibrium growth models. The bell-shaped Mn profile, the inverted Fe-bell, and the shallow, bowl-shaped Mg profile shown in Fig. 3 are normal products of fractionation.

The evidence for disequilibrium growth on garnet comes from the observations that (1) partitioning of Mn between garnet and the matrix phases depends on bulk composition in samples from the same outcrop (Hollister, 1969); (2) partitioning of Mn, Mg, and Fe were observed to depend on grade of the sample by Atherton (1968) and Hollister (1969); the Mn content decreases and the Fe and Mg contents increase as grade increases in rocks with similar bulk composition; and (3) natural zoning profiles suggest that reverse zoning in garnet, identified by increasing Mn concentration, can occur in the center of crystals [e.g., Hollister,

1969, Fig. 7(a)] and near the end of growth (e.g., Hollister, 1969, p. 2478). As discussed above, it is unlikely that diffusion has significantly altered the growth zoning in staurolite-grade, contact metamorphosed rocks and, in any case, cannot explain increasing Mg/Fe ratios with grade.

The variation of Mn partitioning probably can not be explained by differences in bulk composition. Bulk composition can affect partitioning of Mn only through thermodynamic nonideal mixing. Because Mn and Fe mix nearly ideally in garnet (Ganguly and Kennedy, 1974), the large variation of Mn partitioning with bulk composition observed by Hollister can be explained only by disequilibrium processes. Both disequilibrium growth models predict that *Ke* for Mn decreases as overstepping increases (i.e., the Mn content of garnet decreases for a given Mn content of the matrix). At a given temperature of garnet nucleation, the amount of overstepping in Mn-rich samples is greater than in Mn-poor samples because Mn decreases the equilibrium garnet isograd temperature. Consequently, *Ke* for Mn should decrease with increasing bulk Mn content in samples that nucleated garnet at the same temperature. Thus, the disequilibrium growth models explain Hollister's observation that *Ke* decreases with increasing bulk Mn if we assume that garnets in the outcrop nucleated at the same temperature.

The progressive decrease of Mn content and increase of Mg/Fe ratio with increasing grade in rocks of similar composition, found by Hollister and Atherton, can be explained by disequilibrium growth if nucleation occurs at progressively higher temperatures as the ultimate metamorphic grade of a rock increases. It seems reasonable that rocks heated to higher temperatures were probably heated at a faster rate, and the probability that nucleation would be delayed to higher temperatures was greater. Reverse or flattened zoning near the center of grains is consistent with strong disequilibrium and very rapid growth after nucleation, which allows nearly isothermal growth to occur. Reverse zoning near the end of growth may be explained by the declining heating rate to be expected in most metamorphic events as temperatures near their maximum values, but may be due also to slight resorption.

A model of garnet growth in metamorphic rocks that would seem to explain the observed zoning and partitioning trends with grade and bulk composition incorporates the following assumptions: (1) nucleation is delayed to higher temperatures in more rapidly heated rocks; (2) the temperature of disequilibrium nucleation is controlled more by the heating rate than by bulk composition; and (3) disequilibrium is probably greatest at nucleation and decreases near the end of growth. A quantitative calculation of the variable growth rate of garnet necessitated by equilibrium growth during contact metamorphism was shown by Loomis (1982); it illustrates the rate processes that probably cause maximum disequilibrium to be experienced by garnet at nucleation and minimum disequilibrium at the end of crystallization. This model of garnet growth in pelitic rocks is suggested for a single contact metamorphic event; metamorphism in regional terrains and in polymetamorphism will obviously be far more complex. Presumably, the absence in garnet of the very irregular zoning found in plagioclase is due to the lack of convective motions in metamorphic solids.

Our models of disequilibrium growth are not well-enough constrained by experimental data to attempt to distinguish between diffusion- and reaction-controlled growth based on zoning profiles in natural plagioclase and garnet, especially because the major zoning in both models is probably caused by the same process of bulk fractionation. There are, however, several other types of observations that suggest that reaction-controlled processes place limits on the growth of plagioclase and garnet.

Disequilibrium Growth Mechanisms

The previous work on zoning profiles in plagioclase and garnet suggests that disequilibrium growth processes may be important in the development of some crystals. In this section I will review other types of data that indicate the significance of various growth processes in igneous and metamorphic rocks.

The rate of crystal growth may be controlled by (1) environmental factors imposed on the system from outside, (2) intergranular diffusion among crystals, or (3) mechanisms within and on the surface of crystals (interface kinetics and intragranular diffusion). These three classes of growth controls correspond to the three types of crystal growth simulations presented above for plagioclase and garnet. Many researchers have applied models of diffusion-controlled reaction extensively to explain textures and structures in metamorphic rocks (see Loomis, 1979a and 1983b for a bibliography of these works). In contrast, few researchers have investigated the importance of interface kinetics in metamorphic reactions, although some early work on crystallization and nucleation suggested that kinetic factors other than diffusion were important (Kretz, 1966, 1973; Galwey and Jones, 1966). However, the importance of interface kinetic processes in limiting the rate of crystal growth from melts has been the subject of study for many years and has recently been applied to crystallization from magmas (for example: Grey, 1970; Kirkpatrick, 1976, 1977).

Our ability to use assemblage and compositional data from rocks to interpret geological history depends very much on our understanding of reaction processes. The mechanisms of crystal growth that we have investigated above can be related to conceptual models of equilibrium in metamorphic rocks which, in turn, govern the application of thermobarometers and the phase rule. Environmental factors imposed on the system comprise the thermodynamic intensive variables, usually assumed to be temperature, pressure, and the chemical potentials of perfectly mobile components. For example, the rate of a reaction may be limited by the availability of heat or of a perfectly mobile component to drive the reaction, and the system may be considered, from a thermodynamic point of view, to be in chemical equilibrium soon after nucleation. If reactions are controlled principally by environmental factors, then interpretations can be based on the assumption that all crystals in the rock, including inclusions within zoned crystals, were in equilibrium with each other at each stage of development. If mechanism 2, diffusion among crystals, limits the rate of reaction, then the system should be in a state of equilibrium at each small domain in the rock (local equilibrium). If intergranular diffusion controlled the rate of reactions, then touching crystals can be assumed to have been in equilibrium, but separated crystals may not have been. Reaction processes within and on the edge of crystals that could control the rate of reaction (mechanism 3) include intragranular diffusion and interface kinetics. We have already discussed, in terms of Eq. (4), how a zoned crystal can have a rim composition in equilibrium with the rest of the rock and control the rate of reaction through the flux of material arriving at this surface from inside the crystal by diffusion. Interface kinetics can control the rate of reaction in ways ranging from direct limitation of growth or dissolution to selective adsorption of certain components. If interface kinetics controlled reactions, then we can not automatically assume that two touching crystals were in equilibrium, but may be able to assume that two separated crystals were in partitioning equilibrium. Furthermore, rapidly changing conditions that cause disequilibrium reactions may be followed by a sufficiently long period of thermal stability that the matrix of a rock is annealed to an equilibrium state, but the interiors of crystals retain disequilibrium characteristics.

Thermodynamic Systems in Metamorphic Rocks

In broad view, the sizes of thermodynamic systems in metamorphic rocks are ultimately controlled by the rates of intergranular diffusion and infiltration; mineral assemblages change enough over distances of tens of centimeters to indicate that the volumes of equilibration for most components never grow very large. However, the usual assumption in the analysis of metamorphic assemblages is that the gradients of the chemical potentials of most components are small enough in the matrix that a collection of many crystals may be assumed to have experienced the same chemical environment and constitutes a thermodynamic system; in other words, heterogeneous equilibrium in the matrix of the rock can be assumed. Thus, the growth of individual crystals in the assemblage is governed by heterogeneous equilibrium in the matrix even though the growth of the assemblage as a whole (as a segregation, for example) is controlled by intergranular diffusion. This state of heterogeneous equilibrium in the matrix during crystal growth corresponds to the conditions assumed in the reaction-controlled and equilibrium growth models. The alternative assumption is that the gradients of chemical potentials of most components in the matrix are large enough that each crystal experiences a chemical environment different from its neighbors. Each crystal becomes a monomineralic, metasomatic system and heterogeneous equilibrium cannot be assumed. This state of mosaic equilibrium of crystals corresponds to the conditions assumed in diffusion-controlled growth. There is some textural evidence that crystals may act as metasomatic centers defined by their Al content from the association of high-Al minerals with low-Al ones, from the preservation of relic bedding, and from the production of "pressure shadows" low in Al (Carmichael, 1969; Fisher, 1970). However, the success of the many studies of phase compositions and applications of the phase rule that assume heterogeneous equilibrium in the matrix of metamorphic rocks suggests that chemical potential gradients in the matrix of rocks in small groups of crystals are usually small enough to be neglected.

Crystal Morphology and Mineral Compositions

Anisotropy of growth and the euhedral morphology of plagioclase and garnet crystals indicate that crystal structure influences the rate of growth. If the growth of phenocrysts were controlled solely by diffusion in the melt or matrix, the crystals would be spheres, or would have irregular shapes controlled by competition with other crystals for components. The rate of interface-controlled growth of plagioclase from plagioclase melts was measured by Kirkpatrick *et al.* (1979) and extrapolated to natural melts by Loomis (1983a). These calculations suggest that interface processes should limit the rate of plagioclase growth in natural melts also. Kretz (1973) determined that the growth rate of garnet crystals in a high-grade metamorphic rock was not consistent with a diffusion-controlled growth mechanism, but did not depart greatly from the rate expected from an interface-controlled mechanism. The metastable coexistence of reactant phases and the compositional variations of reactant minerals indicate that the rate of disequilibrium reactions in some granulite-grade rocks could not have been controlled by intergranular diffusion (Loomis, 1976, 1979a).

Calculated Growth Rates

Fisher (1978) showed that the growth of a crystal or spherical segregation may pass through three stages: reaction-controlled growth, diffusion-controlled growth, and equilibrium growth. I will attempt to explain the transitions among the three stages qualitatively as follows. When a crystal first nucleates at some overstepped condition, the necessary components for growth are readily available locally and the growth rate will be limited only by interface kinetics. As growth proceeds, the transport of components to and from the crystal through a diffusion halo will become slower and eventually can become the process limiting the growth rate. Eventually, the diffusion halos around crystals will intersect, the entire bulk matrix will become fractionated, and the assemblage will move toward equilibrium.

Fisher attempted to estimate the rate of reaction-controlled growth and dissolution from dissolution data on silica and calcite in aqueous solutions at low temperature. He also assumed that overstepping in metamorphic rocks at nucleation was small. He concluded that the early reaction-controlled growth stage for most metamorphic environments would produce structures too small to observe and that most growth would be controlled by intergranular diffusion or equilibrium. However, there is a very large uncertainty in the interface-controlled growth and dissolution rates that are appropriate for metamorphic rocks, especially for aluminum silicate phases and garnet. For example, Ildefonse and Gabis (1976) measured the diffusion rate of silica in intergranular aqueous solutions at 550°C by reacting pellets of starting materials. They were able to produce reaction rims between many materials that required the transfer of silica and alumina, but found no reaction when corundum should dissolve and were not able to grow aluminum silicates from oxides. Consequently, the dissolution of silica is not a reliable guide to the dissolution or growth of aluminous minerals. As another example, Johannes (1978) found in melting and crystallization experiments in simple igneous systems that (1) the reaction of plagioclase was extremely slow below about 800°C, (2) that plagioclase, once formed, does not react with the residual melt when the temperature was reduced, and that (3) the composition of plagioclase was strongly controlled by surface processes. The experimental difficulties in reacting aluminum silicate phases under metamorphic conditions are well-known from studies of the aluminum silicate system. This evidence of the difficulty of reacting aluminous minerals suggests that the growth and dissolution rates of minerals such as garnet, aluminum silicate phases, and plagioclase could be rate-limiting factors in some metamorphic and igneous rocks.

As reviewed by Fisher, the proportion of the growth history of a crystal governed by each of the three processes depends primarily on the relative rates of reaction-controlled growth, intergranular diffusion, heating, and nucleation (the size of crystals also is a factor). An important variable that can determine whether crystals grow primarily by diffusion-controlled or other mechanisms in metamorphic rocks is nucleation.

Nucleation

We tacitly use nucleation processes to identify the size of metamorphic systems. For example, if a rock is coarse grained with feldspar crystals 1 cm across, but otherwise normal in appearance, it could be considered to be an equilibrium assemblage, and the size of the equilibrating volume is assumed to be much larger than the one feldspar crystal. But if the 1-cm volume is filled with many crystals of the same feldspar, we would suspect a metasomatic or "open" system, and the volume of equilibration would be considered to be less than the size of the feldspar clot. Thus, we expect nuclei of a phase to be somewhat evenly distributed throughout the volume of equilibration and not accidentally grouped together. The suggestion that nuclei should be evenly distributed throughout a rock does not imply that nucleation is strictly controlled by local diffusion; it could be a random process instead (or some combination of the two), as discussed below.

Fisher (1978) points out that an increase in the spacing of crystal nuclei increases the range of conditions in which crystal growth can be controlled by intergranular diffusion. If crystals are closely spaced relative to the ability of intergranular diffusion to transport components in the matrix, the growth rate must be controlled by interface kinetic processes or by equilibrium. It should be possible to identify the mechanism of crystal growth from the distribution and size of crystals. If local diffusion controls growth, competition for constituents should restrict the growth rate of two neighboring crystals but allow solitary ones to become large. Conversely, if several crystals develop in a volume over which diffusion is rapid, the size of a crystal should be independent of the proximity of neighboring crystals. Where careful study of nucleation and growth has been carried out, as for garnets in metamorphic rocks, the unanimous conclusion was that nucleation was random and that the growth of garnet was not controlled by competition for components with neighboring garnets (Jones and Galwey, 1964; Galwey and Jones, 1966; Kretz, 1966, 1973).

The fact that segregation structures related to crystal growth are found in some metamorphic rocks indicates that diffusion-controlled growth may be important in some circumstances. One major factor that determines the relative rate of nucleation and diffusion and, consequently, may determine whether reaction or diffusion processes control growth rate, is undercooling or overstepping. The effect of undercooling or overstepping on diffusion can be estimated from Eqs. (2) and (3). The fractional change of the penetration distance d with change of temperature is

$$\frac{1}{2}(A/RT^2).$$

For reasonable activation energies for diffusion in melts of 20–50 kcal/mole at 900°C, the rate of decrease of d with decreasing temperature is only about 0.5% per degree, and must be less than 100% for 100°C undercooling. In comparison, the data of Fenn (1977) and Swanson (1977) show that the nucleation density of feldspars and quartz in granitic rocks rises rapidly in the first one or two hundred degrees of undercooling by orders of magnitude. Consequently, increasing undercooling in the range less than 100°C should increase the nucleation rate relative to diffusive transport distances. Comparable data are not available for metamorphic rocks. However, we can assume that the activation energy of intergranular diffusion is considerably less than that of Fe–Mg exchange in garnet which indicates that the increase of d with overstepping is considerably less than about 2% per degree. It seems probable that nucleation must increase much faster than this. Thus, we can conclude that increasing undercooling in igneous rocks or overstepping in metamorphic ones will increase the nucleation rate relative to diffusion transport distances. Consequently, the importance of reaction-controlled growth in disequilibrium systems will be emphasized as the amount of disequilibrium increases. In contrast, the most probable origin of diffusion-controlled structures related to crystal growth is in rocks in which near-equilibrium conditions or some other factor has inhibited nucleation, so that only a few nuclei form at widely spaced intervals.

Measurement of Intergranular Diffusion

Both relative and absolute rates of intergranular diffusion have been estimated by many authors from segregations and textures by assuming that the rates of reactions are controlled by the diffusion process. The evidence that interface kinetic processes may control growth and reaction rates was summarized above. Figure 11 illustrates the fact that the existence of diffusion-controlled structures does not guarantee that the rate of reactions are controlled by diffusion. If interface kinetic processes limit the rate of reaction, the rates of intergranular diffusion estimated from structures will be *too slow* by an amount proportional to the extent of interface control.

Thermobarometry

The analysis of disequilibrium growth suggests that the compositions of crystals growing in disequilibrium will differ from the equilibrium value for the composition of matrix minerals and the temperature. For example, if the Mg/Fe ratio in garnet is raised by overstepping, then the calculated temperature using partitioning of Mg and Fe between garnet and biotite (e.g., Ferry and Spear, 1978) will be higher than the actual temperature during growth. The compositional variations of garnet due to overstepping of tens of degrees C predicted by the second partitioning model are probably large enough to produce variations in the calculated temperature on the order of 100°C.

Basic Concepts in Diffusion

There are manifold complexities that can be discussed under the topic of diffusion, many of which have received some attention in the geological literature. In this review, however, I will concentrate on factors that I believe are of greatest practical value in analyzing diffusion zoning in crystals in real rocks. There are four basic steps in formulating a model of diffusion to describe zoning in a crystal: describe the fluxes, establish the geological boundary conditions, apply the continuity equation to derive the diffusion equation, and integrate the equation over time either analytically or numerically. I will concentrate on the first two steps and leave the last two to the many reference works on diffusion (Crank, 1975; Carslaw and Jagger, 1959).

The basis of diffusion is Eq. (1) that describes the flux J of a component as the product of (1) the diffusivity D and (2) the driving chemical force ∇C. The flux equation may also contain other terms if the compositional profile is measured relative to the moving crystal surface or some other reference frame. Other equivalent formulations express the flux in terms of another linear coefficient (L) and the gradient of the chemical potential. In any formulation, the amount of material moved depends on a mobility factor (D or L) and a force factor (∇C, ∇X, or $\nabla \mu$) determined by the geological boundary conditions. We will concentrate on the mobility factor here and then consider the geologically significant boundary conditions in the following section.

The diffusion equation based on Eq. (1) can be formulated for any one component in a crystal, but it is necessary to recognize that diffusion usually affects two or more components simultaneously. For example, the flux of Mg^{2+} in one direction in a pyrope–almandine garnet crystal must be matched by a return flux of Fe^{2+} in the opposite direction. Then, the value of D in the diffusion equation for Mg^{2+} is identical to D in the equation for Fe^{2+}, and both refer to the exchange of Mg and Fe. The value of D in the same equation for Mg^{2+} will be different if the exchange component is Mn^{2+} instead of Fe^{2+}. The exchange reaction that drives diffusion is determined primarily by the boundary conditions that initiate diffusion [∇C in Eq. (1)]. Therefore, the geological boundary conditions indirectly influence the diffusivity of a component through their activation of particular exchange reactions.

The most important diffusion exchange reactions in garnet are Fe–Mn and Fe–Mg. Fe–Mn zoning is inherited from low-grade growth and will be relaxed at high temperature. Diffusion zoning of this type can be induced also during resorption of high-grade garnets if the temperature does not stay high long enough to homogenize the crystal. Fe–Mg diffusion zoning is induced in garnets during high-grade reactions. Estimates of diffusion rates based on relaxation experiments and reactions in metamorphic rocks (Lasaga *et al.*, 1977; Loomis, 1978b; Elphick *et al.*, 1981, 1982) indicate that Fe–Mn exchange should be faster than Fe–Mg exchange diffusion at the same temperature. Exchange diffusion involving Ca may be important in some cases (for example, zoning in garnets in kimberlites). Of course, several components may be exchanged simultaneously and described by multicomponent diffusion theory.

A central problem is determining the dependence of the mobility factor D on temperature, pressure, and the composition of the phase. The temperature dependence of D for an exchange reaction is usually expressed in terms of an activation energy, as I did in Eq. (3). The exponential dependence of diffusivity on temperature has the practical result that the onset of diffusion is rather sudden with rising temperature, as shown in Fig. 4. The dependence of D on composition has been estimated for garnet, but it is usually permissible to assume that D is constant over a small compositional range. If D does depend on composition, the interactions among fluxes can lead to "uphill diffusion" and complex zoning profiles; however, these effects will be most noticeable in the profiles of minor elements. Multicomponent compositional interactions in garnet have been investigated by Anderson and Buckley (1974), Loomis (1978a,b), and Lasaga (1979).

Other potential complications include the specific frame of reference to be used for diffusion and how compositions are represented in the phase. These problems may be significant in some substances, but I have shown that normal mole fractions can be substituted for concentration in diffusion equations for garnet without introducing any appreciable error (Loomis, 1978a).

We can conclude by emphasizing that many problems involving diffusion in garnet can be treated quite accurately by assuming that diffusion occurs by exchange of two components, that mole fractions represent the garnet composition, and that D is an exponential function of temperature.

Diffusion Boundary Conditions: The Petrology of Diffusion

While most analyses of diffusion dwell on the many complexities of multicomponent diffusion, the determination of the geological boundary conditions on diffusion is probably the greatest source of uncertainty in the analysis of natural zoning. I will describe some common boundary conditions in terms of two categories: diffusion induced in homogeneous crystals and relaxation of growth zoning. The next section will draw upon both the growth models above and the conclusions of this section to propose ways in which diffusion during growth modifies zoning profiles.

Induced Diffusion Profiles

Many studies have examined zoning profiles near the edge of garnets in high-grade rocks that appear to have been induced in initially homogeneous crystals. The origin of homogeneous crystals is undoubtedly explained by the great diffusion rates shown in Fig. 4 in high-grade regional and very high-grade contact metamorphic rocks. The zoning profiles preserved on the edges of crystals represent the last stage of diffusion that occurred as the rock cooled to temperatures low enough to effectively stop diffusion. Thus, induced profiles record only the cooling history and latest reaction history of high-grade garnets.

The first process that can change the composition of the edge of a garnet crystal and induce zoning is simple exchange of components with a neighboring crystal. The major requirement for simple exchange is that there be little "leakage" along the interface so that other crystals do not participate also in a multiphase reaction that changes the abundance of the crystal. This situation has been suggested to have occurred in some high-grade rocks during slow cooling. Apparently the grain boundaries of crystals are dry enough and sufficiently annealed that diffusion along them does not occur at a rate much faster than within garnet. Another closed system of this kind is the volume around biotite, cordierite, ilmenite or other inclusions within garnet. One observation that indicates that zoning can be explained by a simple exchange process is that the amount of each component that has diffused out of one crystal is fully accounted for in the adjacent crystal.

Diffusion exchange between crytals is motivated by a change of K_D after the two crystals have come to equilibrium. For example, garnet and biotite at equilibrium at a particular temperature have compositions determined by the bulk composition of the two phases and the value of K_D that controls the distribution of Fe and Mg between them. As the temperature changes, neighboring crystals of garnet and biotite can exchange Fe and Mg across the crystal boundary between them. The composition of each crystal surface during this process will depend not only on the changing value of K_D, but also on the rate of diffusion within each crystal. The rate of diffusion within the crystals is a factor because the apparent bulk compositions that the edges of the crystals sense depend on the flux of material from the interior of the crystals, as shown by Eq. (4). Thus, the description of how the composition at the edge of the garnet should vary through time as the temperature changes requires knowledge of K_D as a function of temperature and the fluxes J as functions of temperature and time. Lasaga *et al.* (1977) have attempted this type of analysis for the exchange of Fe and Mg between garnet and cordierite during cooling of some regional metamorphic rocks. They conclude that diffusivities must be known quite accurately to derive absolute cooling rates by this method.

Another exchange problem is how the composition of inclusions in garnet can change during their thermal history after entrapment. In the case of biotite inclusions, the high diffusion rate in biotite will maintain a homogeneous crystal during diffusion exchange with garnet, while the garnet could become zoned in a shell around the inclusion. At a temperature of about 600 °C, depending on the duration of metamorphism, the thickness of the diffusion shell in the garnet could be only a few micrometers thick and unmeasurable with the microprobe. Consequently, the composition of the homogeneous biotite inclusion that was compared with the interior composition of garnet would be incorrect. If exchange were due to cooling, the calculated temperature would be too high by an amount that depends on how much the biotite was modified by exchange. Assuming that the diffusion shell in the garnet around the inclusion was approximately 4 μm thick, the volume of garnet and the volume of biotite inclusion that exchanged components would be equal if the inclusion were a sphere about 30 μm across. Thus, fairly large inclusions can be modified by undectable diffusion exchange and give false calculated temperatures. The deviation of the calculated temperature from the actual equilibration temperature would be inversely proportional to the size of the inclusion.

If diffusion is not limited to exchange between adjacent crystals, changing conditions can drive heterogeneous reactions that change crystal compositions. The change of the compositions of crystals can be illustrated as a "divariant shift" on phase diagrams. For example, the compositions of garnet and biotite in equilibrium with sillimanite, K–feldspar, and quartz, for a fixed bulk composition, become more Mg-rich with increasing temperature in Fig. 7. This change of composition can also be shown as a shift of the triangle in Fig. 14 to the right with increasing temperature. An important difference between this boundary condition and that based strictly on changes of partitioning is that the garnet must grow or be consumed if its edge composition changes. We must be able to calculate not only how the equilibrium crystal composition of the edge of the crystal changed through time, but also the rate at which the crystal was consumed or grew by the reaction process to interpret diffusion profiles.

A simplification of the divariant shift process is to assume that the changes of conditions responsible for the change of garnet composition occurred rapidly enough that we can make the approximation that the edge of garnet changed instantaneously to a new, final composition. The edge composition can then be assumed to have remained constant during the diffusion process, but the garnet still is consumed by reaction. This approximation has been used extensively in my work on disequilibrium reactions in a high-temperature contact aureole (Loomis, 1975).

Figure 14 shows a phase diagram of the assemblage garnet–biotite–sillimanite–K–feldspar–quartz–water, in which the equilibrium composition of garnet is *Ce*. Assume that a homogeneous garnet crystal of composition *Ci* was formed at the culmination of metamorphism, and subsequent cooling or decrease of pressure caused the equilibrium assemblage phase compositions to shift to their present locations on the phase diagram. At this stage, the garnet is reacting incongruently and diffusing internally. The overall reaction taking place is

$$\text{garnet } (Ci) \rightarrow \text{garnet } (Ce) + \text{biotite} + \text{sillimanite} +/- \text{other phases}, \quad (8)$$

but the composition actually removed from garnet to form nongarnet phases is *Cr* because the product garnet (*Ce*) is retained on the edge of the crystal. The diffusion zoning profile that develops in the garnet is shown also in Fig. 14. The profile is drawn assuming that a state of partial equilibrium prevails in the matrix so that the edge composition of garnet (*Ce*) remains in equilibrium with matrix minerals.

The geological boundary conditions that can be established for this reaction system by chemical analysis of crystals are *Ci, Ce,* and *Cr*. These data, together with Eq. (4) and the diffusion equation, can be used to solve for the velocity of the consumption of the garnet interface (v) as a function of time (Loomis, 1975). An example of the diffusion curves calculated for the edge of garnet as a function of time for a reaction of this type (but a different assemblage) is shown in Fig. 15 (modified from Loomis, 1975, Fig. 10). The profile starts out very steep because the consumption velocity is large. Gradually, the profile becomes broader as the velocity falls. Eventually, the center of the garnet is depleted, the profile flattens to a homogeneous garnet, and the reaction stops. The residual, homogeneous garnet of composition *Ce* is a part of the equilibrium assemblage. In this example, it should be obvious that a major uncertainty in predicting the history of zoning is knowing exactly how *Ce* and *Cr* changed through time.

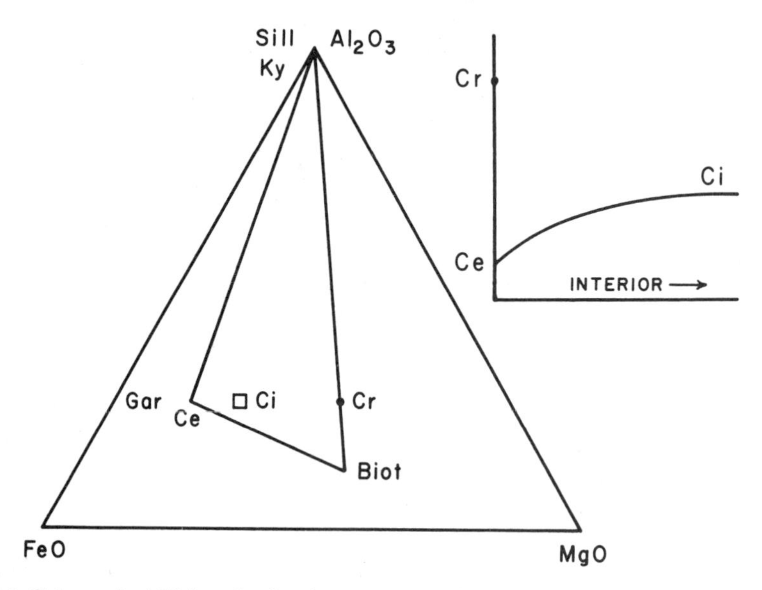

Fig. 14. Schematic AFM projection from quartz and K–feldspar to illustrate disequilibrium reaction of garnet and a schematic zoning profile induced on the edge of garnet. *Ci* is the interior composition of garnet crystals, *Ce* the equilibrium rim composition, and *Cr* the composition of garnet being removed from the crystal during consumption.

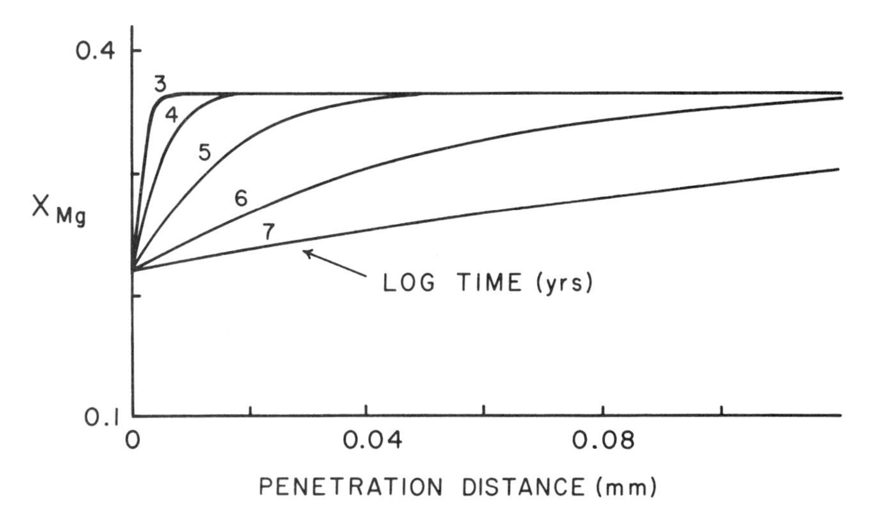

Fig. 15. Simulated diffusion zoning profiles on the edge of a garnet crystal induced by incongruent reaction as a function of the time of reaction (log yrs). The vertical axis is the mole fraction of pyrope in the garnet and the horizontal axis is the distance into the garnet.

An added complication to describing induced diffusion zoning in garnet is diffusion of Mn. A phase diagram like that in Fig. 14 could be drawn to describe the role of Mn, but it is easy to visualize what happens to Mn as garnet is consumed. Mn is strongly partitioned into garnet and most of it will be retained in garnet as garnet is consumed. As Mn builds up at the edge of the crystal, it will diffuse inward, creating a zoning profile. A more subtle complication is that the Mn edge composition of garnet will gradually increase through time as the amount of Mn derived from the consumption of garnet raises the Mn content of the matrix. Thus, the multicomponent composition *Ce* will change through time as the equilibrating bulk composition of the matrix changes. Calculations of the effect of Mn on garnet equilibria are presented by Loomis and Nimick (1982). Despite the complex calculations needed to quantitatively simulate Mn–Fe–Mg induced zoning in garnet, the rising Mn contents of garnet crystals near their edges in high-grade rocks was recognized early as qualitative evidence of consumption of garnet (Chinner, 1962; Evans and Guidotti, 1966; deBethune *et al.*, 1968; and many later papers).

Relaxation of Zoning

Growth of garnet at low- and medium-grade conditions usually produces crystals characterized by strong zoning of Mn and Fe. The zoning in these garnet crystals that grew at low grade may be "relaxed" by diffusion if the garnet is heated further to high-grade conditions. It is necessary to be able to predict the amount of relaxation if we are to use the zoning and compositions of inclusions as evidence of growth conditions.

An ideal, first illustration of the process of relaxation of growth zoning is based on our description of growth zoning in the binary Fe–Mg system shown in Fig. 7. During the hiatus in garnet growth after chlorite disappears, the zoned garnet may react with the matrix assemblage. Let us consider the reaction process in a step by step manner by skipping the first stage of reaction for now and assuming that we have a garnet with inherited zoned interior composition *A-B* and an equilibrium rim composition *J* coexisting with equilibrium matrix biotite of composition *K* at 620°C (Fig. 7). The overall reaction is given by

$$\text{garnet } (A\text{–}B) + \text{biotite } (K) + \text{sillimanite} \rightarrow \text{garnet } (J) +/- \text{ other phases,} \quad (9)$$

the same reaction as (8) but with the sign of reaction coefficients changed, and where *Ci, Ce,* and *Cr* are *A–B, J,* and *K,* respectively. A major difference from the previous example, however, is that Eq. (4) yields a negative velocity, v, meaning that the garnet *grows* as its interior equilibrates with the matrix by diffusion. The velocity of growth will be large at first but decreases to zero as the garnet becomes homogeneous with composition *J*.

The addition of Mn to the system as a bell-shaped zoning profile in garnet will increase the amount of garnet grown because Mn stabilizes garnet (the garnet–biotite loop in Fig. 7 moves down as Mn becomes available to the matrix), but will also cause *J* to vary with time, as discussed above. The general conclusion is that relaxation of growth zoning in garnet by diffusion at high grade should cause an overgrowth of lower Mn, higher Mg garnet on crystals as the zoning profile is flattened.

The next step in the analysis is to examine how the matrix biotite composition *L* generated by garnet growth up to 545°C is changed to *K*. Let us first assume that this reaction happens at 620°C. The easiest way to express the reaction process is in terms of two "component reactions" that add up to the overall reaction (Loomis, 1976). The first one is reaction (9), which produces garnet of the equilibrium rim composition (*J*). The second reaction is

$$\text{garnet } (J) + \text{biotite } (L) \rightarrow \text{biotite } (K) + \text{sillimanite } +/- \text{ other phases,} \quad (10)$$

which consumes the garnet rim to make the equilibrium biotite composition. The overall velocity of the garnet interface is the sum of the growth velocity caused by diffusion relaxation (reaction 9) and the consumption velocity caused by reaction with biotite (reaction 10). Both of these reactions are disequilibrium, irreversible ones. The rate of reaction (9) can be predicted from the rate of diffusion within garnet, but the rate of reaction (10) is controlled by interface reaction processes and intergranular diffusion, and is more difficult to predict.

The final step in our analysis of the relaxation of growth zoning, using the model system in Fig. 7, is to examine the reactions that could occur at about 545°C when chlorite becomes unstable. The predicted equilibrium reaction is garnet (B) + chlorite (F)

$$= \text{biotite } (M) + \text{kyanite} +/- \text{other phases.} \quad (11)$$

If garnet were homogeneous and of composition B, the reaction could proceed at equilibrium as heat were added to the system. But it is impossible for the equilibrium reaction to proceed if the garnet is zoned. There are two possible processes that could occur, in addition to the possibility that nothing happens. The first process occurs in the absence of diffusion in garnet. The edge composition of garnet, which varies between B and A as the reaction proceeds, reacts irreversibly with chlorite to produce some biotite composition and possibly separate crystals of garnet. Secondly, if diffusion occurs within garnet, the interior composition of garnet can react through diffusion with matrix chlorite to produce garnet of the equilibrium rim composition, which then also reacts with chlorite to produce biotite.

The second case can be described in terms of two component reactions as was done for the preceding example at 620°C. The first component reaction is the equilibrium reaction (11). The second is

garnet (A–B) + biotite (M) + kyanite → garnet (B)

$$+/- \text{other phases,} \quad (12)$$

a reaction equivalent to (9). Again, the overall reaction of garnet will depend on the relative rate of consumption by reaction (11) and the rate of production of garnet by reaction (12).

The analysis of garnet reactions in the simple system of Fig. 7 leads to several conclusions of petrological usefulness. The composition of zoned garnet at high-temperature is subject to the effects of a number of disequilibrium reaction processes. Relaxation of growth zoning and reaction with the matrix may cause garnet to continue to grow or to be consumed, depending on the relative rates of disequilibrium reactions. The edge composition of garnet is not necessarily indicative of equilibrium conditions. If diffusion within garnet occurs at a rate fast enough to allow the edge to adjust composition, and if reactions in the matrix are rapid enough to maintain heterogeneous equilibrium, the garnet edge composition does indicate the equilibrium state of the matrix. On the other hand, if garnet does not diffuse internally, the matrix assemblage and garnet rim composition may represent a complex disequilibrium assemblage inherited from lower-grade conditions. The mechanics of reactions involving garnet can be deduced from diffusion zoning in garnet and from the variation of matrix phase compositions as a function of distance from garnet and of extent of reaction. Where this analysis has been done in granulite-grade rocks (Loomis, 1976, 1977, 1979a), it was determined that (1) diffusion in garnet may be a process that limits the rate of reactions that consume garnet, (2) the rate of reactions within the matrix are relatively slow and controlled by interfacial processes, and (3) the absolute rate of reactions can be determined from diffusion zoning in garnet if the rate of diffusion is known. A final point should be emphasized. Only in the exceptional circumstance of very limited mobility in the matrix assemblage can simple exchange of components between crystals be assumed. Most of the realistic reactions that we have discussed that involve diffusion in garnet result in growth or consumption of garnet crystals. It should be clear that the accuracy of zoning models for garnet may benefit much more from a more realistic appraisal of the reaction boundary conditions then from the application of more complex models of diffusion within garnet.

Growth with Diffusion

The growth of garnet from chlorite and biotite at low-grade conditions can reasonably be described without contemporaneous diffusion in garnet. However, growth of garnet at overstepped conditions in high staurolite and higher grade rocks probably involves some diffusion. Overall growth occurs by a combination of the disequilibrium growth mechanisms and relaxation of zoning mechanisms that we discussed above. While growth involving both of these processes can be simulated in theory, it has not been accomplished for garnet in natural systems. I will illustrate the effect of diffusion by considering the end-member case in which internal diffusion in garnet maintains the homogeneity of crystals as they grow. Growth will be assumed to occur by reaction-controlled processes from a matrix that remains homogeneous as it is fractionated. The example will be called the "homogeneous growth" model and can be compared to the reaction-controlled growth model without diffusion above.

If a garnet crystal grows at overstepped conditions, partitioning between the crystal and the matrix must be controlled by a disequilibrium partitioning law. I will assume that the maximum free energy decrease model (partitioning model 1), which I have used to simulate reaction-controlled growth (Loomis, 1982), can be applied here also. Rearrangement of the equations found in that article allows us to solve for the composition of homogeneous garnet as a function of the amount of garnet in the system. The simulations of homogeneous growth shown in Fig. 13 (curves *HG*) indicate that the composition of garnet follows a history with *time* similar to that preserved with *distance* in the zoning of garnets grown without diffusion. In other words, the first, small amount of garnet to form is Mn-rich and Fe-poor like the core composition of zoned garnets, but the composition of homogeneous garnet becomes poorer in Mn and richer in Fe with growth until an equilibrium composition is reached. Concomitant with the changing composition of garnet, the Mn/Fe ratio in matrix phases will decrease as garnet grows.

An example of the significance of models of growth with diffusion is provided by Woodsworth's (1977) analysis of the Mn content of ilmenite inclusions within garnet in cordierite-zone rocks. He found that ilmenite inclusions progressively closer to the center of garnets had increasing Mn contents and interprets this observation as evidence that the garnets originally grew with Mn zoning profiles similar to those found in lower-grade rocks; he proposes that the profiles were subsequently relaxed by diffusion to form the present shallow forms. One observation that suggests that the history may have been more complicated is that

Woodsworth shows staurolite inclusions near the center of garnets, well within his inferred Mn "bell" formed at low-grade growth, while fractionation of Mn was nearly complete before staurolite became stable in his garnet-zone garnets. Thus, the cordierite-zone garnets must have overstepped the garnet isograd and grown at higher-grade conditions.

The homogeneous growth model is an alternative to explain the ilmenite data. If garnet grew at high-grade conditions, diffusion within the crystals could have kept pace with fractionation of the matrix. Then ilmenites trapped early in the growth history would be enriched in Mn relative to inclusions trapped later. The similarity of the predicted trends of Mn content with time suggests that it is probably not possible to distinguish between the two models from inclusion data, especially considering the fractionation effect of the inclusions themselves and the presence of residual zoning in garnet. Thus, garnet could have remained nearly homogeneous during growth and still produced the inclusions observed by Woodsworth.

Other possible models can be advanced to explain the formation of inclusion data and residual zoning in high-grade garnets, but the paucity of data does not warrant their description here. We can conclude that zoning in garnets grown above staurolite-grade probably involves simultaneous diffusion and fractionation of the matrix. The operation of diffusion during growth facilitates the creation of unzoned crystals because the amount of diffusion required to relax the core is reduced if it acts while the crystal is small rather than having to homogenize a large, zoned crystal. The common occurrence of unzoned garnets in high-grade rocks could be due in part to their growth at high-grade conditions as well as to diffusion relaxation of zoning produced during low-grade growth.

Summary and Conclusions

The use of mineral compositions and mineral assemblages to trace geological processes depends on our knowledge of the kinetics of crystallization. It is the composition of crystals that we measure; we infer magmatic composition, fluid composition, and pressure and temperature of formation from an assumed kinetic relationship between the crystal composition and the source material. This relationship is commonly assumed to be complete equilibrium, but it is surely wishful thinking to believe that rocks equilibrate completely and continuously through part of their history and then suddenly freeze up. More likely, the kinetic inhibitions to reequilibration that preserve primary igneous crystals and high-grade metamorphic assemblages also affect the crystallization and prograde metamorphism of these rocks. Compositionally zoned crystals and disequilibrium assemblages contain evidence of how these kinetic processes influence the compositional parameters that we rely upon to interpret the history of rocks. The disequilibrium nature of nucleation and crystal growth is most evident from the interior compositions and inclusions of zoned crystals that have been shielded from the late annealing reactions that affect the matrices of igneous and metamorphic rocks and give them the appearance of being equilibrium systems. Zoned crystals and inclusions can be used to document the $P-T$-fluid composition history of samples and enhance their usefulness in understanding processes of volcanism, plutonism, tectonics, and hydrothermal alteration.

The key to understanding the origin of zoned crystals and to deriving useful geological information from them is knowledge of how components are partitioned between phases. "Normal" zoning in plagioclase and garnet is simply predicted from the unequal partitioning of components coupled with consequent fractionation of the melt or matrix, regardless of whether the growth rate is controlled by intergranular diffusion or interfacial processes. The review of partitioning in this chapter suggests that partitioning under disequilibrium conditions can cause the composition of a crystal to deviate significantly from the equilibrium composition and points to the critical need for more well-designed experiments to empirically constrain models of partitioning. Disequilibrium partitioning severely affects the accuracy of geothermometers and geobarometers. Partitioning models have successfully explained experimentally induced zoning in plagioclase and probably can be used to infer the types of zoning that can be produced in natural crystals. These models suggest that the magnitude of "normal zoning" should be reduced in crystals that nucleate and grow under disequilibrium conditions and predict that reverse and irregular zoning are possible under some natural conditions. It is notable that disequilibrium partitioning may cause a crystal composition to change in a manner opposite to the change of the matrix or melt composition caused by fractionation because partitioning is a function of bulk composition (in addition to nonideal mixing) as well as temperature and pressure. The assumption of Rayleigh fractionation cannot be supported on theoretical grounds for major components.

Growth zoning of plagioclase is very resistant to alteration by diffusion and can be studied in plagioclase phenocrysts from many environments. Natural plagioclase zoning profiles seldom have the smooth character and constantly increasing slope predicted by equilibrium growth simulations. Changes of pressure or even water content do not appear to be capable of explaining the zoning irregularities if plagioclase grows near equilibrium conditions. However, reverse and irregular zoning are readily explained by disequilibrium growth models if plagioclase nucleates (or growth renucleates on resorbed crystals) under disequilibrium conditions. Normal zoning is predicted to occur in plutonic environments where rapid cooling continues after nucleation, where water is lost to the system, or where contemporaneous crystallization of other phases maintain undercooling. Reverse zoning should occur in slowly cooled melts in which residual components accumulate in the melt. These data suggest that plagioclase may show a trend toward reduced normal or increased reverse zoning from the edge of a pluton toward the interior and that granodiorites and diorites may have a greater tendency toward reverse zoning than granites because plagioclase crystallizes without interference from other phases. Sharp changes of plagioclase composition of up to 10% An and oscillatory zoning indicate rapid changes of temperature or melt composition during growth. The commonness of these types of zoning in phenocrysts, together with the occurrence of abraded and broken crystals, suggest that plagioclase phenocrysts in plutonic as well as volcanic environments grow from melts subject to periodic and probably turbulent convective motions. When plagioclase becomes unstable under metamorphic conditions, it seems to react incongruently to form separate crystals of another composition rather than forming zoned crystals, presumably due to the ineffectiveness of intracrystalline diffusion.

Zoning caused by growth without modification by intragranular diffusion can be studied in garnet from contact environments into staurolite grade (around 550 °C). Garnet in rocks from higher-grade conditions or that experienced extended times of metamorphism have Mn–Fe zoning that has been relaxed and garnets from the highest metamorphic grades may be homogenized by diffusion. The general form of growth zoning in garnet is consistent with the predictions of equilibrium growth. The central, major part of the zoning profile is dominated by the fractionation of Mn and by the sympathetic variation of Fe; it will be similar in form regardless of the matrix assemblage because the unequal partitioning of Mn between garnet and other common phases is extreme. The observed variations of partitioning and of garnet core composition with bulk composition and grade of metamorphism can be explained only by disequilibrium growth models; the disequilibrium growth models can also explain reverse and irregular zoning. The data on growth zoning and nucleation of garnet suggest that (1) the amount of overstepping that occurs before most garnets nucleate and grow increases with the heating rate and (2) heating rate rather than bulk composition seems to be the principal factor that controls when nucleation occurs. Nucleation, of course, occurs progressively over a range of oversteppings as the rock is heated. If the growth of garnet is delayed until cordierite-zone conditions in rapidly heated, contact rocks, intragranular diffusion may allow garnet crystals to remain nearly homogeneous during growth, and it can not be assumed that growth zoning as found in lower-grade garnets was formed and subsequently relaxed. The composition of matrix minerals during homogeneous growth will probably change as garnet grows in approximately the same way that they would if garnet grew in equilibrium with the matrix.

Observations that indicate that interface kinetic processes may limit the growth rate of crystals include the unequal dimensions and flat crystal faces of crystals, extrapolations of measured interface-controlled growth rates of plagioclase to natural systems, and interpretations of the growth rate of garnet as a function of size based on relative zoning. Other reaction processes that have been shown to control the rate of reactions in metamorphic rocks are dissolution of reactant crystals and intragranular diffusion in reactant garnet. The inability of intergranular diffusion models to predict reaction rates is most obvious from the coexistence of incompatible aluminum silicate polymorphs in contact in metamorphic rocks. Diffusion gradients in the matrix of rocks can still exist if these processes limit the reaction rate, but the calculated intergranular diffusion rates of components in the matrix will be too slow if it is assumed that the reaction rate is controlled solely by intergranular diffusion.

The model of nucleation and growth supported by this chapter for most rocks is consistent with the commonly used model of thermodynamic equilibrium. It assumes that nucleation of crystals in a heterogeneous assemblage occurs randomly in a volume small enough that most components can be transferred rapidly among crystals. Then partitioning equilibrium of most components in a heterogeneous assemblage can be assumed in many interpretations. In this model, the rate of growth of crystals is limited principally by interface kinetics or equilibrium. If the nucleation density is very sparse with respect to the rate of intergranular diffusion because conditions change very slowly or some other factor inhibits nucleation, then the growth rate of crystal segregations can be controlled principally by intergranular diffusion.

Garnet will probably experience periods of potential disequilibrium growth or resorption even if it grows near equilibrium conditions during slow, prograde metamorphism. During periods when garnet of the previous growth composition is unstable, it may (1) remain inert or react incongruently, as plagioclase does, if internal diffusion is not effective, or (2) react incongruently to form an induced diffusion profile if diffusion is effective above about 600°C. When diffusion is effective, garnet crystals may (1) grow low-Mn rims in response to reactions caused by diffusion relaxation of growth zoning, or (2) develop high-Mn, reverse-zoned rims in response to consumption caused by disequilibrium reactions with matrix minerals, depending on the relative rates of intragranular diffusion and disequilibrium reactions in the matrix. The edge composition of garnet appears to be a better indicator of equilibrium conditions in the matrix than other, homogeneous crystals when diffusion is effective because the edge composition of garnet can be changed by a small amount of reaction, whereas the composition of homogeneous matrix phases require a greater extent of reaction to significantly change composition.

Diffusion zoning in garnet can be used as an indicator of the temperature-time history of metamorphism and of reaction rate. Major uncertainties in analyzing diffusion profiles are (1) how the edge composition changed through time in response to disequilibrium reactions and (2) the rate of consumption or growth of garnet. Most reactions that can induce zoning in garnet involve a change of abundance of this phase that may be determinable from the stoichiometry of disequilibrium reaction. For most geological problems, the errors introduced by assuming that *D*'s of major zoned components are independent of composition and are simple functions of temperature, and by using mole fractions rather than concentrations in the diffusion equations, are much smaller than those introduced by uncertainties in the boundary conditions.

Crystal growth, diffusion, and disequilibrium reactions are subjects that are conceptually difficult to work with. I have changed my mind many times in the course of writing this summary and will not be surprised if the ideas expressed here need to be revised in the future. I believe that theory is severely limited in how far it can take us toward realistic models of growth and equilibration. There is a critical need at this point for carefully designed experiments and detailed studies of rocks with known igneous and metamorphic histories.

Acknowledgments

Many students and colleagues have contributed to the concepts presented in this chapter. I am sure that many of the ideas on disequilibrium that we are presently investigating have been expressed many times in the past literature of geology and have not been adequately acknowledged here. I am especially indebted to S. K. Saxena, J. Ganguly, L. S. Hollister, and R. Strauss for thoughtful reviews of material on which this work is based. NSF contributed financial support for portions of the work, for which I am grateful.

References

Ahrens, T. J., and Schubert, G. (1975) Gabbro-eclogite reaction rate and its geophysical significance, *Reviews Geophys. Space Phys.* **13**, 383–400.

Albarede, F., and Bottinga, Y. (1972) Kinetic disequilibrium in trace element partitioning between phenocrysts and host lava, *Geochim. Cosmochim. Acta* **35**, 141–156.

Anderson, D. E., and Buckley, G. R. (1974) Modeling of diffusion controlled properties of silicates, in *Geochemical Transport and Kinetics,* edited by A. W. Hofmann, B. J. Giletti, H. S. Yoder, Jr., and R. A. Yund, Carnegie Inst. Washington, Washington, D.C.

Atherton, M. P. (1968) The variation in garnet, biotite and chlorite composition in medium grade pelitic rocks from the Dalradian, Scotland, with particular reference to the zonation in garnet, *Contrib. Mineral. Petrol.* **18**, 347–371.

Bollingberg, H. J., and Bryhni, I. (1972) Minor element zonation in an eclogite garnet, *Contrib. Mineral. Petrol.* **36**, 113–122.

Bowen, N. L. (1913) The melting phenomena of the plagioclase feldspar, *Amer. J. Sci.* 4th series **35**, 577–599.

Carmichael, D. M. (1969) On the mechanism of prograde metamorphic reactions in quartz-bearing pelitic rocks, *Contrib. Mineral. Petrol.* **20**, 244–267.

Carpenter, M. A. (1980a) Mechanisms of exsolution in sodic pyroxenes, *Contrib. Mineral. Petrol.* **71**, 289–300.

Carpenter, M. A. (1980b) Composition and cation order variations in a sector-zoned blueschist pyroxene, *Amer. Mineral.* **65**, 313–320.

Carpenter, M. A. (1981) Omphacite microstructures as time-temperature indicators of blueschist- and eclogite-facies metamorphism, *Contrib. Mineral. Petrol.* **78**, 441–451.

Carslaw, H. S., and Jaeger, J. C. (1959) *Conduction of Heat in Solids,* Clarendon Press, Oxford.

Chinner, G. A. (1962) Almandine in thermal aureoles, *J. Petrol.* **3**, 316–340.

Crank, J. (1975) *The Mathematics of Diffusion,* Oxford, London.

Day, H. W., and Brown, V. M. (1980) Evolution of perthite composition and microstructure during progressive metamorphism of hypersolvus granite, Rhode Island, USA, *Contrib. Mineral. Petrol.* **72**, 353–366.

deBethune, P., Laudron, D., Martin, H., and Theunissin, K. (1968) Grenats zonés de la zone du Mont Rose (Valle Anzasca, Prov. de Novara, Italie), *Bull. Suisse Mineral. Petrog.* **48**, 437–454.

deGroot, S. R., and Mazur, P. (1969) *Non-equilibrium Thermodynamics,* North-Holland, Amsterdam.

Delaney, J. R., Muenow, D. W., and Graham, D. G. (1978) Abundance and distribution of water, carbon, and sulfur in the glassy rims of submarine pillow basalts, *Geochim. Cosmochim. Acta* **42**, 581–594.

Downes, M. J. (1974) Sector and oscillatory zoning in calcic augites from M. Etna, Sicily, *Contrib. Mineral. Petrol.* **47**, 187–196.

Dowty, E. (1976) Crystal structure and crystal growth. II. Sector zoning in minerals, *Amer. Mineral.* **66**, 460–469.

Dowty, E. (1977) The importance of adsorption in igneous partitioning of trace elements, *Geochim. Cosmochim. Acta* **41**, 1643–1646.

Elphick, S. C., Ganguly, J., and Loomis, T. P. (1981) Experimental study of Fe–Mg interdiffusion in aluminum silicate garnet, *Amer. Geophys. Union Trans.* **62**, 411 (abstr.).

Elphick, S. C., Ganguly, J., and Loomis, T. P. (1982) Experimental study of Fe–Mn interdiffusion in aluminsilicate garnet, *Geol. Soc. Amer.* Abs. with Pgm. **14**, 483 (abstr.).

Evans, B. W., and Guidotti, C. V. (1966) The sillimanite-potash feldspar isograd in western Maine, U.S.A., *Contrib. Mineral. Petrol.* **12**, 25–62.

Fenn, P. M. (1977) Nucleation and growth of alkali feldspar from hydrous melts, *Can. Mineral.* **15**, 135–161.

Ferry, J. M., and Spear, F. S. (1978) Experimental calibration of the paritioning of Fe and Mg between biotite and garnet, *Contrib. Mineral. Petrol.* **66**, 113–117.

Fisher, G. W. (1970) The application of ionic equilibria to metamorphic differentiation: an example, *Contrib. Mineral Petrol.* **29**, 91–103.

Fisher, G. W. (1978) Rate laws in metamorphism, *Geochim. Cosmochim. Acta* **42**, 1035–1050.

Fitts, D. D. (1962) *Nonequilibrium Thermodynamics,* McGraw-Hill, New York.

Galwey, A. K., and Jones, K. A. (1966) Crystal size frequency distribution of garnets in some analysed metamorphic rocks from Mallaig, Inverness, Scotland, *Geol. Mag.* **103**, 143–153.

Ganguly, J. (1977) Compositional variables and chemical equilibrium in metamorphism, in *Energetics of Geological Processes,* edited by S. K. Saxena and S. Bhattacharji, Springer-Verlag, Berlin/New York/Heidelberg.

Ganguly, J., and Kennedy, G. C. (1974) The energetics of natural garnet solid solution. I. Mixing of the aluminosilicate end-members, *Contrib. Mineral. Petrol.* **48**, 137–148.

Ghent, E. D. (1975) Temperature, pressure and mixed volatile equilibria attending metamorphism of staurolite-kyanite bearing assemblages, Esplanade Range, British Columbia, *Geol. Soc. Amer. Bull.* **86**, 1654–1660.

Ghent, E. D., Robbins, D. B., and Stout, M. Z. (1979) Geothermometry, geobarometry, and fluid compositions of metamorphosed calc-silicates and pelites, Mica Creek, British Columbia, *Amer. Mineral.* **64**, 874–885.

Grey, N. H. (1970) Crystal growth and nucleation in two large diabase dikes, *Can. J. Earth Sci.* **7**, 366–375.

Grove, T. L. (1982) Use of exsolution lamellae in lunar clinopyroxenes as cooling rate speedometers: an experimental calibration, *Amer. Mineral.* **67**, 251–268.

Gurney, J. J., Harris, J. W., and Rickard, R. S. (1979) Silicate and oxide inclusions in diamonds from the Finsch kimberlite pipe, in *Kimberlites, Diatremes, and Diamonds:*

Their Geology, Petrology, and Geochemistry, 1, edited by F. R. Boyd and H. O. A. Meyer, Amer. Geophys. Union, Washington, D.C.

Harkins, E., and Hollister, L. S. (1977) Sector zoning of clinopyroxene from a weakly metamorphosed diabase, *Amer. Mineral.* **62**, 390–394.

Harte, B., Gurney, J. J., and Harris, J. W. (1980) The formation of peridotite suite inclusions in diamonds. *Contrib. Mineral. Petrol.* **72**, 181–190.

Harte, B., and Henley, K. J. (1966) Occurrence of compositionally zoned almanditic garnets in regionally metamorphosed rocks, *Nature* **210**, 689–692.

Hollister, L. S. (1966) Garnet zoning: an interpretation based on the Rayleigh fractionation model, *Science* **154**, 1647–1651.

Hollister, L. S. (1969) Contact metamorphism in the Kwoiek area of British Columbia: an end member of the metamorphic process, *Geol. Soc. Amer. Bull.* **80**, 2456–2493.

Hollister, L. S. (1970) Origin, mechanism and consequences of composition sector-zoning in staurolite, *Amer. Mineral.* **55**, 742–766.

Hollister, L. S., and Gancarz, A. J. (1971) Compositional sector-zoning in clinopyroxene from the Narce area, Italy, *Amer. Mineral.* **56**, 959–979.

Hopper, R. W., and Uhlmann, D. R. (1974) Solute redistribution during crystallization at constant velocity and constant temperature, *J. Crystal Growth* **21**, 203–213.

Ildefonse, J-P., and Gabis, V. (1976) Experimental study of silica diffusion during metasomatic reactions in the presence of water at 550°C and 1000 bars, *Geochim. Cosmochim. Acta* **40**, 297–303.

Johannes, W. (1978) Melting of plagioclase in the system Ab–An–H_2O and Qz–Ab–An–H_2O at P_{H_2O} = 5 kbars, an equilibrium problem, *Contrib. Mineral. Petrol.* **66**, 295–303.

Jones, K. A., and Galwey, A. K. (1964) A study of possible factors concerning garnet formation in rocks from Ardara, Co. Donegal, Ireland, *Geol. Mag.* **101**, 76–93.

Katchalsky, A., and Curran, P. F. (1967) *Nonequilibrium Thermodynamics in Biophysics,* Harvard, Cambridge.

Kirkpatrick, R. J. (1975) Crystal growth from the melt: A review, *Amer. Mineral.* **60**, 798–814.

Kirkpatrick, R. J. (1976) Towards a kinetic model for magma crystallization, *J. Geophys. Res.* **81**, 2565–2571.

Kirkpatrick, R. J. (1977) Nucleation and growth of plagioclase, Makaopuhi and Alae lava lakes, Kilauea volcano, Hawaii, *Geol. Soc. Amer. Bull.* **88**, 78–84.

Kirkpatrick, R. J., Klein, L., Uhlmann, D. R., and Hays, J. F. (1979) Rates and processes of crystal growth in the system anorthite-albite, *J. Geophys. Res.* **84**, 3671–3676.

Korzhinskii, D. S. (1959) *Physio-chemical Basis of the Analysis of the Paragenesis of Minerals,* Consultants Bureau, Inc., New York.

Kretz, R. (1966) Grain-size distribution for certain metamorphic minerals in relation to nucleation and growth, *J. Geol.* **74**, 147–173.

Kretz, R. (1973) Kinetics of the crystallization of garnet at two localities near Yellowknife, *Can. Mineral.* **12**, 1–20.

Kuo, L.-C., and Kirkpatrick, R. J. (1982) Pre-eruption history of phyric basalts from DSDP legs 45 and 46: evidence from morphology and zoning patterns in plagioclase, *Contrib. Mineral. Petrol.* **79**, 13–27.

Larsen, L. M. (1981) Sector zoned aegirine from the Ilimaussaq alkaline intrusion, south Greenland, *Contrib. Mineral. Petrol.* **76**, 285–291.

Lasaga, A. C. (1979) Multicomponent exchange and diffusion in silicates, *Geochim. Cosmochim. Acta* **43**, 455–469.

Lasaga, A. C., Richardson, S. M., and Holland, H. D. (1977) The mathematics of cation diffusion and exchange between silicate minerals during retrograde metamorphism, in *Energetics of Geologic Processes,* edited by S. K. Saxena and S. Bhattacharji, Springer-Verlag, New York.

Lofgren, G. (1972) Temperature induced zoning in synthetic plagioclase feldspar, in *The Feldspars,* edited by W. S. MacKenzie, University of Manchester Press, Manchester.

Loomis, T. P. (1972a) Contact metamorphism of pelitic rock by the Ronda ultramafic intrusion, southern Spain, *Geol. Soc. Amer. Bull.* **83**, 2449–2474.

Loomis, T. P. (1972b) Diapiric emplacement of the Ronda high-temperature ultramafic intrusion, southern Spain, *Geol. Soc. Amer. Bull.* **83**, 2475–2496.

Loomis, T. P. (1972c) Coexisting aluminum silicate phases in contact metamorphic aureoles, *Amer. J. Sci.* **272**, 933–945.

Loomis, T. P. (1974) Tertiary mantle diapirism, orogeny, and plate tectonics east of the Strait of Gibraltar, *Amer. J. Sci.* **275**, 1–30.

Loomis, T. P. (1975) Reaction zoning of garnet, *Contrib. Mineral. Petrol.* **52**, 285–305.

Loomis, T. P. (1976) Irreversible reactions in high-grade metamorphic rocks, *J. Petrol.* **17**, 559–588.

Loomis, T. P. (1977) Kinetics of a garnet granulite reaction, *Contrib. Mineral. Petrol.* **62**, 1–22.

Loomis, T. P. (1978a) Multicomponent diffusion in garnet: I. Formulation of isothermal models, *Amer. J. Sci* **278,** 1099–1118.

Loomis, T. P. (1978b) Multicomponent diffusion in garnet: II. Comparison of models with natural data, *Amer. J. Sci.* **278,** 1119–1137.

Loomis, T. P. (1979a) A natural example of metastable reactions involving garnet and sillimanite, *J. Petrol.* **20**, 271–292.

Loomis, T. P. (1979b) An empirical model for plagioclase equilibrium in hydrous melts, *Geochim. Cosmochim. Acta* **43**, 1753–1759.

Loomis, T. P. (1981) An investigation of disequilibrium growth processes of plagioclase in the system anorthite–albite–water by methods of numerical simulation, *Contrib. Mineral. Petrol.* **76**, 196–205.

Loomis, T. P. (1982) Numerical simulation of disequilibrium growth processes of garnet in chlorite-bearing, aluminous pelitic rocks, *Canad. Mineral.* **20**, 411–423.

Loomis, T. P. (1983a) Numerical simulations of crystallization processes of plagioclase in complex melts: The origin of major and oscillatory zoning in plagioclase, *Contrib. Mineral. Petrol.*, **81,** 219–229.

Loomis, T. P. (1983b) Metamorphic Petrology, in *1979–1982 U.S. National Report to IUGG,* edited by D. E. James, Amer. Geophys. Union, Washington D.C.

Loomis, T. P., and Nimick, F. B. (1982) Equilibrium in Mn–Fe–Mg aluminous pelitic compositions and the equilibrium growth of garnet, *Canad. Mineral.* **20**, 393–410.

Loomis, T. P., and Welber, P. W. (1983) Crystallization processes in the Rocky Hill granodiorite pluton, California: An interpretation based on compositional zoning of plagioclase, *Contrib. Mineral. Petrol.,* **81,** 230–239.

Maaloe, S. (1976) The zoned plagioclase of the Skaergaard intrusion, east Greenland, *J. Petrol.* **17**, 398–419.

McCallister, R. H., and Nord, G. L., Jr. (1981) Subcalcic diopsides from kimberlites: Chemistry, exsolution microstructures, and thermal history, *Contrib. Mineral. Petrol.* **78**, 118–125.

McDowell, S. D. (1978) Little Chief granite porphyry: Feldspar crystallization history, *Geol. Soc. Amer. Bull.* **89**, 33–49.

Mueller, R. F. (1967) Mobility of elements during metamorphism, *J. Geol.* **75,** 565–582.

Nakumura, Y. (1973) Origin of sector-zoning of igneous clinopyroxenes, *Amer. Mineral.* **58**, 986–990.

Prigogine, I.. and Defay, R. (1954) *Chemical Thermodynamics,* Longman, London.

Pringle, G. J., Trembath, L. T., and Pajari, G. E., Jr. (1974) Crystallization history of a zoned plagioclase, *Mineral Mag.* **39**, 867–877.

Rosenfeld, J. L. (1969) Stress effects around quartz inclusions in almandine and the piezothermometry of coexisting aluminum silicates, *Amer. J. Sci.* **267**, 317–351.

Rutter, J. W., and Chalmers, B. (1953) A prismatic substructure formed during the solidification of metals, *Can. J. Phys.* **31**, 15–39.

Swanson, S. E. (1977) Relation of nucleation and crystal-growth rate to the development of granitic textures, *Amer. Mineral.* **62**, 966–978.

Thompson, A. B., Tracy, R. J., Lyttle, P. T., and Thompson, J. B., Jr. (1977) Prograde reaction histories deduced from compositional zonation and mineral inclusions in garnet from the Gassetts schist, Vermont, *Amer. J. Sci.* **277**, 1152–1167.

Thompson, J. B. Jr. (1959) Local equilibrium in metasomatic processes, in *Researches in Geochemistry,* vol. 1, edited by P. H. Abelson, John Wiley & Sons, New York.

Tracy, R. J., Robinson, P., and Thompson, A. B. (1976) Garnet composition and zoning in the determination of temperature and pressure of metamorphism, central Massachusetts, *Amer. Mineral.* **61**, 762–775.

Trzcienski, W. E., Jr. (1977) Garnet zoning- product of a continuous reaction, *Canad. Mineral.* **15**, 250–256.

Vance, J. A. (1962) Zoning in igneous plagioclase: Normal and oscillatory zoning, *Amer. J. Sci.* **260**, 746–760.

Vance, J. A. (1965) Zoning in igneous plagioclase: Patchy zoning, *J. Geol.* **73**, 636–651.

Walker, D., Kirkpatrick, R. J., Longi, J., and Hays, J. F. (1976) Crystallization history of lunar picritic basalt 12002: Phase-equilibrium and cooling-rate studies, *Geol. Soc. Amer. Bull.* **87**, 646–656.

Watson, E. B. (1976) Glass inclusions as samples of early magmatic liquid: determinative method and application to a South Atlantic basalt, *J. Volcanol. Geotherm. Res.* **1**, 73–84.

Wiebe, R. A. (1968) Plagioclase stratigraphy: A record of magmatic conditions and events in a granite stock, *Amer. J. Sci.* **266**, 670–703.

Woodsworth, G. J. (1977) Homogenization of zoned garnets from pelitic schists, *Can. Mineral.* **15**, 230–242.

Yund, R. A., and Ackermand, D. (1979) Development of perthite microstructures in the Storm King granite, N.Y., *Contrib. Mineral. Petrol.* **70**, 273–280.

Yund, R. A., and Tullis, J. (1980) The effect of water, pressure, and strain on Al/Si order–disorder kinetics in feldspar, *Contrib. Mineral. Petrol.* **72**, 297–302.

Chapter 2
Exsolution and Fe^{2+}–Mg Order–Disorder in Pyroxenes

S. K. Saxena

Introduction

In recent years there has been significant development in our understanding of exsolution phenomena in minerals through (a) the use of petrography (Robinson *et al.*, 1977) and transmission electron microscopy (Champness and Lorimer, 1976; see other articles in Wenk, 1976), and (b) the use of kinetic experiments and theory (Yund and McCallister, 1970; McCallister and Yund, 1977; Lasaga and Kirkpatrick, 1981; Putnis and McConnell, 1980; McConnell, 1975). The references quoted above are only examples of the extensive literature available on exsolution phenomena. Buseck *et al.* (1980) have reviewed the results on pyroxenes. In all these studies there has been little mention of the Mg–Fe^{2+} order–disorder in pyroxenes that must continue simultaneously with the intercrystalline processes of ion exchanges leading to exsolution (cf. Kretz, 1982b). As discussed extensively by Putnis and McConnell (1980), exsolution, like all other reactions, is a time-dependent transformation and is best displayed on a time–temperature–transformation (TTT) diagram. These may be considered as kinetic phase diagrams. Mg–Fe^{2+} order–disorder in silicates is also a time-dependent process and the results are best displayed on TTT diagrams (Seifert and Virgo, 1975; Ganguly, 1982). Is it possible then to plot both kinds of results, i.e., percent transformation of an intercrystalline ion-exchange reaction between two crystals, e.g., the host augite and pigeonite lamellae and the Mg–Fe^{2+} order–disorder in the exsolving or other coexisting pyroxenes?

As has been demonstrated by Seifert and Virgo (1975) and Ganguly (1982), who used Mueller's (1969) kinetic model, the estimation of the cooling rates of the host rock containing orthopyroxene and/or amphiboles can be effectively constrained if kinetic rates of order–disorder and the site occupancy (e.g., in M1 and M2 in pyroxenes) of Fe–Mg are known for the minerals. As discussed by Ganguly (1982), the information on the cooling history of the rock during the time when the crystal maintains equilibrium with falling temperature cannot be obtained from the order–disorder data. Cooling rates determined using this method are truly applicable over a small part of the history of the rock—generally close to the temperature at which a determined site occupancy was quenched. However, if this cooling rate could be considered as applicable throughout, we may be able to use this information in understanding processes of exsolution occurring in coexisting minerals in the same rock. On the other hand, if independent transformation rates are available for exsolving crystals, we may combine the two types of informations on the same TTT diagram and achieve a greater understanding of the petrogenetic history of the rock.

This chapter presents the possible relationships between exsolution and Fe–Mg order–disorder in pyroxenes. For both these phenomena, experimental results are meager. The final results in this chapter, therefore, are based on assumed data found by comparison with natural observations. Most numbers and quantities are for the purpose of illustrating possible relationships and should not be taken as definitive. This chapter, in essence, provides a model for experimental studies on pyroxenes involving both exsolution and order–disorder.

Abbreviations and Symbols

Opx	orthopyroxene
Cpx	clinopyroxene
Pig	pigeonite
X_{Fe}	atomic fraction of Fe^{2+} (Fe/(Fe + Mg)) in a crystal or on a crystallographic site
X_{Mg}	atomic fraction of Mg (Mg/(Mg + Fe)) in a crystal or on a crystallographic site
(Fe), (Mg), (Ca)	number of ions in a pyroxene totalling 2
η	cooling rate constant
K_D	distribution coefficient for either intracrystalline or intercrystalline reactions
G^{Ex}	excess Gibbs free energy of mixing
W_{ij}	simple mixture parameter for mixing of i and j
T	absolute temperature
R	gas constant
G	total Gibbs free energy of a solution (molar)

From Homogeneous to Heterogeneous Equilibria

As has been discussed amply by Yund and McCallister (1970), Champness and Lorimer (1976), and Buseck *et al.* (1980), exsolution and formation of one pyroxene phase in another may be initiated either by nucleation or by spinodal decomposition. Nucleation may be heterogeneous, taking place on dislocations, preexisting grain boundaries, etc., or it may be homogeneous. Incipient lamellae may also form by spinodal decomposition. Once formed, growth and coarsening of the lamellae may be considered through the transfer reaction (Kretz, 1982a)

$$CaMgSi_2O_6 \rightleftharpoons CaMgSi_2O_6 \tag{a}$$

and the ion-exchange reaction

$$\text{Fe–Cpx} + \text{Mg–Opx/Pig} \rightleftharpoons \text{Mg–Cpx} + \text{Fe–Opx/Pig}. \qquad \text{(b)}$$

A third ion-exchange reaction,

$$\text{Fe–M2} + \text{Mg–M1} \rightleftharpoons \text{Fe–M1} + \text{Mg–M2}, \qquad \text{(c)}$$

continues during exsolution—beginning when the crystal is homogeneous at high temperature and continuing in both phases when the nucleation of the new pyroxene phase takes place and/or the spinodal decomposition is initiated. Reaction (a), requiring the diffusion of Ca through the host crystal, may be the slowest kinetically. We note that Ca diffusion must be involved in either the formation of the nucleus or in the compositional fluctuation leading to spinodal decomposition. Similarly, reaction (b) must be operative both at the initial stage and during the growth and coarsening of the lamellae.

Figure 1 shows a schematic relationship between the cooling rate of the rock,

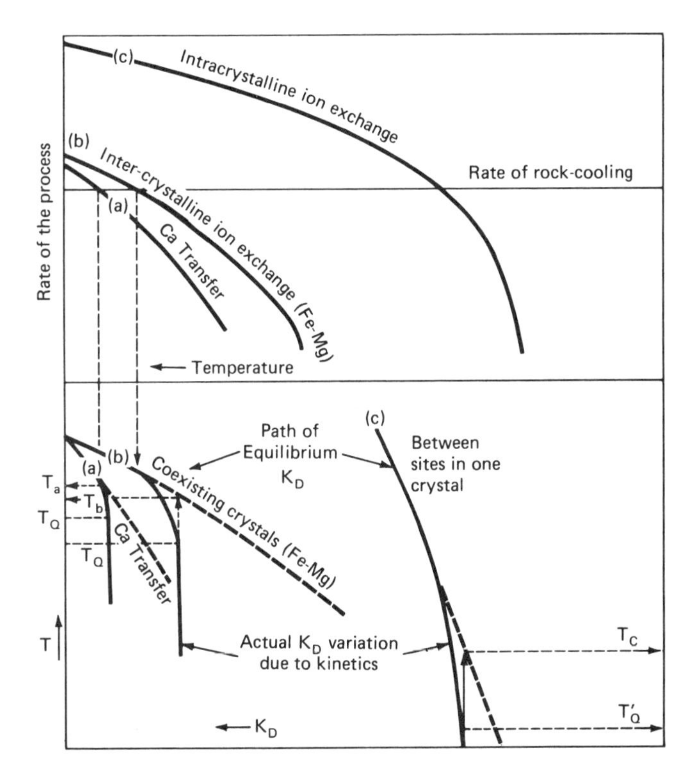

Fig. 1. Relationship between intercrystalline and intracrystalline ion exchange reactions. The top part of the figure shows the relative rates of (a) Ca transfer, (b) Fe–Mg exchange between crystals, (c) Fe–Mg exchange between M1 and M2 sites and of the cooling of the rock. The bottom part of the figure shows the closure reaction temperatures as $T_C < T_b < T_a$. T_Q, T'_Q are quenching temperatures. See text for further explanation.

which is assumed constant with temperature, and the three rate processes (a), (b), and (c). Note that although the cooling rate of the rock is assumed constant, the temperature of the rock itself is a variable and the reaction rates are functions of this temperature change. We note that as the rate of the process approaches the cooling rate of the rock, the transfer or ion-exchange starts falling behind its equilibrium value and finally quenches at a certain temperature T_Q when the rate of the process is much lower than the rate of cooling. T_a and T_b represent the measured temperatures usually obtained from compositions of coexisting pyroxenes (Kretz, 1982a)—T_a representing the transfer reaction equilibrium and T_b representing the Fe–Mg exchange equilibrium. T_c is the temperature of apparent equilibrium for the ion-exchange between M1 and M2 sites in one of the pyroxenes. The rates of the three reactions are functions of the composition of the original crystal. This is also true about the processes (nucleation/spinodal) which bring about the existence of the embryonic lamellae.

Exsolution in Enstatite–Diopside Solid Solution

To understand further the relationship among the three reactions, we begin by first considering the binary system enstatite–diopside in which only the transfer reaction is operative. In this compositional series, the structural differences between enstatite and diopside require that the solutions be considered separately as ortho- and clinopyroxenes. Orthoenstatite with a fictive endmember orthodiopside forms the orthopyroxene solid solution while diopside forms the clinopyroxene solid solution with clinoenstatite. The excess Gibbs free energies of the two solutions are given by

$$\begin{aligned} G^{\mathrm{Ex}}\ (\mathrm{Opx}) &= W^{\mathrm{Opx}} X_{\mathrm{En}}^{\mathrm{Opx}} X_{\mathrm{Di}}^{\mathrm{Opx}}, \\ G^{\mathrm{Ex}}\ (\mathrm{Cpx}) &= W_{12} X_{\mathrm{En}}^{\mathrm{Cpx}} (X_{\mathrm{Di}}^{\mathrm{Cpx}})^2 + W_{21} (X_{\mathrm{En}}^{\mathrm{Cpx}})^2 X_{\mathrm{Di}}^{\mathrm{Cpx}}. \end{aligned} \tag{1}$$

From Lindsley *et al.* (1981) for the W's, we have

$$\begin{aligned} W^{\mathrm{Opx}} &= 25 \mathrm{kJ \cdot mol^{-1}} \\ W_{12} &= 31.216 - 0.0061\mathrm{P}\ \mathrm{kJ \cdot mol^{-1}} \quad (\mathrm{P\ in\ kbar}), \\ W_{21} &= 25.484 + 0.0812\mathrm{P}\ \mathrm{kJ \cdot mol^{-1}} \quad (\mathrm{P\ in\ Kbar}). \end{aligned} \tag{2}$$

From these thermochemical data, it is possible to calculate the spinodal in the Cpx solution. The spinodal region is characterized by a negative value of the second derivative of the total Gibbs free energy of the solution (see a lucid discussion by Grover, 1977). The spinode for the Cpx solution is given by

$$\left(\frac{\partial^2 G_T}{\partial X_{\mathrm{Di}}^2}\right)_{P,T} = \frac{RT}{X_{\mathrm{En}} X_{\mathrm{Di}}} - 2[W_{12}(3X_{\mathrm{Di}} - 1) + W_{21}(3X_{\mathrm{En}} - 1)]. \tag{3}$$

Figure 2 shows the calculated spinodal in the Cpx solution. The Cpx solvus at 1 atmosphere is adopted from Lindsley *et al.* (1981). The significance of the spinodal in Cpx has been extensively discussed by Busek *et al.* (1980). Inside this

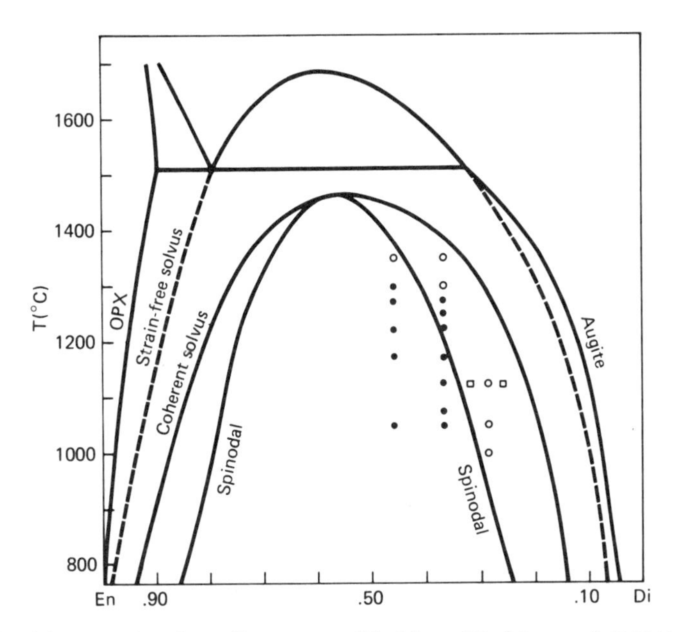

Fig. 2. Diopside–enstatite phase diagram, modified from Lindsley *et al.* (1981), showing the calculated spinodal from this work. For comparision, the data from McCallister and Yund (1977) has been plotted, the filled circles showing unmixing after heating experiments. The open circles represent no unmixing and the squares represent unmixing after prolonged heating.

compositional-temperature field coherent spinodal decomposition is the main process of exsolution.

Experimental data of McCallister and Yund (1977) on heated pyroxenes are shown in Fig. 2. These authors heated the pyroxenes for various lengths of time and studied the results by transmission electron microscopy. All the filled circles show the pyroxenes that exsolved by heating between 19 and 186 hours. The calculated spinodal is in fair agreement with the experiments. In the spinodal calculation, no account has been taken of the pigeonite to protoenstatite to rhombic enstatite transformations and some discrepancy between experiments and the calculated spinodal may be attributed to this.

In the region between the coherent spinodal and the calculated solvus (the strain-free solvus), exsolution is initiated and developed through nucleation and growth. In the region enclosed by the coherent solvus, exsolution could occur by homogeneous nucleation while outside but within the solvus it may occur by heterogeneous nucleation. The coherent solvus is drawn schematically in Fig. 2.

The progress of any kinetic process may be studied through TTT diagrams. These diagrams are plots of the isolines for percentage transformation of reactants to products (or for site-occupancies when studying order–disorder) on time–temperature coordinates. The kinetics of the transfer reaction (including formation of embryonic lamellae through nucleation or spinodal process) is a function of the original composition of the crystal. It is instructive to consider the relationship between the cooling of a rock and possible kinetic phase relations in crystals of various compositions schematically. In the binary system, we consider the cooling of a crystal with 54.0% diopside. In McCallister and Yund's (1977) experiments exsolution was noted first at 1320°C. In all runs between 1250 and 1320°C, exsolution was noted when the durations were between 19 and 24 hours. For the purpose of illustration, let us assume that a rock with such a crystal cools at a rate such that the temperature varies between 1310 and 1290°C during 20 hours. Cooling rates of rocks could be modelled with either the asymptotic equation $1/T = 1/T + \eta\Delta t$ or the exponential equation $T = T\exp(-\eta t)$. Using the asymptotic model, the value of η that will give us the required duration of heating at 1300°C is 6.736E-9 per minute. Figure 3 shows a possible kinetic relationship for this crystal. The figure, which is partly based on the McCallister–Yund data, shows that exsolution, in this case perhaps by the spinodal mechanism, begins at 1200°C in about 16–17 hours. With both increasing and decreasing temperatures, the time required for exsolution to begin increases, as shown by curve (a) in Fig. 3. The "finish" curve (b) schematically drawn shows the time required for achieving equilibrium in the transfer reaction (a).

In cooling from about 1600 to 1300°C, our model crystal passes through regions where nucleation and growth is possible (see Fig. 2). According to the cooling rate chosen, the crystal cools from 1600 to 1300°C in approximately 70 days (Fig. 3). If either heterogeneous nucleation (beginning at 1540°C) or homogeneous nucleation (beginning at 1440°C) is possible in such a period, the crystal will have a more complex history than depicted above. If there was no interruption by any other process, the model crystal during cooling from 1300°C will develop lamellae recognizable through electron microscopy but perhaps not through use of the petrologic microscope.

Let us consider the possibility that the cooling rate for the model crystal changes after reaching 1300°C. This may happen in nature if, for example, the system loses considerable heat in the beginning, reaches the upper crust, and cools slowly, perhaps because it is still over a hot spot. If the new cooling curve (c′ in Fig. 3) either touches or intersects the "finish" curve (b), the crystal reaches equilibrium in the transfer reaction (a) and the two grown phases (host + lamellae) have compositions lying on the solvus. In the model crystal this happens at 1210°C with the two compositions $Di_{0.05}\,En_{0.95}$ and $Di_{0.88}\,En_{0.12}$ (from Fig. 2). The two crystals should now cool through a region of exsolution by heterogeneous nucleation. This history, however, cannot be followed through the kinetic phase relations in Fig. 3, since they are applicable only to the model crystal with 54 diopside. If heterogeneous nucleation and growth do take place in the new crystals, we will end up with new exsolution textures within each lamella.

The kinetic phase relations change in an important way with crystal compositions. As may be seen in Fig. 2, a crystal with critical composition of mixing passes directly from the region of heterogeneous nucleation to the region of spinodal decomposition. The rates of diffusion also change with various degrees of saturation. Figure 4 shows possible kinetic relations in other model crystal with 68% Di matching another crystal heated by McCallister and Yund (1977). In their exper-

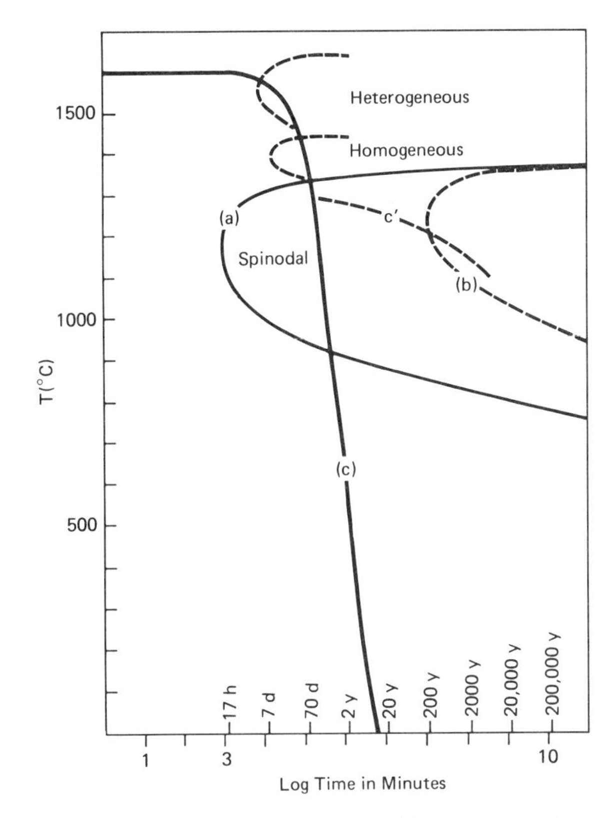

Fig. 3. Schematic TTT diagram for the Cpx with 54% Di. Curve (c) represents cooling curve for the rock while curves (a) and (b) represent beginning of the spinodal decomposition and attainment of equilibrium in Ca transfer reaction respectively. Possible beginnings of heterogeneous and homogeneous nucleation are also shown. See text for discussion.

iment (T = 1125°C, 696.5 hours), the crystal developed regular widely spaced lamellae. In Fig. 4, the cooling curve is drawn with an η value of 2.45E-10 per minute such that the crystal is held between 1135 and 1115°C for 700 hours during cooling. The crystal cools through the region of heterogeneous nucleation beginning at 1490°C in about 3 years, through the region of homogeneous nucleation beginning at 1325°C in about 17 years and finally through the region of spinodal decomposition beginning at 1050°C in about 10 years. Strictly, if the cooling crystal is affected by any one mechanism, the succeeding cooling history must be shown on differing diagrams for the two newly generated compositions which may be in equilibrium or disequilibrium with respect to the transfer reaction (a). If our second model crystal does not develop significant heterogeneous nucleation and enters the field of homogeneous nucleation, there will be two sets of lamellae, most likely in disequilibrium or metastable equilibrium, each of which may further decompose by other mechanisms depending on the composition (see Fig. 2). The Ca-rich lamellae will be in the region of either homogeneous or het-

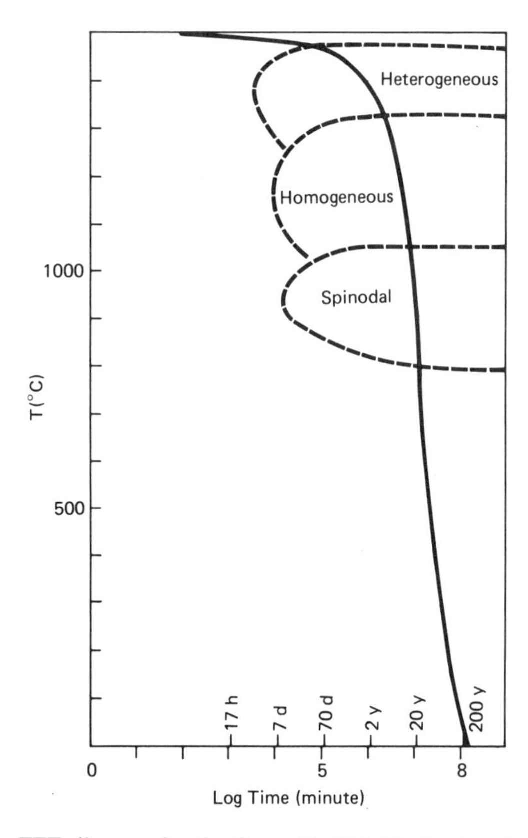

Fig. 4. Schematic TTT diagram for the Cpx with 64% Di. See text for discussion.

erogeneous nucleation while the Ca-poor lamellae will most probably be in the region of spinodal decomposition. If the cooling rate is slow enough for the lamellae to equilibrate in the region of homogeneous nucleation, exsolution of individual lamellae would be possible only by heterogeneous nucleation.

Exsolution in Ternary Mg–Fe–Ca Pyroxenes

For ternary pyroxenes, we must consider the results of an exsolution process through the kinetic rates and attainment of equilibrium in the three reactions (a), (b), and (c). There are data on the equilibrium relations, but except for (c) in orthopyroxene, there is little information on the kinetics of the reactions (cf. Grove, 1982). Nord and McCallister (1979) found that in a heated Cpx of $Wo_{0.25}$ $En_{0.31}$ $Fs_{0.44}$ spinodal decomposition was detectable in 10 hours of heating at 1000 °C and in 2000 hours at 800 °C. The effect of Fe on the rate of the transfer reaction (if other kinetic conditions are similar) may be judged (for spinodal mechanism) by comparing this time with the time required for the binary system. This could be roughly estimated as 7 days at 1000 °C from Fig. 3.

The presence of Fe^{2+} in the pyroxenes introduces Fe^{2+}–Mg order–disorder in each of the two coexisting pyroxene phases formed by exsolution. The relationship between the process of exsolution and Fe^{2+}–Mg order–disorder has received little consideration in the literature. When a subcalcic augite cools, we have ordering

$$\mathrm{FeM1} + \mathrm{MgM2} \rightleftharpoons \mathrm{FeM2} + \mathrm{MgM1}. \qquad (c')$$

When an exsolution process is initiated, we form coexisting pyroxenes with the Ca-poor pyroxene being richer in Fe than the Ca-rich pyroxene. Kretz (1982a) demonstrated that in coexisting pyroxenes (Mg:Fe)M1 in pigeonite $\simeq$ (Mg:Fe)M1 in augite, thus confirming Mueller's (1962) postulate. Therefore (Fe:Mg)M2 in pigeonite $>$ (Fe:Mg)M2 in augite. To extract Fe from Ca-rich regions of the crystal the ordering process (c′) must be effective and the reaction (c′) may be viewed as promoting exsolution. Thus we may consider

$$\text{Ca (pigeonite region)} + \text{Fe,Mg (augite region)} \rightleftharpoons \text{Ca (augite region)} + \text{Fe,Mg (pigeonite region)}.$$

If the M1 sites remain unchanged as the reaction progresses (see Kretz, 1982b), the Fe–Mg exchange equilibrium can be achieved by

$$\text{Mg–M2 (pigeonite)} + \text{Fe–M2 (augite region)} \rightleftharpoons \text{Fe–M2 (pigeonite region)} + \text{Mg–M2 (augite region)}.$$

The intracrystal equilibria would be satisfied in both phases. The two processes, could, of course, occur simultaneously, so it would not be necessary for the Mg atom to move from augite to pigeonite and then back to augite.

To understand the kinetic relationships among the cooling of the crystal and the three reactions (a), (b), and (c), let us first briefly review the equilibrium relations in coexisting pyroxenes and the intracrystalline equilibrium in orthopyroxene. Due to lack of data, no distinction will be made between orthopyroxene and pigeonite and both will be considered as Ca-poor pyroxenes.

Equilibrium Relations in Coexisting Pyroxenes

Although there is extensive literature on coexisting pyroxenes in natural assemblages (e.g., Davidson, 1968), experimental studies are relatively few. The latter have been recently reviewed by Kretz (1982a) and Lindsley (1982). Since the purpose of this chapter is only to illustrate possible equilibrium and kinetic relationships in pyroxenes, Kretz's (1982a) results, because of their simplicity and easy algorithm, will be used here. According to Kretz (1982a) the distribution coefficient for Fe^{2+}–Mg fractionation between coexisting Ca-rich and Ca-poor pyroxenes (orthopyroxene and pigeonite not distinguished) is given by (reaction (b))

$$\ln K_D = (1130/T) - 0.505, \qquad (4)$$

where

$$K_D = (X_{\mathrm{Fe}}/X_{\mathrm{Mg}})^{\mathrm{Opx}}(X_{\mathrm{Mg}}/X_{\mathrm{Fe}})^{\mathrm{Cpx}}.$$

This equation is applicable to a composition $X_{Fe}^{Opx} \approx 0.3$. For the transfer reaction (a), Kretz (1982a) finds for $T < 1080\,^{\circ}C$:

$$T\,(^{\circ}K) = 1000/(0.054 + 0.608\,X_{Fe}^{Cpx} - 0.304\ln(1 - 2X_{Ca})), \tag{5}$$

where X_{Fe} is Fe/(Fe + Mg) and X_{Ca} is Ca/(Ca + Fe + Mg). The temperature of the transfer reaction equilibrium may generally be different from the equilibrium temperature of the Fe–Mg exchange equilibrium.

Intracrystalline Equilibrium Relations in Pyroxenes

Equilibrium distribution of Mg–Fe in orthopyroxene on the two sites is considered on the basis of the ion-exchange reaction:

$$\text{Mg(M1)} + \text{Fe(M2)} \rightleftharpoons \text{Fe(M1)} + \text{Mg(M2)}. \tag{c}$$

Reaction (c), while written in analogy with heterogeneous ion-exchange reaction involving two separate phases, represents a homogeneous reaction where the site M1 may not be independent of the site M2. The cooperative effect may be taken care of by the following formulation of the activity coefficient of an endmember (Thompson, 1969; Sack, 1980); e.g., for ferrosilite, we have

$$\gamma_{FS} = \gamma_{Fe}^{M1}\gamma_{Fe}^{M2}\,[\exp\,(X_{Mg}^{M1})(X_{Mg}^{M2})\,\Delta G_d^{\circ}/RT], \tag{6}$$

where ΔG_d° represents the standard free energy change for the reciprocal reaction

$$\begin{aligned}\text{Mg(M1)Mg(M2)} + \text{Fe(M1)Fe(M2)} \rightleftharpoons\; &\text{Mg(M1)Fe(M2)}\\ &+ \text{Mg(M2)Fe(M1)}.\end{aligned} \tag{d}$$

Saxena and Ghose (1971) and Saxena (1973) used a nonideal expression describing the solution on each site but ignored ΔG_d°. Following Thompson (1969) and Sack (1980), for equilibrium in (c), we may write:

$$\begin{aligned}RT\ln K_D = &-\Delta G_c^{\circ} + W^{M2}(1 - 2X_{Fe}^{M2})\\ &- W^{M1}(1 - 2X_{Fe}^{M1}) + \Delta G_d^{\circ}(X_{Fe}^{M2} - X_{Fe}^{M1}),\end{aligned} \tag{7}$$

where K_D is $(X_{Fe}^{M1}X_{Mg}^{M2}/X_{Mg}^{M1}X_{Fe}^{M2})$.

The constants ΔG_c°, W^{M1}, W^{M2}, and ΔG_d° may all be determined from isothermal site occupancy data of Saxena and Ghose (1971) and Virgo and Hafner (1969). Unfortunately, multiple regression analysis (including stepwise procedure) generally results in large residuals if ΔG_d° is forced as a parameter. The best results seem to be those adopted by Ganguly (1982):

$$\ln K_c = 0.1435 - 1562/T, \tag{8}$$

$W^{M1} = 1524$; $W^{M2} = 1080$ cal/Mol. Since these results will be used to calculate equilibrium temperatures and not activities of the endmembers, the fact that ΔG_d° may not be actually zero is of little consequence in the present work. Equilibrium temperatures for cation distribution between M1 and M2 sites may be calculated from

$$T = \frac{3104 - 1080(1 - 2X_{Fe}^{M2}) + 1524(1 - 2X_{Fe}^{M1})}{0.2851 - 1.987 \ln K_D}. \quad (9)$$

Temperature estimates of the experimental data of Saxena and Ghose (1971) using Eq. (9) result in the following errors for the pyroxenes in the compositional range of X_{Fe} from 0.181 to 0.72:

578 ± 20°C for the data at 600°C,
716 ± 20°C for the data at 700°C,
830 ± 25°C for the data at 800°C.

The progressive increasing difference between estimated and experimental temperatures may be traced to the accommodation of Virgo and Hafner's (1969) data at 1000°C by Ganguly (1982). Obviously, it is necessary to redetermine the equilibrium site occupancy data particularly at the high- and low-iron end compositions using X-ray techniques. In the meantime Eq. (9) may be used to estimate relative temperatures of cation order–disorder equilibrium.

Possible Kinetic Phase Diagram of an Exsolved Pyroxene

The relationships among compositions of coexisting lamellae and the compositions of the M1 and M2 sites in a pyroxene from Skaergaard (sample 4430, Brown, 1957; Nobugai *et al.*, 1978) were estimated by Kretz (1982b). These relationships are explored here further using TTT diagrams. Specifically, we consider the exsolution of pigeonite with the following compositions (from table 1, Kretz, 1982b):

Pig grain $Ca_{0.087}Mg_{0.461}Fe_{0.452}$,
Opx host $Ca_{0.028}Mg_{0.485}Fe_{0.485}$,
Cpx lamellae $Ca_{0.437}Mg_{0.376}Fe_{0.186}$.

For coexisting augite grain ($Ca_{0.356}Mg_{0.381}Fe_{0.263}$), Kretz (1982b) showed that exsolution of pigeonite crystals may have continued to 750°C, after which the lamellae did not grow further but Fe–Mg exchange between host and lamellae continued down to ~540°C. As is evident by Kretz's (1982b) discussion, there are several difficulties in the interpretation of the compositions of the exsolved pyroxenes and the results are only possible estimates.

The Cpx lamellae in the pyroxene grain may grow until equilibrium in the transfer reaction (a) has been reached. Assuming this was indeed the case for the Skaergaard pigeonite grain, we may calculate the equilibrium temperature for reaction (a) from (5). However, this requires X_{Fe}^{Cpx} at the temperature of transfer equilibrium which is unknown. This may be a problem if T is being calculated from the composition of the augite lamellae in pigeonite host. Because the augite lamellae are small in volume, the exchange reaction could decrease X_{Fe}^{Cpx} significantly. The problem is not serious if the composition of the augite host is used because here the volume of augite may be much larger than the volume of the pigeonite lamellae and the exchange produces very slight change in the augite composition.

The calculation of equilibrium temperature for Fe–Mg exchange (b) by using

Eqs. (4) presents certain other problems. The formulation of these equations is based on data on naturally coexisting pyroxenes in which different degrees of equilibrium might have been approached in reactions (a) and (b). In the present case a fixed content of Ca (~0.874) has to be used at all temperatures. The method used in determining the temperature of Fe–Mg exchange equilibrium is outlined below.

As long as the intercrystalline Fe–Mg exchange (reaction (b)) continues, we are assured that the intracrystalline reaction (c) of Fe–Mg ordering within each phase will also continue. In such a situation, if we invoke Mueller's (1962) postulate (see also Kretz, 1982a; Ganguly, 1982), namely that the equivalent crystallographic sites in different crystals will contain very similar Fe/Mg ratio, we have

$$(X_{\mathrm{Fe}}^{\mathrm{M1}})^{\mathrm{Opx}} \simeq (X_{\mathrm{Fe}}^{\mathrm{M1}})^{\mathrm{Cpx}}.$$

Since orthopyroxene host crystal is large relative to the exsolving augite lamellae, the composition of the host does not change significantly with decreasing temperature. As such $X_{\mathrm{Fe}}^{\mathrm{M1}}$ in Opx can be estimated at different temperatures by solving the equations:

$$2X_{\mathrm{Fe}}^{\mathrm{Opx}} = (X_{\mathrm{Fe}}^{\mathrm{M1}} + X_{\mathrm{Fe}}^{\mathrm{M2}})\mathrm{Opx}, \tag{10}$$

$$RT \ln K_D = \left[(0.1435 - 1562/T) + \frac{1080}{RT}(1 - 2X_{\mathrm{Fe}}^{\mathrm{M2}}) - \frac{1524}{RT}(1 - 2X_{\mathrm{Fe}}^{\mathrm{M1}}) \right], \tag{11}$$

$$K_D = (X_{\mathrm{Fe}}^{\mathrm{M1}}/X_{\mathrm{Mg}}^{\mathrm{M1}})(X_{\mathrm{Mg}}^{\mathrm{M2}}/X_{\mathrm{Fe}}^{\mathrm{M2}}). \tag{12}$$

By equating the calculated $X_{\mathrm{Fe}}^{\mathrm{M1}}$ in Opx with $X_{\mathrm{Fe}}^{\mathrm{M2}}$ in Cpx, it is possible to calculate the Cpx composition as a function of decreasing temperature through the equations for interrelating the $X_{\mathrm{Fe}}^{\mathrm{M1}}$ and other site occupancies in Cpx. Data on site occupancies in equilibrated Cpx are meager and the following two equations are based on the data of McCallister *et al.* (1976):

$$\ln K_D\,(675\,^{\circ}\mathrm{C}) = -0.561 - 1.733\,(R^{3+}) - 2.343(\mathrm{Ca}), \tag{13}$$

$$\ln K_D\,(927\,^{\circ}\mathrm{C}) = -1.235 - 0.520\,(R^{3+}) - 0.718(\mathrm{Ca}) \tag{14}$$

where

$$K_D = (X_{\mathrm{Fe}}^{\mathrm{M1}} X_{\mathrm{Mg}}^{\mathrm{M2}} / X_{\mathrm{Mg}}^{\mathrm{M1}} X_{\mathrm{Fe}}^{\mathrm{M2}}) \text{ in Cpx}$$

and R^{3+} is the sum of octahedrally coordinated ions (Al^{3+}, Ti^{4+}, Cr^{3+} and Fe^{3+}) and (Ca) the sum of Ca, Mn, and Na ions. Equations (13) and (14), assuming that the coefficients are linear with temperature, yield

$$\ln K_D = -[(7324(\mathrm{Ca}) - 3039)/T] + 5.383(\mathrm{Ca}) - 3.767 \tag{15}$$

(ignoring R^{3+} ions in the present discussion). Equation (15) is obviously very approximate and applicable only over a limited range of temperature (650 ~ 950 °C) and of composition (iron-poor and close to those used by McCallister *et al.*, 1976). Despite these problems, Eq. (15) is used here to continue with the development of an approximate cooling model for the Skaergaard pyroxene.

The composition of the exsolving Cpx lamellae, now may be obtained by first determining the X_{Fe}^{M1} in Opx through the use of Eqs. (10)–(12), then using this X_{Fe}^{M1} value as X_{Fe}^{M1} in Cpx and then by solving Eq. (15) and

$$X_{Fe}^{M2} = 1 - [X_{Fe}^{M1}/(1 - X_{Fe}^{M1})K_D]^{-1} \tag{16}$$

simultaneously. We then have

$$(Fe) = 2X_{Fe}^{Cpx} = X_{Fe}^{M1} + X_{Fe}^{M2} \tag{17}$$

and

$$(Mg) = 2X_{Fe}^{Cpx} = (1 - (Ca) - X_{Fe}^{M2})+(1 - X_{Fe}^{M1}), \tag{18}$$

where (Fe), (Mg), and (Ca) are number of cations in Cpx per 2-cation basis.

Figure 5 shows the changing composition of Cpx lamellae with temperature. The calculated Fe^{2+} content of Cpx approaches the chemically determined value of 0.372 (per 2 cation) at a temperature of 720°C. The calculated Mg content also approaches the required 0.752 in the Cpx lamellae at 720°C. This temperature, therefore, represents the Fe–Mg exchange equilibrium temperature. If we used Eq. (9) directly, the estimated temperature would be 628°C. Because of the difficulties with Eq. (15), it may be that the equilibrium temperature lies somewhere in the range of 628 and 720°C. At the temperature at which reaction (a) ceases, the lamellae also stop growing. This temperature, if calculated directly by using Kretz's (1982a) Eq. (5), is 860°C. The present value of X_{Fe}^{Cpx} used in this calculation must have been different at the equilibrium temperature of reaction (a). This estimate, therefore, may be wrong by several degrees. We may determine this temperature more accurately (if the equations to be used are based on precise data) by combining Eqs. (15)–(18) and Eq. (5). In this method, we determine X_{Fe}^{Cpx} at a fixed content of (Ca) (in our case 0.874) at different temperatures. The combinations of X_{Fe}^{Cpx} and T are substituted in

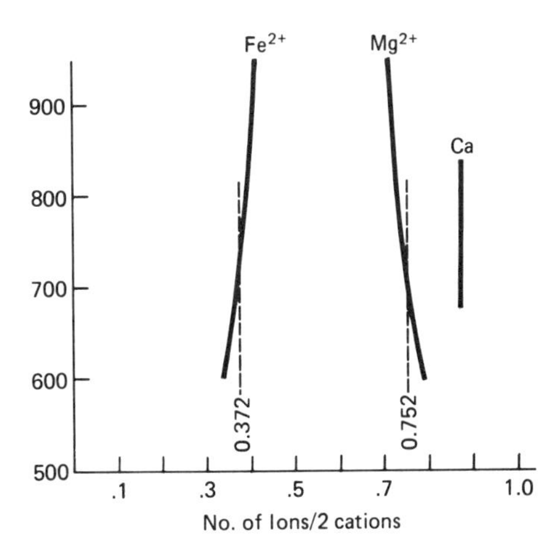

Fig. 5. Calculation of the equilibrium temperature for reaction (a). See text.

$$(\text{Ca}) = (1 - (\exp(-1000/T - 0.054 - 0.608X_{\text{Fe}}^{\text{Cpx}})/0.304))/2 \quad (19)$$

until the required (Ca) is found. This method results in a composition $Ca_{0.437}$ $Fe_{0.197}$ $Mg_{0.366}$ at 840°C, which, in view of the probable errors, is not significantly different from Kretz's estimate.

From the above discussion and calculations, it appears that the pigeonite crystal ($Ca_{0.087}$ $Mg_{0.461}$ $Fe_{0.452}$) cooled and exsolved into pigeonite host ($Ca_{0.028}$ $Fe_{0.485}$ $Mg_{0.485}$, the composition might have been slightly different) and augite ($Ca_{0.437}$ $Mg_{0.361}$ $Fe_{0.201}$) lamellae. If local equilibrium was reached in this process, the temperature at which the transfer reaction ceased in this crystal was 830°C. Elsewhere in the rock, e.g., in the augite crystal studied by Nobugai *et al.* (1978) and discussed by Nobugai and Morimoto (1979) and Kretz (1982b), the reaction continued to lower temperatures perhaps to ~750°C. In the pigeonite crystal under discussion here, Fe–Mg exchange between pigeonite host and augite lamellae ceased at approximately ~674°C (628 to 720°C). At some temperature above or below the pigeonite host inverted to Opx ($Ca_{0.028}$ $Mg_{0.485}$ $Fe_{0.485}$). During cooling from 830 to 674°C the augite lamellae changed in composition from $Ca_{0.437}$ $Mg_{0.366}$ $Fe_{0.197}$ to $Ca_{0.437}$ $Mg_{0.376}$ $Fe_{0.186}$.

The above information is largely similar to that provided and discussed by Kretz (1982b). Little can be added to it without kinetic data on the various reactions. While there is virtually no information on kinetics of intercrystalline exchanges or reactions in pyroxenes, there is some data on the intracrystalline Fe–Mg exchange in orthopyroxene. For the purpose of demonstrating kinetic relationships among intracrystalline and intercrystalline reactions, we assume that the site occupancy data in the natural Opx ($Ca_{0.028}$ $Fe_{0.485}$ $Mg_{0.485}$) host crystal are available. By comparison with Opx from other plutonic rocks, we adopt a value of $X_{\text{Fe}}^{\text{M2}}$ as 0.883, which corresponds to a T of about 340°C.

This little added information, which is not difficult to obtain by X-ray crystal structural refinement of the host Opx, can lead to several important results, which are presented and discussed below.

Fe^{2+}–Mg Ordering in Host Opx and Cooling Rate of the Rock

Let us consider the kinetics of Fe^{2+}–Mg exchange over M1 and M2 sites in orthopyroxene. The ion-exchange reaction (e) can be considered as a result of the forward (disordering) and backward (ordering) reactions, so that a change in the atomic fraction of Fe in M2 is given by

$$-dX_{\text{Fe}}^{\text{M2}}/dt = \vec{K}X_{\text{Mg}}^{\text{M1}}X_{\text{Fe}}^{\text{M2}} - \overleftarrow{K}X_{\text{Fe}}^{\text{M1}}X_{\text{Mg}}^{\text{M2}},$$

where $\vec{K}$ and $\overleftarrow{K}$ are the specific rate constants for disordering and ordering, respectively, and X represent site occupancy. At equilibrium $dX_{\text{Fe}}^{\text{M2}}/dt = 0$ and Eq. (5) reduces to

$$(X_{\text{Fe}}^{\text{M1}}X_{\text{Mg}}^{\text{M2}})/(X_{\text{Mg}}^{\text{M1}}X_{\text{Fe}}^{\text{M2}}) = \vec{K}/\overleftarrow{K} = \vec{K}^{\circ}/\overleftarrow{K}^{\circ}K_{\gamma} = K_{D}, \tag{20}$$

where $\vec{K}^{\circ}$ and $\overleftarrow{K}^{\circ}$ are now rate constants representing some standard state of reference. K_{γ} takes care of the nonideality of the two sites.

Mueller's (1967, 1969) integrated expression relating the change of site occupancy with time under isothermal conditions would be

$$-C_0\vec{K}\Delta t = \frac{1}{(b^2 - 4ac)^{1/2}}\left|\ln\frac{\pm(2aX_{\text{Fe}}^{\text{M2}} + b) \mp (b^2 - 4ac)^{1/2}}{(2aX_{\text{Fe}}^{\text{M2}} + b) + (b^2 - 4ac)^{1/2}}\right|_{X_{\text{Fe}(t\phi)}^{\text{M2}}}^{X_{\text{Fe}(t)}^{\text{M2}}}, \tag{21}$$

where C_0 is the total number of sites per unit volume of the crystal, Δt is the change in time and

$$\begin{aligned} a &= 0.5(1 - K_D^{-1}), \\ b &= 0.5 - X_{\text{Fe}} + K_D^{-1}(X_{\text{Fe}} + 0.5), \\ c &= -K_D^{-1}X_{\text{Fe}}, \\ X_{\text{Fe}} &= \tfrac{1}{2}(X_{\text{Fe}}^{\text{M1}} + X_{\text{Fe}}^{\text{M2}}). \end{aligned}$$

The upper signs in the numerator of the bracket hold when

$$2\,a\,X_{\text{Fe}}^{\text{M2}} + b > (b^2 - 4\,ac)^{1/2}$$

and the lower signs hold for the opposite condition.

To determine the rate constant $\vec{K}$, data on site occupancy at a fixed temperature and as a function of time are required. The function $C_0\vec{K}\Delta t$ can be determined from such data using Eq. (21). Conversely, if the rate constants are known the changing isothermal site occupancy can be determined as a function of time.

Seifert and Virgo (1975) used the TTT plot to calculate the cooling path that matches with the site occupancy in the natural mineral. As explained by Ganguly (1982), through this method we are calculating the time required by a crystal to attain its cation ordering under isothermal conditions and assuming that this time is the same as in a continuously cooling system. In practice we assume that the cooling behavior of the rock is simulated by either the asymptotic model

$$\left(\frac{1}{T}\right) = \left(\frac{1}{T_0} + \eta t\right) \tag{22}$$

or the exponential model

$$T = T\exp(-\eta t), \tag{23}$$

where T_0 is the crystallization temperature, η the rate constant, and Δt the time. A value of η is chosen by trial and error, such that the cooling curve generated is tangential to the natural site occupancy curve calculated by Eq. (21) and plotted on a temperature–time diagram. A refinement of the η value can generally be achieved by calculating site occupancies at certain temperature intervals between the crystallization temperature and the closure temperature for times permitted by the cooling curve. The η value is finally adjusted such that the calculated $X_{\text{Fe}}^{\text{M2}}$ value below the closure temperature matches with the $X_{\text{Fe}}^{\text{M2}}$ in the natural sample.

Figure 6 is a TTT diagram showing a possible history of cooling of the host rock containing the exsolved pyroxenes. The cooling curve for the rock is obtained from the site occupancy in the Opx host (assumed $X_{Fe}^{M2} = 0.883$). The rate constant used for Opx is from the review of Besancon's (1981) data by Ganguly (1982) and is given by

$$C_0\vec{K}(T) = 6.4761 \times 10^{11} \exp(-30549/T) \text{ min } \vec{K}^{-1}.$$

The resulting rate constant η for the cooling rock using asymptotic model is 1×10^{-7} per year. This has been calculated by the method proposed by Ganguly (1982). The cooling rate is strictly applicable in the vicinity of the quenching temperature. In the following discussion, it is assumed that the same cooling rate prevailed throughout the history of the rock.

Some useful estimates of kinetic rates of diffusion of Ca^{2+}, Fe^{2+}, and Mg^{2+} required for the growth of augite lamellae may be obtained from the results shown in Fig. 6. In Fig. 7 the data are replotted on a TTT diagram. The process of exsolution (which includes nucleation; in this case perhaps heterogeneous) requiring diffusion of Mg^{2+} and Fe^{2+} mostly from the M2 site and replacing Ca^{2+} was initiated slightly below the equilibrium temperature (~1100°C; augite $Ca_{0.356}$

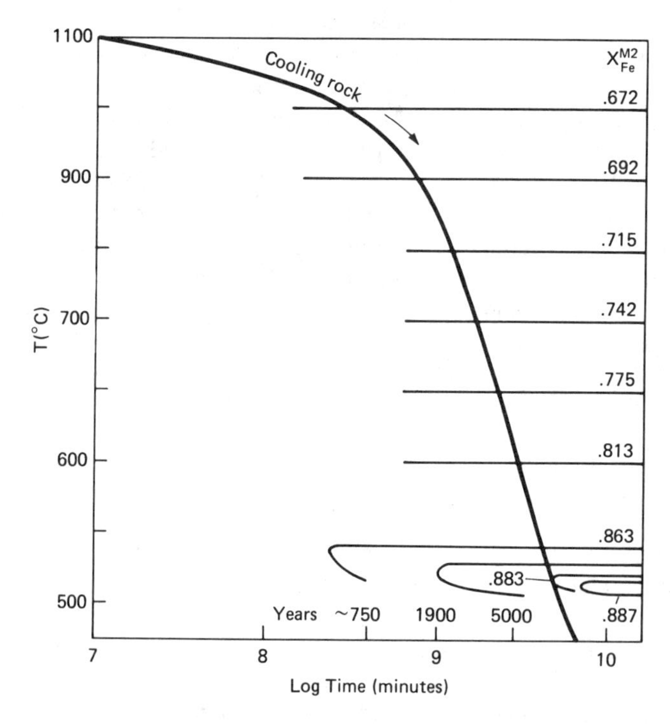

Fig. 6. Estimation of the cooling rate of a rock from the site occupancy data of orthopyroxene. The cooling behavior is assumed to follow the asymptotic model for which the rate constant (η) is calculated as 1×10^{-7}/year.

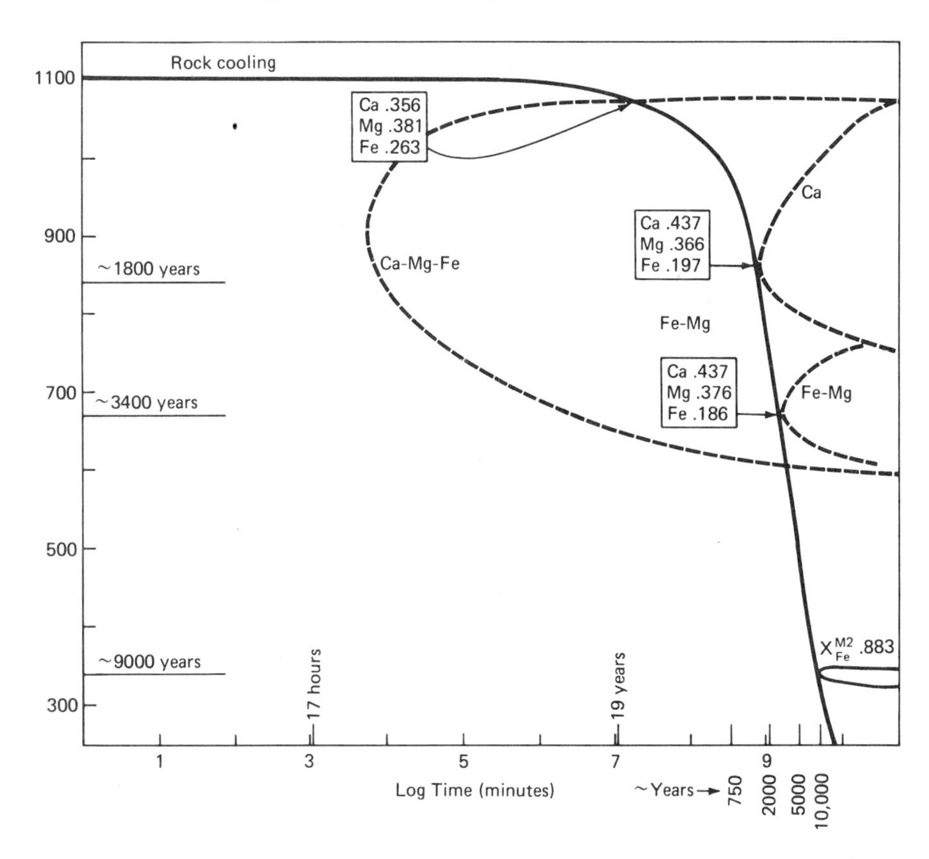

Fig. 7. Kinetic compositional relationships in an exsolving pyroxene with a rate as determined in Fig. 6. The dashed curve Ca–Mg–Fe marks the beginning of the exsolution when diffusion of all species is possible. There are two "finish" curves marked as Ca and Fe–Mg to which the cooling curve is tangential. These curves signify the end of any effective diffusion of the Ca and the Fe–Mg components, respectively.

$Mg_{0.381}$ $Fe_{0.263}$; pigeonite $Ca_{0.087}$ $Mg_{0.461}$ $Fe_{0.452}$). This is schematically shown to occur in 20 years with temperature reaching about 1070°C. The composition of the embryonic lamellae of augite at 1070°C may be assumed to be the same as that of the exsolving augite crystals coexisting with the exsolving pigeonite. During the first ~1800 years the rock cooled from 1100 to 840°C. In this period Ca^{2+}–Mg^{2+}–Fe^{2+} exchange diffusion and the growth of the lamellae continued. From here on the Ca^{2+} diffusion became too sluggish but the Fe^{2+}–Mg^{2+} exchange between the Opx/Pig host and the augite ($Ca_{0.437}$ $Mg_{0.366}$ $Fe_{0.197}$) continued for another 1600 years until the temperature of ~675°C was reached. No further intercrystalline exchanges took place but the intracrystalline Fe^{2+}–Mg^{2+} exchange within the host Opx continued for another 5600 years. Similar ordering of augite must have continued but there are no kinetic data to get this information.

Discussion and Conclusions

Although Fig. 7 is drawn schematically in many respects and on the basis of an assumed site occupancy in Opx, the results may not be far from the real situation. The curve representing Ca–Mg–Fe exchange diffusion does not make a distinction among processes of heterogeneous or homogeneous nucleation or spinodal decomposition. Close to the solvus temperature, homogeneous nucleation will not be effective because it requires some degree of supercooling. While the beginnings of the diffusion processes cannot be estimated on the basis of such data, the endings of these reactions are well constrained because the cooling curve for the rock must be the same irrespective of the method of its determination. This could not, of course, be valid for rocks with complicated cooling history, e.g., polymetamorphic rocks.

In conclusion, petrographic, electron microscopic and X-ray structural refinement studies of different cyrstals of pyroxenes from a rock can yield important information on the petrogenetic history of a rock which cannot be obtained through any single method of study. Site-occupancy determinations in orthopyroxene or pigeonite coexisting with exsolved Ca-pyroxene should be obtained for many different kinds of igneous and metamorphic rocks. As yet only limited kinetic data on Mg–Fe order–disorder in orthopyroxene (for medium compositions in Fe:Mg) are available (see Ganguly, 1982), but this situation should improve considerably in the near future. One discouraging aspect of the determination of the cooling rate of a rock through site occupancies is the large error associated with it. This was discussed by Ganguly (1982) and is evident in Fig. 6, which shows a change in site occupancy of X_{Fe}^{M2} from 0.883 to 0.887 could make a difference of several thousand years. Therefore a study of several crystals from the same rock may be required for a good result.

Acknowledgments

Thanks are due to Ralph Kretz and J. Ganguly for suggestions for improvements and to Dr. G. Rossi, C.N.R. at the University of Pavia, and Professor A. Dal Negro, University of Padova, for financial support and warm hospitality.

References

Besancon, J. R. (1981) Rate of cation ordering in orthopyroxenes, *Amer. Mineral.* **66**, 965–973.

Brown, G. M. (1957) Pyroxenes from the early and middle stages of fractionation of the Skaergaard intrusion, East Greenland, *Mineral. Mag.* **31**, 511–543.

Buseck, P. R., Nord, Jr., G. L., and Veblen, D. R. (1980) Subsolidus phenomena in pyroxenes, *Pyroxenes,* edited by C. T. Prewitt, *Reviews in Mineralogy,* Vol. 7, pp. 117–211. Mineralogical Soc. of America, Washington, D.C.

Champness, P. E., and Lorimer, G. W. (1976) Exsolution in silicates, in *Electron Microscopy in Mineralogy,* edited by H.-R. Wenk, pp. 174–204, Springer-Verlag, New York.

Davidson, L. R. (1968) Variation in ferrous iron-magnesium distribution coefficients of metamorphic pyroxenes from Quairading, Western Australia, *Contrib. Mineral. Petrol.* **19**, 239–259.

Ganguly, J. (1982) Mg–Fe order–disorder in ferromagnesian silicates II. Thermodynamics, kinetics and geological applications, in *Advances in Physical Geochemistry,* Vol. 2, pp. 58–100, Springer-Verlag, New York.

Grove, T. L. (1982) Use of exsolution lamellae in lunar clinopyroxenes as cooling rate speedometers: an experimental calibration, *Amer. Mineral.* **67**, 251–268.

Grover, J. (1977) Chemical mixing in multicomponent solutions, in *Thermodynamics in Geology,* edited by D. G. Fraser, p. 68, Reidel, Dordrecht, Holland/Boston.

Kretz, R. (1982a) Transfer and exchange equilibria in a portion of the pyroxene quadrilateral as deduced from natural and experimental data, *Geochim. Cosmochim. Acta* **46**, 411–422.

Kretz, R. (1982b) Redistribution of Ca, Mg, and Fe during pyroxene exsolution; potential rate-of-cooling indicator, in *Advances in Physical Geochemistry,* Vol. 2, pp. 101–115, Springer-Verlag, New York.

Lasaga, A. C., and Kirkpatrick, R. J. (Eds.) (1981) Kinetics of geochemical processes, in *Reviews in Mineralogy,* Vol. 8, Mineral. Soc. Amer., pp. 1–398. Mineralogical Soc. of America, Washington, D.C.

Lindsley, D. H. (1982) A two pyroxene thermometer. Abstr. *Lunar and Planetary Science* **13**, 435–436.

Lindsley, D. H., Grover, J. E., and Davidson, P. M. (1981) The thermodynamics of the Mg Si O–CaMgSi O Join: A review and an improved model, in *Advances in Physical Geochemistry,* Vol. 1, edited by R. C. Newton, A. Navrotsky, and B. J. Wood. Springer-Verlag, New York.

McCallister, R. H., and Yund, R. A. (1977) Coherent exsolution in Fe-free pyroxenes, *Amer. Mineral.* **62**, 721–726.

McCallister, R. H., Finger, L. W., and Ohashi, Y. (1976) Intracrystalline Fe-Mg equilibria in three natural Ca-rich clinopyroxenes. *Amer. Mineral.* **61**, 671–676.

McConnell, J. D. C. (1975) Microstructures of minerals as petrogenetic indicators, *Ann. Rev. Earth Planetary Sci.* **3**, 1–123.

Mueller, R. F. (1962) Energetics of certain silicate solid solutions, *Geochim. Cosmochim. Acta,* **26**, 265–275.

Mueller, R. F. (1967) Model for order–disorder kinetics in certain quasi-binary crystals of continuously variable composition, *J. Phys. Chem. Solids* **28**, 2239–2243.

Mueller, R. F. (1969) Kinetics and thermodynamics of intracrystalline distributions, *Mineral. Soc. Amer. Spec. Pap.* **2**, 83–93.

Nobugai, K., and Morimoto, N. (1979) Formation mechanism of pigeonite lamellae in Skaergaard augite, *Phys. Chem. Minerals* **4**, 361–371.

Nobugai, K., Tokonami, M., and Morimoto, N. (1978) A study of subsolidus relations of the Skaergaard pyroxenes by analytical election microscopy, *Contrib. Mineral. Petrol.* **67**, 111–117.

Nord, G. L., and McCallister, R. H. (1979) Kinetics and mechanism of decomposition in Wo En Fs clinopyroene (abstr.), *Geol. Soc. Amer. Annual Meeting* **11**, 488.

Putnis, A., and McConnell, J. D. C. (1980) *Principles of Mineral Behaviour,* pp. 1–257, Blackwell Scientific Publications, Oxford.

Robinson, P., Ross, M., Nord, Jr., G. L., Smyth, J. R., and Jaffe, H. W. (1977) Exsolution lamellae in augite and pigeonite: fossil indicators of lattice parameters at high temperatures and pressures. *Amer. Mineral.,* **62**, 857–873.

Sack, R. A. (1980) Some constraints on the thermodynamic mixing properties of Fe–Mg orthopyroxenes and olivine, *Contr. Mineral. Petrol.* **71**, 257–269.

Saxena, S. K. (1973) *Thermodynamics of Rock-Forming Crystalline Solutions,* Springer-Verlag, New York.

Saxena, S. K., and Ghose, S. (1971) Mg–Fe order–disorder and the thermodynamics of the orthopyroxene crystalline solution, *Amer. Mineral.* **56**, 532–559.

Seifert, F. A., and Virgo, D. (1975) Kinetics of the Fe–Mg order–disorder reaction in anthophyllites: Quantitative cooling rates, *Science* **188**, 1107–1109.

Thompson, J. B. (1969) Chemical reactions in crystals, *Amer. Mineral.* **54**, 341–375.

Virgo, D., and Hafner, S. S. (1969) Fe–Mg disorder in heated orthopyroxenes, *Miner. Soc. Amer. Spec. Pap.* **2**, 67–81.

Wenk, H. R. (Ed.) (1976) *Electron Microscopy Mineralogy,* Springer-Verlag, New York.

Yund, R. A., and McCallister, R. H. (1970) Kinetics and mechanisms of exsolution, *Chem. Geol.* **6**, 5–30.

Chapter 3
Geospeedometry: An Extension of Geothermometry

Antonio C. Lasaga

Introduction

One of the basic aims of petrology is to relate observed mineral assemblages to their previous temperature–pressure history. To assess the previous thermal history, we have relied on an understanding of thermodynamics and phase equilibria. The assumption is made that the various minerals and their chemical compositions measured today will reveal a temperature and pressure which are geologically significant. This procedure is quite valid if equilibrium is achieved and preserved over significant portions of the mineral assemblage. However, in this latter case, by the very nature of equilibrium, effects of any previous or posterior processes are completely lost. An important point is that, by the use of thermodynamics, the dynamics or history of a rock is immediately subjugated to a secondary role.

The geologic history of the set of ion exchange geothermometers commonly used covers quite a range of temperatures (a few hundred °C–1400°C) and pressures (0–50 kbar). A very widely used set, especially with ferromagnesian minerals, relies on Fe–Mg or Ca–Mg exchange. Some of these ion-exchange geothermometers are given in Table 1.

Note that the geologic terranes in Table 1 cover a wide range of possible thermal histories. Therefore, the P–T-time paths of the potential exchange geothermometers span a significant portion of the terrestial P–T space.

The thermodynamic background for the use of geothermometers has been discussed in many works (e.g., Wood and Fraser, 1978; Wood, 1977; Newton, 1977; Navrotsky, 1976; Saxena, 1972, 1973). In this chapter we will assume the thermodynamics of the relevant minerals is known. Our concern is with the next critical step, that is, relating the P–T–t history of the assemblage to the P and T obtained today by geothermometric methods. The fundamental quantity now is not the ΔG° or ΔH° of the exchange reaction or the activity–composition relations, but the *kinetic* response of the minerals. In stressing the geologic process, geothermometry is extended to geospeedometry.

The ideal case in geothermometry is obtained when two adjacent minerals are unaltered and exhibit essentially no chemical composition zoning throughout any one crystal (Fig. 1). One implication in this case is that an unzoned crystal can

Table 1. Some ion-exchange geothermometers.

Mineral pair	Ions	Applications
Garnet–Clinopyroxene	Fe^{2+}/Mg^{2+}	Eclogites, kimberlites, gneissic terranes, blueschist terranes
Garnet–Biotite	Fe^{2+}/Mg^{2+}	Pelitic metamorphism
Garnet–Cordierite	Fe^{2+}/Mg^{2+}	Pelitic metamorphism
Clino–Orthopyroxene	Ca^{2+}/Mg^{2+}	Mafic granulite gneisses
Olivine–Clinopyroxene	Ca^{2+}/Mg^{2+}	Garnet lherzolites Spinel lherzolites
Olivine–Spinel	Fe^{2+}/Mg^{2+}	Peridotites

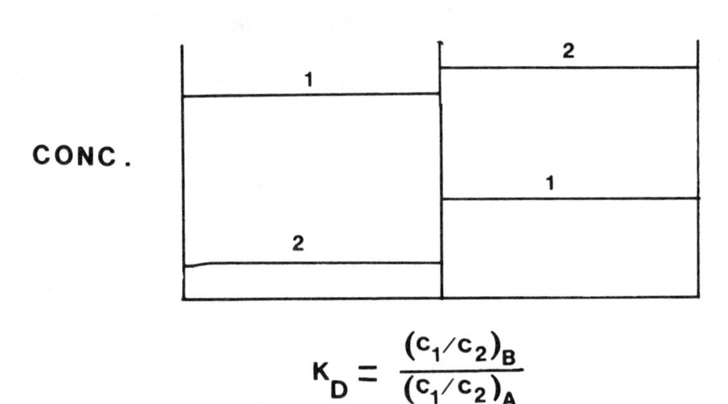

$$K_D = \frac{(c_1/c_2)_B}{(c_1/c_2)_A}$$

DOES K_D REFLECT PEAK TEMPERATURE?

Fig. 1. Ideal case for an ion exchange geothermometer. K_D is the effective equilibrium constant.

be used as evidence that equilibrium between minerals A and B was achieved and preserved at some point. In this ideal case, the concentrations of exchangeable components 1 and 2 are used to obtain the value for the exchange equilibrium constant:

$$K_D = \frac{(c_1/c_2)_A}{(c_1/c_2)_B}. \tag{1}$$

If, for example, the pressure is independently known (or bracketed), the value of K_D will correspond to some temperature T, which can be obtained from thermodynamics (including activity–composition relations). c in Eq. (1) can refer to either mol % or mole fraction. Another common assumption in cases such as that shown in Fig. 1 is that the temperature obtained reflects the "peak" temperature, T_0. This peak temperature is either the temperature at which one or both of the crystals were first grown in equilibrium or the temperature at which they came initially in contact and at which they were held for a "substantial" period of time. Of course, the final assumption (and the one of interest here!) is that the profiles were "frozen in," so that we can recover today the value of T_0.

Less ideal cases have crystals that exhibit compositional zoning, especially at their common boundaries, as is shown in Fig. 2. In these cases, a common assumption is that the compositions of the cores of the crystals can be used to obtain an original peak temperature, T_0. The interpretation of the zoning is then ascribed to some later effect.

The next section will set up the necessary tools for answering questions about geospeedometry. The general results are then given in the third section, and the applications are introduced in the final section.

Ion-Exchange Geospeedometry

Consider two minerals, A and B, and two exchangeable components, 1 and 2. The system is initially in equilibrium at temperature, T_0, as in Fig. 1. The pair of minerals will then traverse a P–T–t path from P_0, T_0 at $t = 0$ to 1 bar, 298°C, as the thin section sits under our microscope. Once the minerals are thrown out of equilibrium due to changing P–T conditions, the response will depend on kinetics. The immediate change occurs at the mutual boundary (which is set at $x = 0$ in this paper) between A and B. Because K_D is changing, the concentrations of 1 and 2 at the mutual boundary will have to change to satisfy

$$\frac{c_{1A}(0,t)/c_{2A}(0,t)}{c_{1B}(0,t)/c_{2B}(0,t)} = K_D(t), \tag{2}$$

where $c_{i\phi}$ is the mol % of i in phase ϕ.

These changes will induce diffusive fluxes in each mineral. Therefore, within

GEOTHERMOMETERS

Non-ideal case:

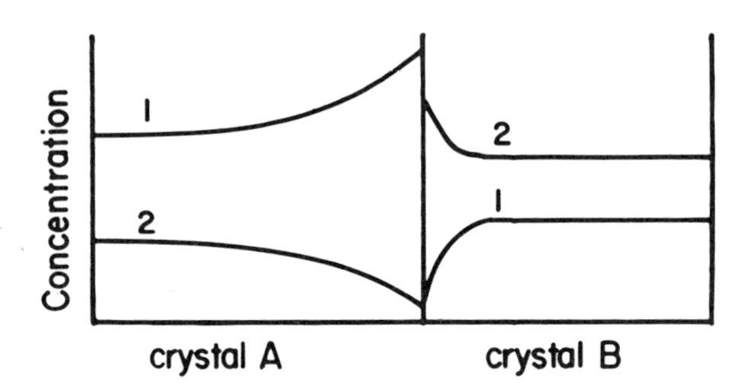

Can cores be used to obtain peak temperatures ?

What information is in the zoning?

Fig. 2. Nonideal case for an ion exchange geothermometer. There is zoning at mutual boundaries.

each mineral, the diffusion equation must be applied:

$$\frac{\partial c_{1A}}{\partial t} = D_A(t) \frac{\partial^2 c_{1A}}{\partial x^2}, \tag{3}$$

$$\frac{\partial c_{1B}}{\partial t} = D_B(t) \frac{\partial^2 c_{1B}}{\partial x^2}. \tag{4}$$

At this point, the concentrations of 1 or 2 will be functions of both distance from the interface, x, and time, t. D_A and D_B are binary interdiffusion, not tracer diffusion, coefficients (see Lasaga, 1979; Anderson, 1981), and are the same for both components.

Equations (3) and (4) assume that the interdiffusion coefficients, D_A and D_B, are not significantly dependent on composition (and hence x) over the relevant changes in composition. There are also analogous equations for component 2.

$$\frac{\partial c_{2A}}{\partial t} = D_A(t) \frac{\partial^2 c_{2A}}{\partial x^2}, \tag{5}$$

$$\frac{\partial c_{2B}}{\partial t} = D_B(t) \frac{\partial^2 c_{2B}}{\partial x^2}. \tag{6}$$

Finally, the flux of components 1 and 2 at the boundary ($x = 0$) between the two minerals must conserve mass. If there is no leakage of ions out of the system (e.g., by grain boundaries), then the conservation of mass requires that the flux out of mineral A must equal the flux into mineral B and vice versa. Therefore, using Fick's first law to compute the diffusive flux:

$$-D_A(t)\left.\frac{\partial c_{1A}}{\partial x}\right|_{x=0} = -D_B(t)\left.\frac{\partial c_{1B}}{\partial x}\right|_{x=0}, \tag{7}$$

$$-D_A(t)\left.\frac{\partial c_{2A}}{\partial x}\right|_{x=0} = -D_B(t)\left.\frac{\partial c_{2B}}{\partial x}\right|_{x=0}. \tag{8}$$

If the units used in Eqs. (7) and (8) are mol %, there are some corrections necessary to take into account molar volumes; however, these are usually very small (see Appendix A). In the cases discussed here only binary exchange is treated. Therefore, as the relative ratio of cations change in a mineral the sum of their concentrations is constrained by stoichiometry, i.e.,

$$c_{1A}(x,t) + c_{2A}(x,t) = \text{constant} = c_{1A}^I + c_{2A}^I, \tag{9}$$

$$c_{1B}(x,t) + c_{2B}(x,t) = \text{constant} = c_{1B}^I + c_{2B}^I. \tag{10}$$

In this case, we need only worry about one component, e.g., component 1.

The role of the thermal history of a mineral assemblage in the compositional evolution of a geothermometer is quantified through the time dependence of the diffusion coefficients and of the exchange equilibrium constant, K_D. Diffusion coefficients vary strongly with temperature (Lasaga, 1981; Freer, 1981):

$$D(T) = Ae^{-(Ea/RT)}, \tag{11}$$

where A is the pre-exponential factor and E_a the diffusion activation energy in the Arrhenius expression.

If the temperature varies with time as in most geologic processes, then D will also vary with time:

$$D(t) = Ae^{-[Ea/RT(t)]}.$$

It is often convenient to rewrite the last equation using the value of D at some initial temperature, T_0:

$$D(t) = D_0 \exp\left\{\frac{E_a}{R}\left[\frac{1}{T(t)} - \frac{1}{T_0}\right]\right\}, \tag{12}$$

where D_0 is the value of D at T_0.

The exponential term in Eq. (12) indicates that D will have a strong time dependence if the activation energy is high. This strong time dependence makes Eqs. (5) and (6) cumbersome. We can remove this time variation dependence from the diffusion equations (3) or (4) by the introduction of a *compressed time scale* (Lasaga *et al.*, 1977; Dodson, 1973). To do this we write $D(t)$ as

$$D(t) = D_0 d(t), \tag{13}$$

where $d(t)$ is the dimensionless factor that scales D_0 to the correct value of D at time t. Then the compressed time t' is defined by

$$t' \equiv \int_0^t d(\tau)\, d\tau \tag{14}$$

or using Eq. (12)

$$t' = \int_0^t \exp\left\{ -\frac{E_a}{R}\left[\frac{1}{T(\tau)} - \frac{1}{T_0}\right]\right\} d\tau. \tag{15}$$

There are some important results of viewing geothermometry from the t' point of view. First, the diffusion equations simplify considerably. Realizing from (14) that

$$dt' = d(t)\, dt,$$

it follows that if

$$\frac{\partial c}{\partial t} = D(t)\,\frac{\partial^2 c}{\partial x^2}$$

or

$$\frac{\partial c}{\partial t} = D_0 d(t)\,\frac{\partial^2 c}{\partial x^2}$$

then

$$\frac{1}{d(t)}\frac{\partial c}{\partial t} = D_0\,\frac{\partial^2 c}{\partial x^2}$$

or

$$\frac{\partial c}{\partial t'} = D_0\,\frac{\partial^2 c}{\partial x^2}. \tag{16}$$

Thus, using t' we can reduce the problem to a constant-D diffusion equation. Any compositional re-adjustments during the thermal history of the pair of minerals will occur as long as t' varies. The crucial point is that t', unlike t, cannot generally keep on increasing but will rather reach a limiting finite value. We have now reached the quantitative representation of the whole concept of preservation of previous compositions. If t', on the one hand, could reach very high values and equilibrium could be re-established at later times, all rocks observed today will not have preserved *any* of their previous thermal history. That is the essence of thermodynamics (i.e., equilibrium does *not* depend on path). If t', on the other hand, can only achieve very small values (even for $t \rightarrow \infty$), then there will be no "time" available to change a profile and the original composition will be preserved.

To illustrate the behavior of t', let us take the most common problem of a near-linear temperature decrease:

$$T = \frac{T_0}{1 + (st/T_0)} \approx T_0 - st, \tag{17}$$

where s is the initial cooling rate.

In this case, inserting Eq. (17) into (12) and (15), we have

$$D(t) = D_0 e^{-\gamma t} \tag{18}$$

and

$$t' = \frac{1}{\gamma}[1 - e^{-\gamma t}], \tag{19}$$

where

$$\gamma \equiv \frac{E_a s}{RT_0^2}. \tag{20}$$

Equation (19) clearly illustrates that while $t' \approx t$ for small times ($t \ll 1/\gamma$), for longer times t' will never get any larger than $1/\gamma$, even as $t \to \infty$! As a consequence, the original problem is reduced to an equivalent problem where diffusion goes on at constant D (e.g., constant T) but is only allowed to go on for t' time units and then is abruptly terminated. Obviously, the maximum value of t' will determine the net effect of the thermal history on the mineral assemblage. For example, if $E_a = 50$ kcal/mol, $s = 2°$C/my and $T_0 = 1000$ K, then Eq. (20) yields $\gamma = 0.05$ my^{-1}. The upper limit of t' in this case is 20 million years, no matter how much normal time increases. This "compression" of the normal time scale due to the thermal history is illustrated in Fig. 3.

The temperature dependence of the equilibrium constant is given by standard thermodynamics

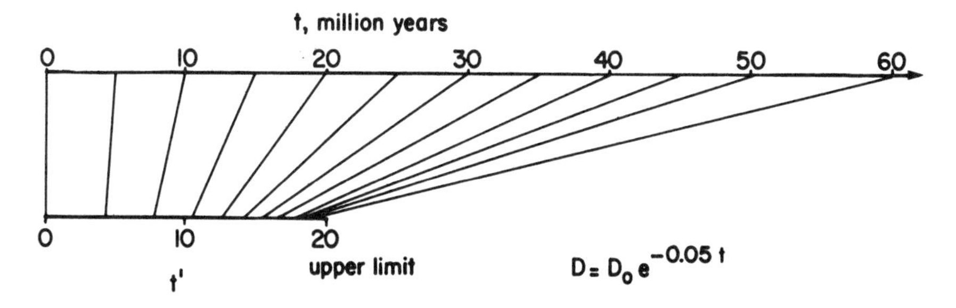

Fig. 3. Illustration of the compression of normal time, t, by the t' transformation. The example is based on typical values of γ as discussed in the text. Note that for small t, t' and t are essentially the same. However, t' cannot reach values greater than 20 million years.

$$K_D = K_D^0 \exp\left\{\frac{\Delta H^\circ}{R}\left[\frac{1}{T} - \frac{1}{T_0}\right]\right\}, \tag{21}$$

where K_D^0 is the value at temperature T_0 and ΔH° is the standard enthalpy change for the exchange reaction. If K_D is given in terms of concentration, ΔH° will also contain a composition dependent term due to activity coefficients. If we use the temperature–time relationship in Eq. (17), then

$$K_D(t) = K_D^0 e^{-\beta' t}, \tag{22}$$

where

$$\beta' \equiv \frac{\Delta H^\circ s}{RT_0^2}. \tag{23}$$

To solve the diffusion equation,

$$\frac{\partial c}{\partial t'} = D_0 \frac{\partial^2 c}{\partial x^2},$$

we need a boundary condition at $x = 0$ [i.e., the concentration, $c(0,t)$], of the component in *one* of the minerals at the interface between the two minerals. For the time being let us assume that we know $c(0,t)$ for any component in our system. First, we should stress that if the two crystals were homogeneous initially and there were only boundary changes at $x = 0$ due to the thermal history, the resulting profiles would *only* depend on x (not on the other directions, e.g., y or z). In fact, if no material was exchanged along the other sides of a mineral, then the full set of equations (y and z now dropping out) would be

$$c(x,y,z,t) = c(x,t) = c(x,t'),$$

$$\frac{\partial c}{\partial t'} = D_0 \frac{\partial^2 c}{\partial x^2}, \tag{24a}$$

$$c(x,0) = c^I, \tag{24b}$$

$$c(0,t) = f(t), \qquad \left.\frac{\partial c}{\partial x}\right|_{x=a} = 0, \tag{24c}$$

where $f(t)$, the boundary concentration, is assumed known, a is the crystal length in the x-direction, and c^I is the initial concentration. We can make Eqs. (24) dimensionless by use of the new variables

$$t'_r \equiv \frac{D_0}{a^2} t', \qquad x'_r = \frac{x}{a}, \tag{25}$$

so that

$$\frac{\partial c}{\partial t'_r} = \frac{\partial^2 c}{\partial x_r'^2} \tag{26}$$

and

$$c(0,t'_r) = c(0,t(t'_r)), \tag{27}$$

$$\left.\frac{\partial c}{\partial x'_r}\right|_{x'_r = 1} = 0, \tag{28}$$

$$c(x'_r,0) = c^I. \tag{29}$$

The purpose of introducing t'_r and x'_r is that we will then be able to categorize the behavior of many geothermometers with just a few parameters.

We must now return to the boundary condition, $c(0,t)$. If $c(0,t)$ did not change from c^I with time, there would be no profile changes. Hence, it is ultimately the changes in $c(0,t)$ caused by thermal changes in K_D that drive the evolution of a geothermometer. The detailed work by Lasaga *et al.* (1977) and Lasaga (1979) have shown that given the K_D in Eq. (22), the function $c(0,t)$ is reasonably well approximated by

$$c_{1A}(0,t) = c^I_{1A}e^{-\beta t}, \tag{30}$$

where

$$\beta = \frac{\Delta H^\circ s\sqrt{D^0_B/D^0_A}}{RT_0^2\{\sqrt{D^0_B/D^0_A}\,[1 + (c^I_{1A}/c^I_{2A})] + [(c^I_{1A}/c^I_{1B}) + (c^I_{1A}/c^I_{2B})]\}}. \tag{31}$$

Note that β is different but related to β' in Eq. (23). $c(0,t)$ in (30) now specifically refers to the concentration of component 1 in crystal A. D^0_A, D^0_B are the initial interdiffusion coefficients of minerals A and B. $c^I_{i\phi}$ refers to the initial (homogeneous) concentration of component i (1, 2) in phase ϕ (A or B). Similar formulas, of course, apply to the other $c_{i\phi}(0,t)$.

Note that it is the boundary concentration as a function of t'_r that is of interest in solving (26). Therefore, we rewrite Eq. (30) using Eq. (19) as

$$c(0,t') = c^I_{1A}\,(1 - \gamma t')^{\beta/\gamma}$$

or using (25) as

$$\begin{aligned} c(0,t'_r) &= c^I_{1A}\left(1 - \frac{a^2}{D_0}\gamma t'_r\right)^{\beta/\gamma}, \\ c(0,t'_r) &= c^I_{1A}\,(1 - \gamma' t'_r)^{\beta/\gamma}, \end{aligned} \tag{32}$$

where

$$\gamma' \equiv \frac{a^2}{D_0}\gamma. \tag{33}$$

Let us now comment on some powerful generalizations from these equations. Equations (30) and (31) are very useful. They show that the rate of change of the boundary concentration depends not just on the cooling rate s and the enthalpy of exchange, ΔH°, but also on the ratio of the diffusion coefficients and on the initial composition of both minerals. It is interesting to note that the rate of change is slowest for the mineral with the fastest diffusion coefficient [this follows from Eqs. (7) and (8)]. In other words, if D^0_B/D^0_A is very small, then β will be small.

Obviously, whether a particular component in a mineral is considered to be diffusing fast or slow is only meaningful *relative* to the adjacent mineral. Therefore, garnet may be the mineral with the rapidly varying boundary only if other minerals such as biotite or cordierite with much faster diffusion coefficients are exchanging with it. However, garnet may be the mineral with the slowly varying boundary in the case of exchange with clinopyroxenes (see further below). Lasaga (1979) has discussed further the implications of Eq. (31).

The next very important observation stems from the fact that the original problem of the thermal evolution of an ion-exchange geothermometer has been reduced to that of solving Eq. (26) with boundary conditions given by (28), (29), and (32). Note from (19) and (25) that t'_r can only reach values of

$$t'_r \leq \frac{D_0}{a^2\gamma} = \frac{1}{\gamma'}.$$

Therefore, x'_r always takes values between 0 and 1, while t'_r takes values between 0 and $1/\gamma'$. Furthermore, the solution to (26) only depends (aside from (28) and (29)) on the condition

$$c(0,t'_r) = c^I_{1A}(1 - \gamma' t'_r)^{\beta/\gamma}.$$

The net result is that the entire evolution of the geothermometer is characterized by just two dimensionless parameters, γ' and β/γ. This simplification is the reason for the use of these dimensionless variables. The next section discusses this dependence of the evolution on γ' and β/γ.

Solutions—Thermal Evolution

The solution to Eq. (26) that satisfies (28), (29), and (32) is

$$c(x'_r,t'_r) = \sum_{n=0}^{\infty} \exp\left[-\frac{(2n+1)^2\pi^2}{4} t'_r\right] A_n \sin\left[\frac{(2n+1)\pi}{2} x'_r\right], \tag{34}$$

where

$$A_n \equiv \frac{(2n+1)\pi e^{\alpha_n} c^I_{1A}}{\gamma'} \int_{1-\gamma' t'_r}^{1} e^{-\alpha_n u} u^{\beta/\gamma} du + \frac{4}{(2n+1)\pi} c^I_{1A} \tag{35}$$

and

$$\alpha_n \equiv \frac{(2n+1)^2\pi^2}{4\gamma'}. \tag{36}$$

For each value of β/γ and γ' there will be a final frozen profile ($t \to \infty$) given by $c(x'_r)$ when t'_r is set equal to $1/\gamma'$ in Eq. (34). The results are shown in Figs. 4 and 5. The calculations reported here used a $c^I_{1A} = 10$ mol %. However, the choice of c^I_{1A} is immaterial since, as can be seen from Eq. (35) for A_n, the entire solution for $c(x'_r, t'_r)$ is simply multiplied by c^I_{1A}; thus c^I_{1A} does *not* affect the shape of the profile. In this sense the results in Figs. 4 and 5 are quite general.

It is important to clarify what the profiles in Figs. 4 and 5 mean. Initially, the

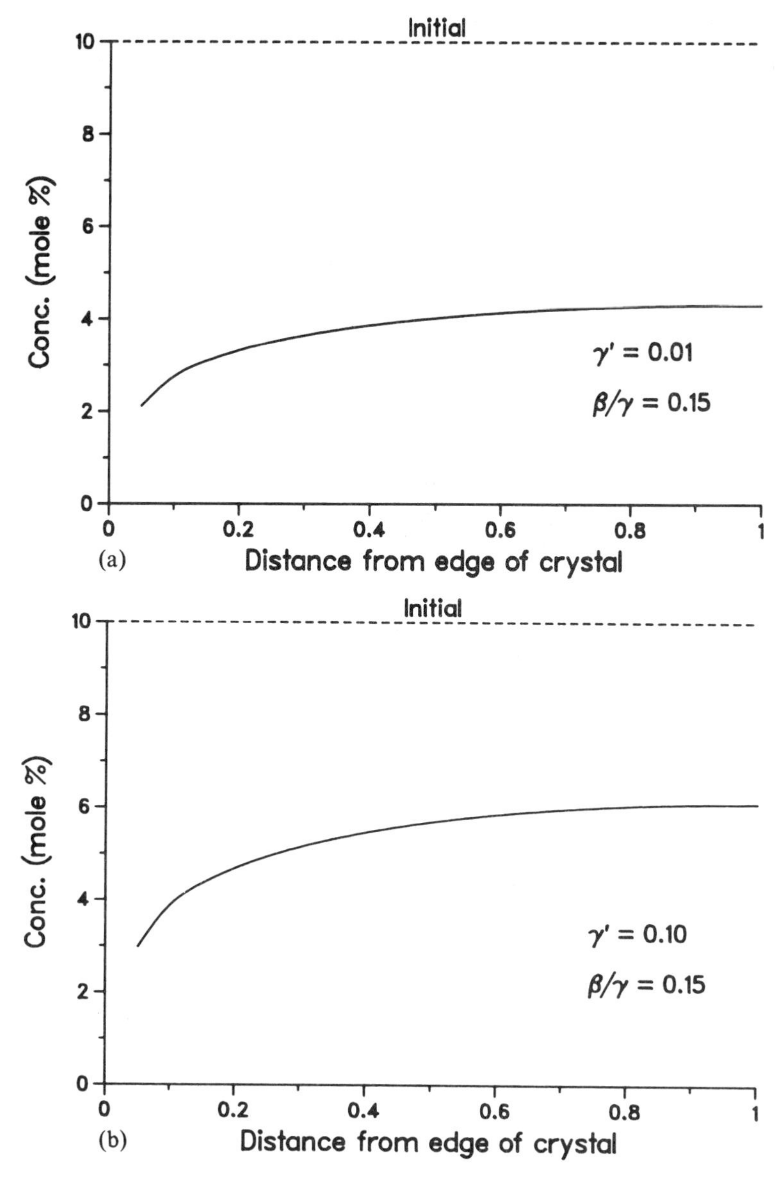

Fig. 4. General curves for the kinetic response of the mineral with the *slow* diffusion in a geothermometer (high $\beta/\gamma \simeq 0.15$). (a) slow cooling, $\gamma' = 0.01$. Note the extensive re-equilibration and the generally homogeneous profile. (b) $\gamma' = 0.1$. Still a rather homogeneous profile with re-equilibration. (c) $\gamma' = 1.0$. Here the approach to the "final" profile is illustrated: ——— — curve is for $t_r' = 0.9$, which would correspond to a time interval of 73 million years if $D_0 = 10^{-17}$ cm^2/sec and $a = 1000$ μm. ——— · curve is for $t_r' = 0.995$, which would correspond to a time interval of 168 million years (same D_0 and a). ——— curve is for $t_r' = 1.0$, which would correspond to a time interval of infinity. (d) $\gamma' = 10.0$. Note that the core of the profile is now essentially unchanged from the initial composition. (e) $\gamma' = 100$ (fast cooling). This corresponds to the ideal case shown in Fig. 1. (f) $\gamma' = 1000$.

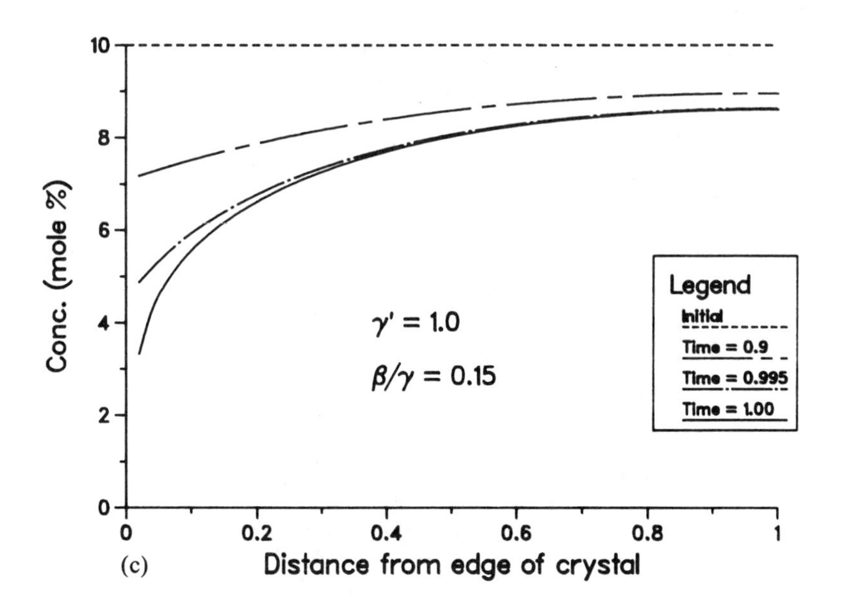

(c)

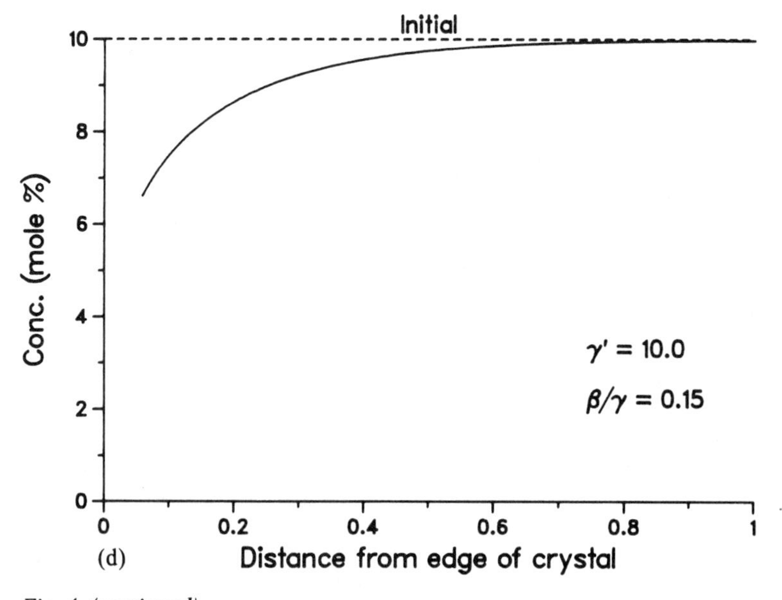

(d)

Fig. 4. (continued)

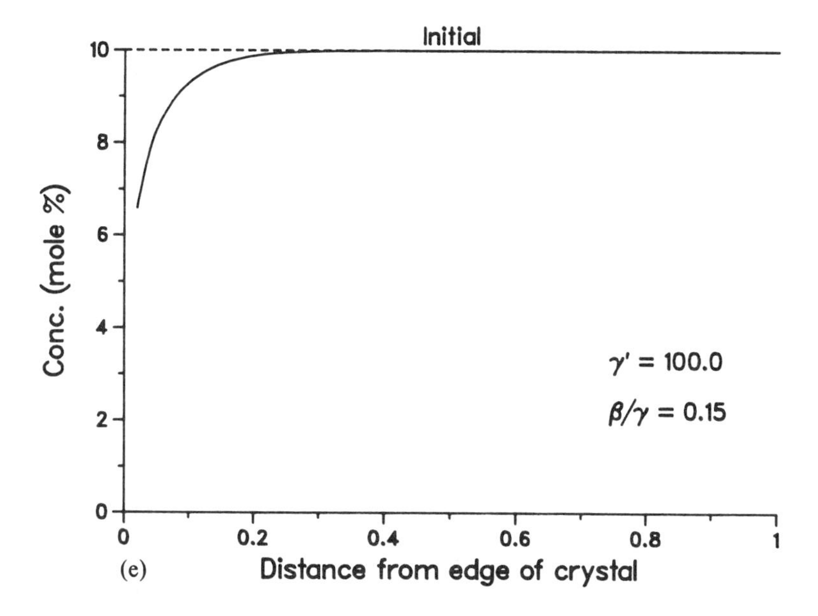

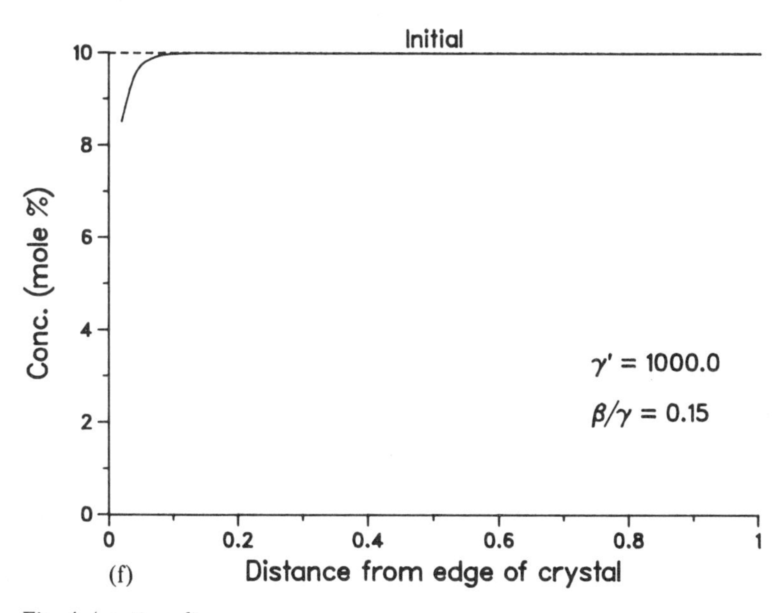

Fig. 4. (continued)

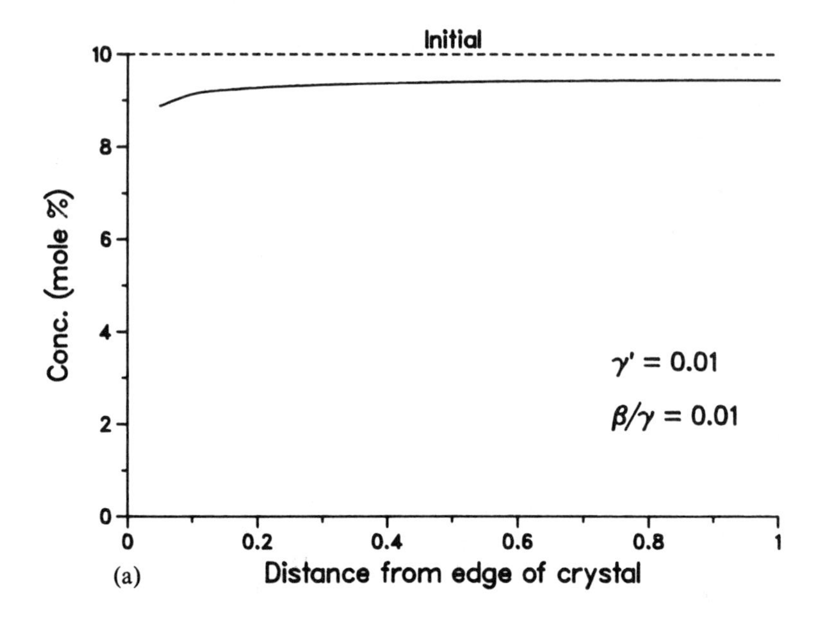

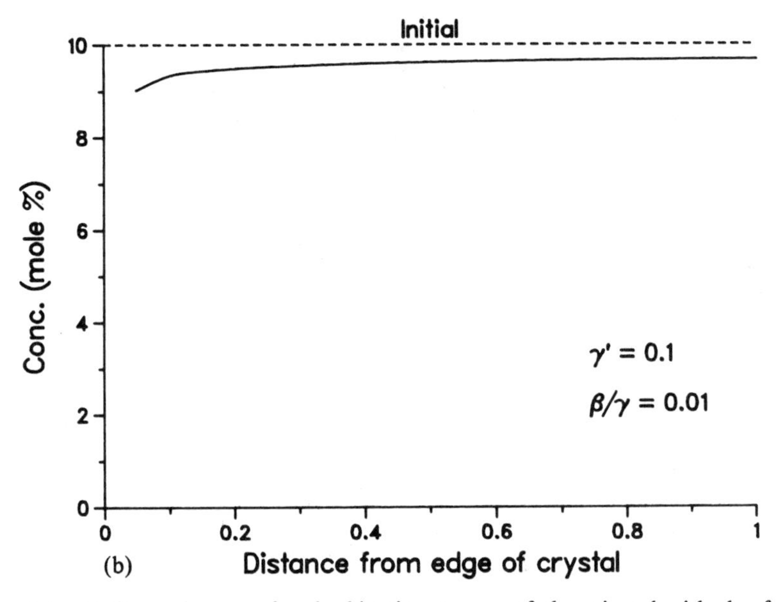

Fig. 5. (a)–(d) General curves for the kinetic response of the mineral with the *faster* diffusion coefficient i.e., $\beta/\gamma \simeq 0.01$. All else is the same as in Fig. 4. Note that in general the profiles are very homogeneous and close to the initial composition.

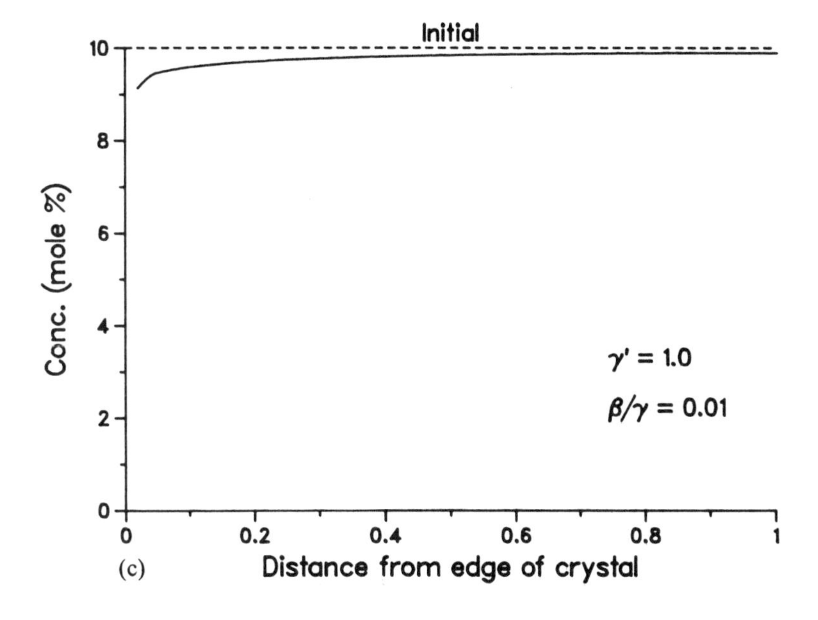

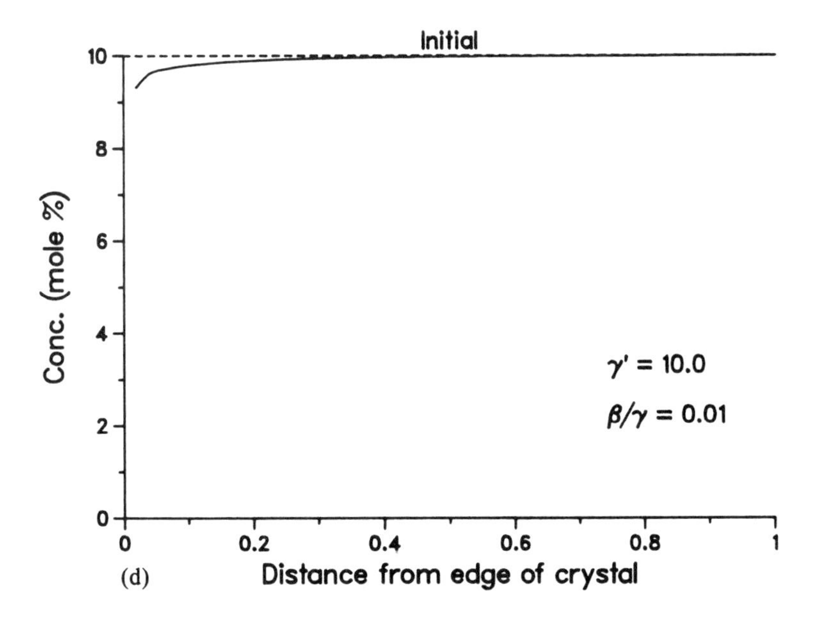

Fig. 5. (continued)

composition is homogeneous for each crystal. As the thermal history progresses, the composition of each crystal re-adjusts. This re-adjustment continues in a manner prescribed by Eq. (34). However, the whole point of introducing t' (or t'_r) is that this kinetically controlled re-adjustment cannot continue indefinitely because the kinetics become extremely slow. Mathematically, this result is manifested in the fact that t' or t'_r has a maximum value that it can attain, even as regular time goes on indefinitely. Therefore, if we were to take a snapshot of the composition through time, there would be variations in the composition profile at first, then a slowing of the variations, and beyond a certain point no more variation. It is this *final* profile that is obtained by setting $t'_r = 1/\gamma'$. This final profile is plotted in Figs. 4 and 5. Of course, this is also the profile that would be obtained by analysis of the crystal as it finally emerges on the earth's surface.

Figure 4 uses a value of $\beta/\gamma = 0.15$ and Fig. 5 uses a value of $\beta/\gamma = 0.01$. These two values are representative of the minerals with slow diffusion and fast diffusion respectively as we will show. In fact, we can use (20) and (31) to write β/γ as

$$\beta/\gamma = \frac{\Delta H^\circ}{E_a} \frac{\sqrt{D_B^0/D_A^0}}{\{\sqrt{D_B^0/D_A^0}\,[1 + (c_{1A}^I/c_{2A}^I)] + [(c_{1A}^I/c_{1B}^I) + (c_{1A}^I/c_{2B}^I)\}}. \quad (37)$$

Note that β/γ is *not* dependent on the value of the rate of change of temperature, s, and thus is more an intrinsic property of the pair of exchanging minerals than it is a property of the thermal history. On the other hand, γ' is very dependent on the thermal history (see below).

If $D_B^0 \gg D_A^0$ and $c_{2A}^I > c_{1A}^I$ then Eq. (37) simplifies to

$$\beta/\gamma \approx \frac{\Delta H^\circ}{E_a}. \quad (38)$$

Table 2 lists some of the reaction enthalpies for several exchange geothermometers. Table 3 lists some relevant diffusion data (see Appendix B for a discussion of the diffusion data). It can be surmised that for typical values of ΔH° and E_a, Eq. (38) yields

$$\beta/\gamma \approx 0.1 \text{ to } 0.2.$$

Therefore, $\beta/\gamma = 0.15$ is a reasonable number for minerals with slow diffusion (where (38) applies). On the other hand, if D_B^0/D_A^0 is small (fast diffusion in A) then

$$\beta/\gamma \approx \frac{\Delta H^\circ}{E_a}\left(\frac{D_B^0}{D_A^0}\right)^{1/2}. \quad (39)$$

For the type of values of D_B° and D_A° (see Table 3) β/γ can be reduced by 1 or 2 orders of magnitude (e.g., $D_B^0/D_A^0 = 10^{-2}$–10^{-4}). Therefore, we chose $\beta/\gamma = 0.01$ as typical for this case.

Table 2. Some enthalpies for ion-exchange reactions.

Mineral pair	Ions	$\Delta H°$ (kcal/mol)	Reference
Clinopyroxene–garnet	Fe–Mg	9.2	Roheim and Green (1974)
Clinopyroxene–garnet	Fe–Mg	5.6	Mori and Green (1978)
Clinopyroxene–garnet	Fe–Mg	3–5	Dahl (1980)
Garnet–biotite	Fe–Mg	4.2	Ferry and Spear (1978)
Garnet–cordierite	Fe–Mg	6.0	Lasaga *et al.* (1977)
Garnet–cordierite	Fe–Mg	5.4	Thompson (1976)
Olivine–clinopyroxene	Ca–Mg	3–4	Adams and Bishop (1982)
Ilmenite–clinopyroxene	Fe–Mg	3–6	Bishop (1980)

The value of γ' is given from (20) and (33) as

$$\gamma' \equiv \frac{E_a s a^2}{D_0 R T_0^2}. \tag{40}$$

Note that D_0 is a function of T_0. Given an Arrhenius equation for D, as in Table 3,

$$D = Ae^{-Ea/RT},$$

and also assuming some values for the cooling rate s and the crystal size, a, we are able to compute γ' as a function of T_0. We will return to Eq. (40) below.

Figures 4 and 5 summarize the possible scenarios allowed for the thermal evolution of the geothermometer. If the mineral has a relatively slow diffusion coefficient, β/γ is high (~0.15); the possible evolutions are given in Fig. 4. As may be expected, low values of γ' yield higher t_r' values and hence more re-equilibration. For example, if $\gamma' = 0.01$ [Fig. 4(a)] the final profile (at $t \to \infty$) changes drastically from the initial profile. In this case, not only has the profile changed

Table 3. Diffusion data.

Mineral	D (cm^2/sec)	Workers
Garnet	6.11 exp $(-82.2/RT)$	Freer (1981)
Garnet	0.023 exp $(-88.3/RT)$	Elphick *et al.* (1981)
Garnet	0.0275 exp $(-70.0/RT)$	Lasaga *et al.* (1977)
Pyroxene	0.0039 exp $(-86.2/RT)$	Brady and McCallister (1983)
Pyroxene	659 exp $(-80.0/RT)$	Sanford and Huebner (1979)
Pyroxene	1.306 exp $(-80.0/RT)$	McCallister *et al.* (1978)
Olivine	0.03 exp $(-61.0/RT)$[a]	Buening and Buseck (1973)
	8×10^{-7} exp $(-31.7/RT)$[b]	Buening and Buseck (1973)
Spinel	0.02 exp $(-86.1/RT)$	Lindner and Akerstrom (1958)

[a] $T > 1125°C$.

[b] $T < 1125°C$.

but the low value of γ' means that the final profile is also relatively flat. This is an important observation because it is precisely this homogeneity that is used as evidence for equilibrium. However, the profile in Fig. 1(a) is certainly a nonequilibrium profile! Therefore, values of $\gamma' \leq 0.01$ will make the geothermometer less applicable. Furthermore, the kinetic analysis shows that one must be careful to check that $\gamma' \gg 0.01$, if homogeneity is to be used as a criterion for equilibrium (see applications below).

If $\gamma' = 0.10$ [Fig. 4(b)] there is still considerable change in the core concentration of the crystal but there is now more pronounced zonation. This continues until at $\gamma' = 10.0$ [Fig. 4(d)] the core region of the crystal has not changed much while the outer edge of the crystal contains severe zoning. At this point ($\gamma' = 10$), the usual assumption about the use of the core concentration to obtain T_0 is indeed warranted. Finally, depending on the spatial resolution of the analysis, a $\gamma' =$ 100–1000 will yield a final profile that is very close to the initial profile. This would then correspond to the ideal case shown in Fig. 1.

The situation is quite different in the case of a mineral with *relatively* fast diffusion, i.e., low β/γ values (Fig. 5). Now all profiles show very slight zoning and only for $\gamma' \leq 0.01$ is the deviation of the final profile from the initial profile significant.

The major point from Fig. 5 is that if the interdiffusion rates of the adjacent minerals are quite disparate, the success of the geothermometer will hinge in large part on the kinetics of the mineral with the slow diffusion. This major result also makes it easier to characterize the kinetic behavior of geothermometers.

Applications

It is clear from the previous section that the crucial mineral in a geothermometric pair is that which has the slowest diffusion coefficient (high β/γ). Table 3 presents diffusion data for the more ubiquitous of these types of minerals. To apply our general results we need merely compute values of γ' from Eq. (40) and use Fig. 4! As stated earlier, given an Arrhenius expression and the cooling rate and size of the crystal we can compute γ' as a function of the initial temperature, T_0. Alternatively, we can compute the values of T_0 for which γ' will obtain a particular value, i.e., we can compute iso-γ' temperatures.

The thermal history, as represented by the parameter s, is obviously of central importance in the kinetic discussion. Useful estimates of s may be obtained in the case of tectonically driven uplift. Bottinga and Allegre (1976) have discussed some simple models. For our purposes (Fig. 6) we may imagine a slab moving toward the surface due to erosion and uplift at a velocity v_z. If k is the thermal conductivity of the slab, the temperature (ignoring lateral variations) would obey the usual conduction–convection equation

$$\frac{\partial T}{\partial t} = k \frac{\partial^2 T}{\partial z^2} + v_z \frac{\partial T}{\partial z}. \tag{41}$$

If we assume steady state has been reached ($\partial T/\partial t = 0$) and that (a) $T = 0$ at the surface, $z = 0$, and (b) $T = T_m$ at some maximum depth, z_m, then the solution to (41) is

$$T(z) = T_m \frac{(1 - e^{-(v_z/k)z})}{(1 - e^{-(v_z/k)z_m})} . \tag{42}$$

But if some material was at z_m at time $t = 0$, the same material would be at $z_m - v_z t$ at time t. Therefore, the temperature of this material at time t would be

$$T(t) = T(z_m - v_z t) = T_m \frac{(1 - e^{-(v_z z_m/k)} e^{(v_z^2/k)t})}{(1 - e^{-(v_z/k)z_m})} . \tag{43}$$

Equation (43) can be differentiated to obtain s:

$$s \equiv \frac{dT}{dt} = -\frac{T_m v_z^2}{k} \frac{e^{-(v_z/k)z_m} e^{(v_z^2/k)t}}{(1 - e^{-(v_z/k)z_m})} . \tag{44}$$

For example, if $v_z = 0.1$ cm/yr, $z_m = 30$ km, $T_m = 700°$C and $k = 0.01$ cm^2/sec, then Eq. (44) would yield $s = 14°$C/my at $t = 0$ and, as the surface is approached, s increases to 36°C/my. Reasonable values of v_z, z_m, and T_m generally yield s values between 10°C/my and 100°C/my. Furthermore, the variation in s is not significant (less than a factor of two) over changes of temperature of several hundred degrees from T_m.

The previous section indicated that a nearly frozen profile is obtained when $\gamma' = 100$. Table 4 gives the iso-γ' temperatures (for $\gamma' = 100$) obtained from the Arrhenius expressions in Table 3, where the cooling rate was taken as $s = 10°$C/my and the crystal size at $a = 1000$ μm. Note that Eq. (40) shows that these iso-γ' temperatures are unchanged if for example $s = 1000°$C/my and $a = 100$ μm or $s = 0.1°$C/my and $a = 10000$ μm, etc.

The meaning of these iso-γ' temperatures is as follows. For any T_0 that is less

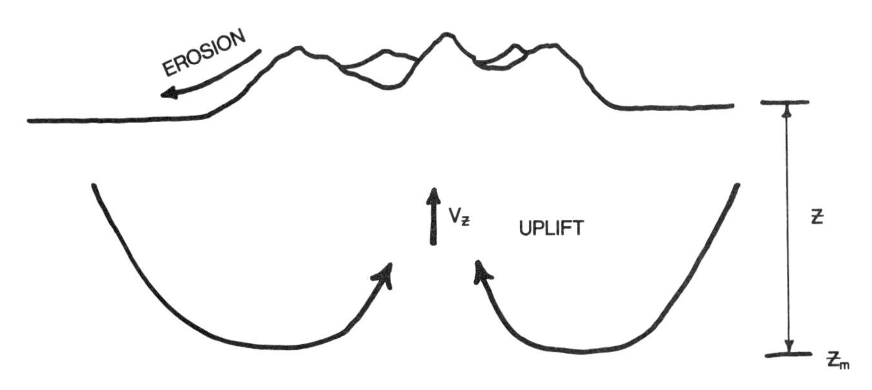

Fig. 6. Uplift model used in the text. The material moves up at a rate of v_z cm/year due to the combined effects of erosion and uplift.

than the one in Table 4, the initial T_0 will be preserved by the geothermometer. For T_0 greater than those in Table 4 more zoning will be evident.

Table 5 gives the iso-γ' temperatures for a $\gamma' = 10$, i.e., where there is zoning but the cores of the minerals are not affected. Again, if T_0 is less than this T_0, we are guaranteed at least preservation of T_0 in the composition of the cores of the minerals. $\gamma' = 10$ *is probably the best transition point between workable and non-workable geothermometers.*

Tables 6 and 7 give iso-γ' temperatures for $\gamma' = 1.0$ and 0.1; the latter γ' value represents the case where we have an "equilibrium-like" nearly homogeneous profile but no preservation of T_0!

The applicability of a geothermometer obviously depends on the thermal history. This can be illustrated by Tables 8–10, where the T_0 values required to produce $\gamma' = 10$ are given as a function of the cooling rate, s.

Pyroxenes exhibit a wide T_0 range, in part reflecting the diffusion of different ions. The data of McCallister *et al.* (1978) are for tracer calcium (^{45}Ca) diffusion in diopside; the data of Brady and McCallister (1983) are for homogenization of fine-scale pigeonite lamellae in subcalcic diopside (see Appendix B). The latter represent Ca–Mg interdiffusion and are more appropriate for the model, because the D's in the model are interdiffusion coefficients.

Table 8 shows the iso-γ' temperatures for pyroxenes using the diffusion data of Brady and McCallister (1983). Pyroxenes have been widely used as upper mantle geothermometers and geobarometers. The kinetic problems have been recently discussed by Dahl (1979). Table 8 can be used to analyze the significance of eclogite and garnet lherzolite temperatures.

For slow cooling rates of less than 10°C per million years, temperatures of 900–1000°C can be considered frozen equilibrium temperatures, i.e., either formation or recrystallization at the temperature obtained from the garnet–pyroxene geothermometer. However, the preservation of higher temperatures, 1100–1300°C, would require very fast transport from the mantle, i.e., fast cooling rates.

Table 4. Iso-γ' temperatures.[a]

Mineral	D	T_0 (°C)
Garnet	$6.11 \exp(-82.2/RT)$	690
Garnet	$0.023 \exp(-88.3/RT)$	907
Garnet	$0.0275 \exp(-70.0/RT)$	665
Pyroxene	$0.0039 \exp(-86.2/RT)$	935
Pyroxene	$659 \exp(-80.0/RT)$	580
Pyroxene	$1.306 \exp(-80.0/RT)$	700
Olivine	$8 \times 10^{-7} \exp(-31.7/RT)$	315
Spinel	$0.02 \exp(-86.1/RT)$	885

$\gamma' = 100$; $s = 10$°C/my; $a = 1000$ μm.

[a] All sources of data are as in Table 3.

Table 5. Iso-γ' temperatures.[a]

Mineral	D	T_0 (°C)
Garnet	$6.11 \exp(-82.2/RT)$	745
Garnet	$0.023 \exp(-88.3/RT)$	980
Garnet	$0.0275 \exp(-70.0/RT)$	720
Pyroxene	$0.0039 \exp(-86.2/RT)$	1010
Pyroxene	$659 \exp(-80.0/RT)$	620
Pyroxene	$1.306 \exp(-80.0/RT)$	755
Olivine	$8 \times 10^{-7} \exp(-31.7/RT)$	370
Spinel	$0.02 \exp(-86.1/RT)$	955

$\gamma' = 10.0$; $s = 10$°C/my; $a = 1000$ μm.

[a]All sources of data are as in Table 3.

Table 6. Iso-γ' temperatures.[a]

Mineral	D	T_0 (°C)
Garnet	$6.11 \exp(-82.2/RT)$	800
Garnet	$0.023 \exp(-88.3/RT)$	1050
Garnet	$0.0275 \exp(-70.0/RT)$	790
Pyroxene	$0.0039 \exp(-86.2/RT)$	1100
Pyroxene	$659 \exp(-80.0/RT)$	665
Pyroxene	$1.306 \exp(-80.0/RT)$	810
Olivine	$8 \times 10^{-7} \exp(-31.7/RT)$	430
Spinel	$0.02 \exp(-86.1/RT)$	1035

$\gamma' = 1.0$; $s = 10$°C/my; $a = 1000$ μm.

[a]All sources of data are as in Table 3.

Table 7. Iso-γ' temperatures.[a]

Mineral	D	T_0 (°C)
Garnet	$6.11 \exp(-82.2/RT)$	865
Garnet	$0.023 \exp(-88.3/RT)$	1155
Garnet	$0.0275 \exp(-70.0/RT)$	860
Pyroxene	$0.0039 \exp(-86.2/RT)$	1200
Pyroxene	$659 \exp(-80.0/RT)$	715
Pyroxene	$1.306 \exp(-80.0/RT)$	880
Olivine	$8 \times 10^{-7} \exp(-31.7/RT)$	500
Spinel	$0.02 \exp(-86.1/RT)$	1125

$\gamma' = 0.1$; $s = 10$°C/my; $a = 1000$ μm.

[a]All sources of data are as in Table 3.

Table 8. Pyroxene iso-γ' temperatures.[a]

s (°C/my)	T_0 (°C)
0.01	805
0.10	865
1.0	935
10.0	1010
100.0	1100
1000.0	1200
10000.0	1320

$\gamma' = 10.0$; $a = 1000 \mu m$.

[a]Diffusion data of Brady and McCallister (1983).

Alternatively, a re-calibration of the geothermometer may be needed. We will return to this example below.

Tables 4 and 5 give garnet T_0 temperatures that suggest that even high-grade metamorphic rocks will retain their garnet composition, at least in the core, if the cooling is not too slow. Therefore, we have a kinetic justification for the various garnet geothermometers.

Table 9 shows similar results for garnet using the diffusion data of Lasaga *et al.* (1977) and of Elphick *et al.* (1981). The data of Lasaga *et al.* (1977) yield temperatures which are much lower than the corresponding ones for pyroxene. However, the applicability of garnet-containing geothermometers in metamorphic terranes is again generally justified (e.g., Ferry, 1980; Tracy *et al.*, 1976). Table 9 shows that if for some reason the cooling paths were slower (e.g., 0.01 °C/my), then $\gamma' = 10$ only at 565 °C. In this case there will be re-equilibration of even the cores down to amphibolite facies range.

Of course, these numbers are based on a crystal size of 1 mm. For smaller

Table 9. Garnet iso-γ' temperatures.

s (°C/my)	T_0[a]	T_0 (°C)[b]
0.01	565°C	785
0.10	610°C	840
1.0	665°C	907
10.0	720°C	980
100.0	790°C	1050
1000.0	860°C	1155
10000.0	945°C	1260

$\gamma' = 10.0$; $a = 1000\ \mu m$.

[a]Diffusion data of Lasaga *et al.* (1977).

[b]Diffusion data of Elphick *et al.* (1981).

crystals the γ' numbers would change according to Eq. (40). The quadratic dependence on "a" means that small garnets, e.g., 100 μm, would reach $\gamma' = 10$ if $s = 10^\circ$C/my only at $T_0 = 610^\circ$C rather than 720°C. The same remarks apply to the other tables.

The data of Lasaga *et al.* (1977) would indicate that garnet would not be as useful at higher temperatures, e.g., in upper mantle geothermometry. However, in this case, garnet may be the mineral with fast diffusion (e.g., when compared to pyroxene) and one of our results is that the critical T_0 is the T_0 of the mineral with the slowest diffusion. Table 8 has indicated that use of pyroxene pushes the acceptable T_0 to higher values (although still not covering the entire range).

Table 10 gives the results for olivine. It should be noted that the diffusion data of Buening and Buseck have a kink (change in mechanism) at around 1125°C. Therefore, the appropriate D vs. T curve has been used in Table 10. It is also important to realize that D varies as $f_{O_2}^{1/6}$ in olivine (Buening and Buseck, 1973). Therefore, f_{O_2} values other than 10^{-12} bar will change the numbers accordingly. It is clear that olivine is absolutely unacceptable as a geothermometer in plutonic environments (see also Smith and Roden (1981)). However, as we raise the value of s to those of extrusive events, the T_0 for olivine finally climbs to acceptable levels. For temperatures greater than 1000°C, the cooling rate would have to be greater than 10°C/year, if olivine is to be acceptable. However, it should be repeated that this constraint applies only if olivine is the mineral with slow diffusion.

The spinel ($MgAl_2O_4$) T_0 is similar to those obtained for pyroxene. Therefore the spinel is a good candidate for geothermometry of ultramafic rocks.

It is probably instructive to close with a more detailed example of the power and simplicity in the use of γ' and Figs. 4 and 5. Recently there has been interest in deciphering the meaning of the temperatures obtained for garnet–clinopyroxene pairs, expecially in garnet lherzolites found in crustal rocks (see Carswell and

Table 10. Olivine iso-γ' temperatures.[a]

s (°C/my)	T (°C)
1.0	315
10^1	370
10^2	430
10^3	500
10^4	585
10^5	695
10^6	830
10^7	1005
10^8	1235

$\gamma' = 10.0$; $a = 1000$ μm.

[a]Data of Buening and Buseck (1973) and $f_{O_2} = 10^{-12}$ bars.

Gibb, 1980; Harte *et al.*, 1981; Harte and Freer, 1982; Dahl, 1979). The kinetic questions that arise are amenable to the treatment presented here. For calculations, we have used the revised K_D of Mori and Green (1978):

$$K_D \equiv \frac{(\mathrm{Fe/Mg})_{\mathrm{cpx}}}{(\mathrm{Fe/Mg})_{\mathrm{gar}}} \tag{45}$$

where

$$K_D = 3.287\ e^{-5600/RT}. \tag{46}$$

The diffusion data of Brady and McCallister (1983) for clinopyroxene and of Elphick *et al.* (1981) and Lasaga *et al.* (1977) for garnet (see Table 3) have been used. The initial compositions were assumed in equilibrium at 1200°C:

$$c_{\mathrm{Mg}}^{\mathrm{gar}} = 22 \text{ mol }\%, \quad c_{\mathrm{Fe}}^{\mathrm{gar}} = 15 \text{ mol }\%,$$
$$c_{\mathrm{Mg}}^{\mathrm{cpx}} = 20 \text{ mol }\%, \quad c_{\mathrm{Fe}}^{\mathrm{cpx}} = 6.62 \text{ mol }\%.$$

The garnet size is taken as 0.6 mm and the clinopyroxene crystal as 0.3 mm. How would this geothermometer respond to various thermal histories?

The previous equations were rewritten as

$$\frac{\partial c_A}{\partial t_r'} = \frac{\partial^2 c_A}{\partial x'^2}, \qquad \frac{\partial c_B}{\partial t_r'} = \frac{D_B(t)}{D_A(t)} \frac{\partial^2 c_B}{\partial x'^2},$$
$$x' = 0 \text{ to } 1, \qquad x' = 0 \text{ to } b/a,$$

and

$$\begin{aligned} \frac{c_A(c_B^{\mathrm{tot}} - c_B)}{c_B(c_A^{\mathrm{tot}} - c_A)} &= K_D^0 e^{-\beta t} \\ &= K_D^0 (1 - \gamma' t_r')^{\beta/\gamma}, \\ D_A(t_r') \left.\frac{\partial c_A}{\partial x'}\right|_{x'=0} &= D_B(t_r') \left.\frac{\partial c_B}{\partial x'}\right|_{x'=0}, \\ \left.\frac{\partial c_A}{\partial x'}\right|_{x'=1} &= 0, \qquad \left.\frac{\partial c_B}{\partial x'}\right|_{x'=b/a} = 0, \end{aligned}$$

where b is the length of mineral B and the other variables are defined as before. c_A now stands for c_{1A} and c_B for c_{1B}. (Component 1 was taken as Fe and mineral A as clinopyroxene, B as garnet.) c_A^{tot} and c_B^{tot} are the sums in Eqs. (9) and (10) (e.g., 26.62 and 37 mol %). These equations were solved *exactly* by a general finite difference program (program is available from author). The results are shown for each set of garnet diffusion data in Figs. 7 and 8.

Returning to our general scheme, clinopyroxene is the slow diffusing mineral in this pair. Therefore we should compute γ' for cpx. Using $a = 300\ \mu$m, the results are shown in Table 11. If $s = 10$°C/my, $\gamma' = 0.009$ at 1200°C. Therefore, it is clear that extensive re-equilibration will take place as is shown in Figs. 7 and 8. Note the slight change in the garnet profiles, which is predicted by Fig. 5. We had stated earlier that γ' would have to reach 10 before the cores of the

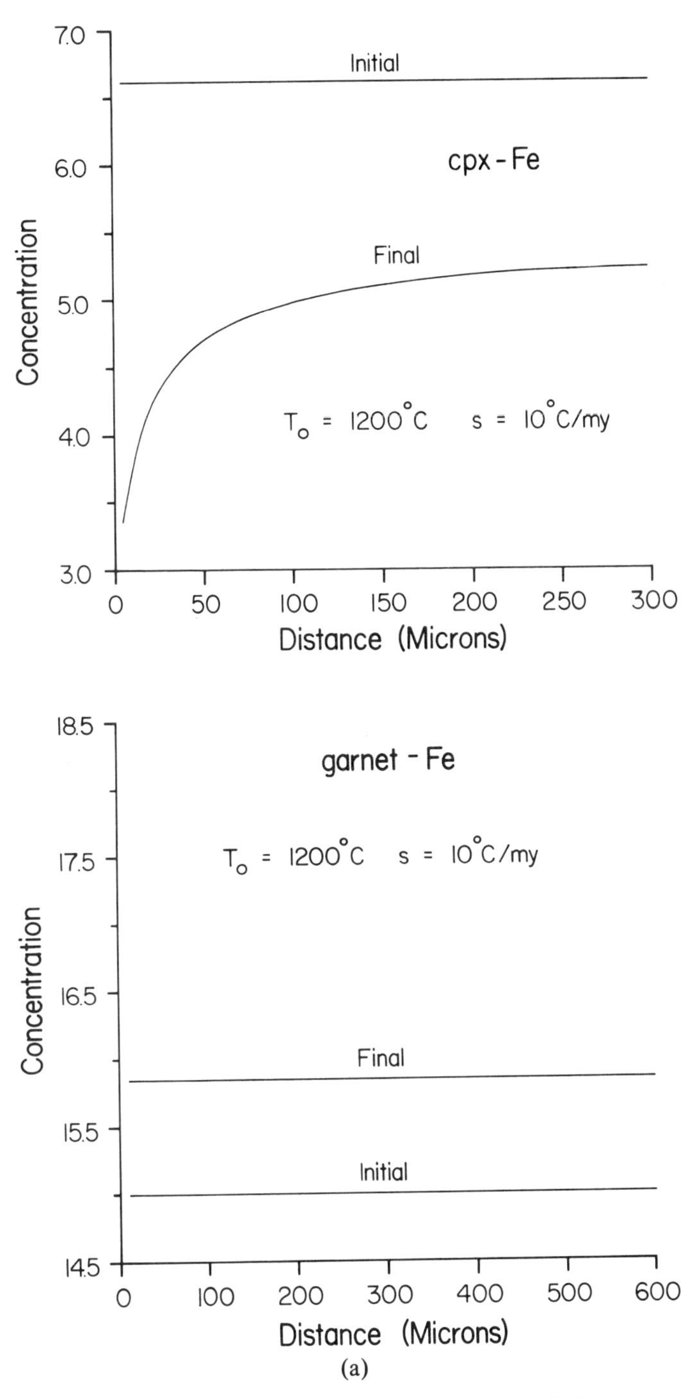

(a)

Fig. 7. (a) Final Fe composition profiles (in mol %) in garnet and clinopyroxene after re-equilibration with $s = 10^\circ C/my$. The initial profiles are as discussed in the text. Diffusion data of Lasaga *et al.* (1977) are used for garnet. Note the very homogeneous profile obtained for the garnet and the more heavily zoned and re-equilibrated clinopyroxene.

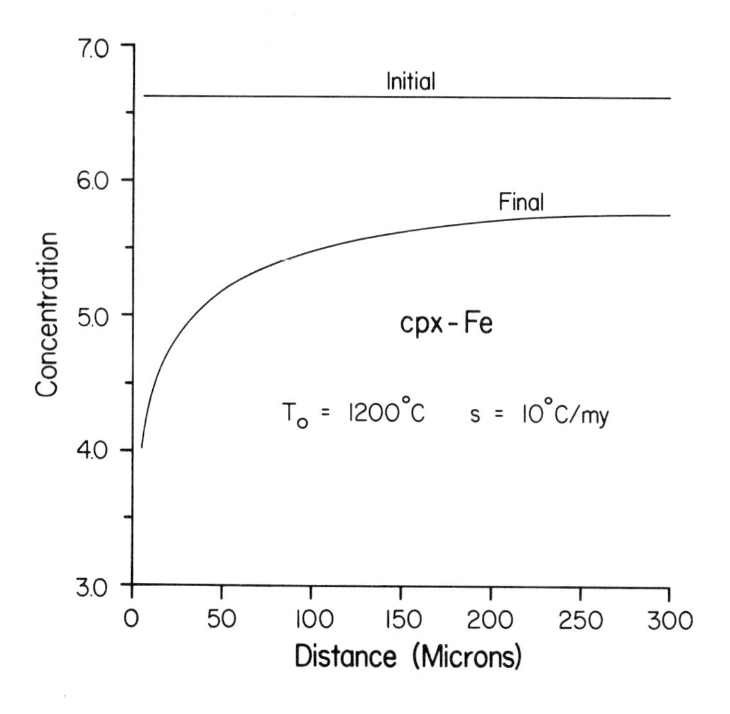

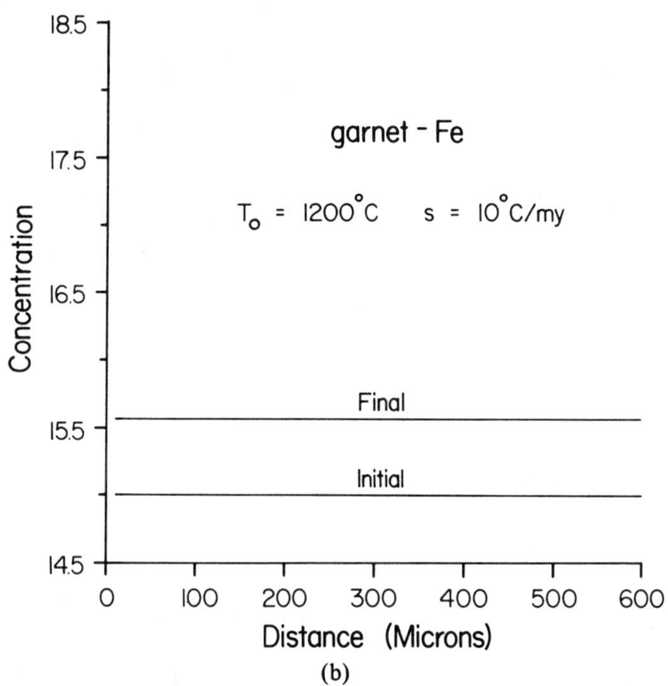

(b)

Fig. 7. (b) Same as (a) except that now $s = 100\,^\circ\mathrm{C/my}$. Note the much less re-equilibration of the clinopyroxene.

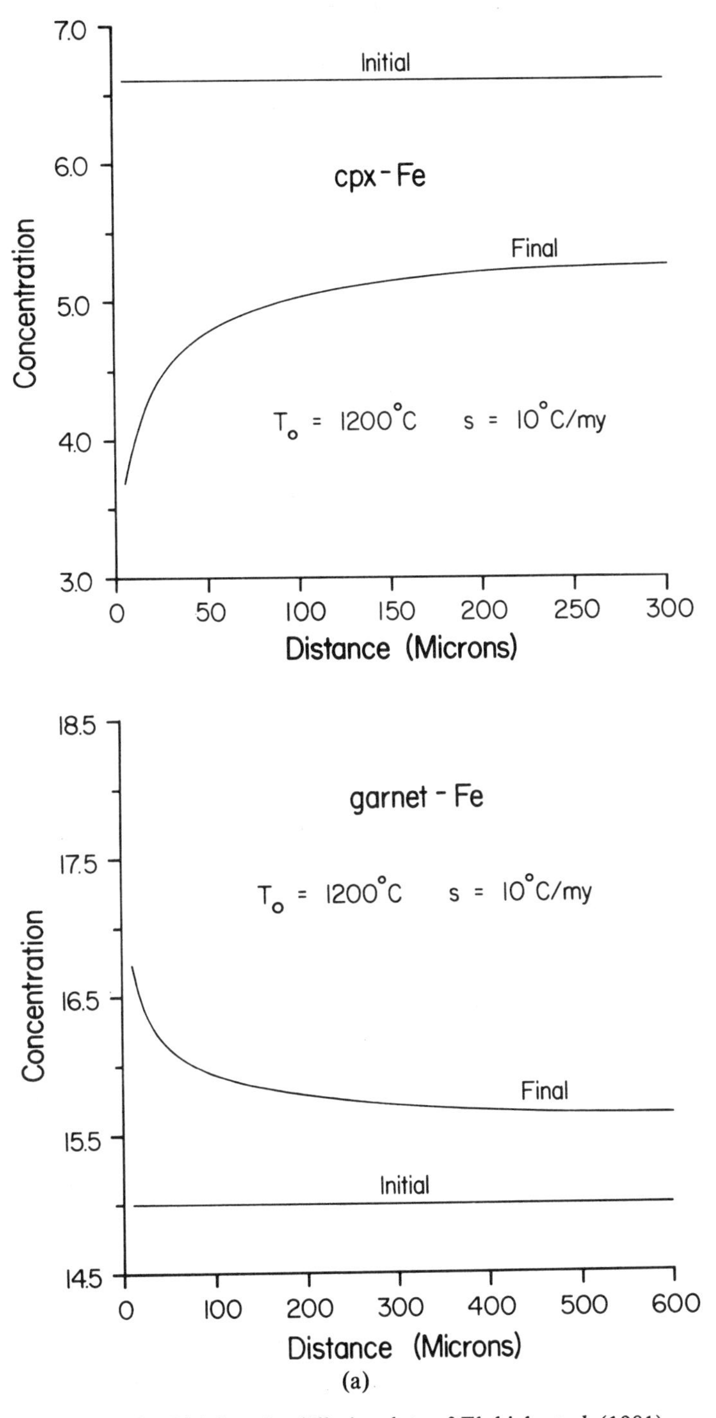

Fig. 8. (a) Same as Fig. 7(a) but the diffusion data of Elphick *et al.* (1981) are used for garnet.

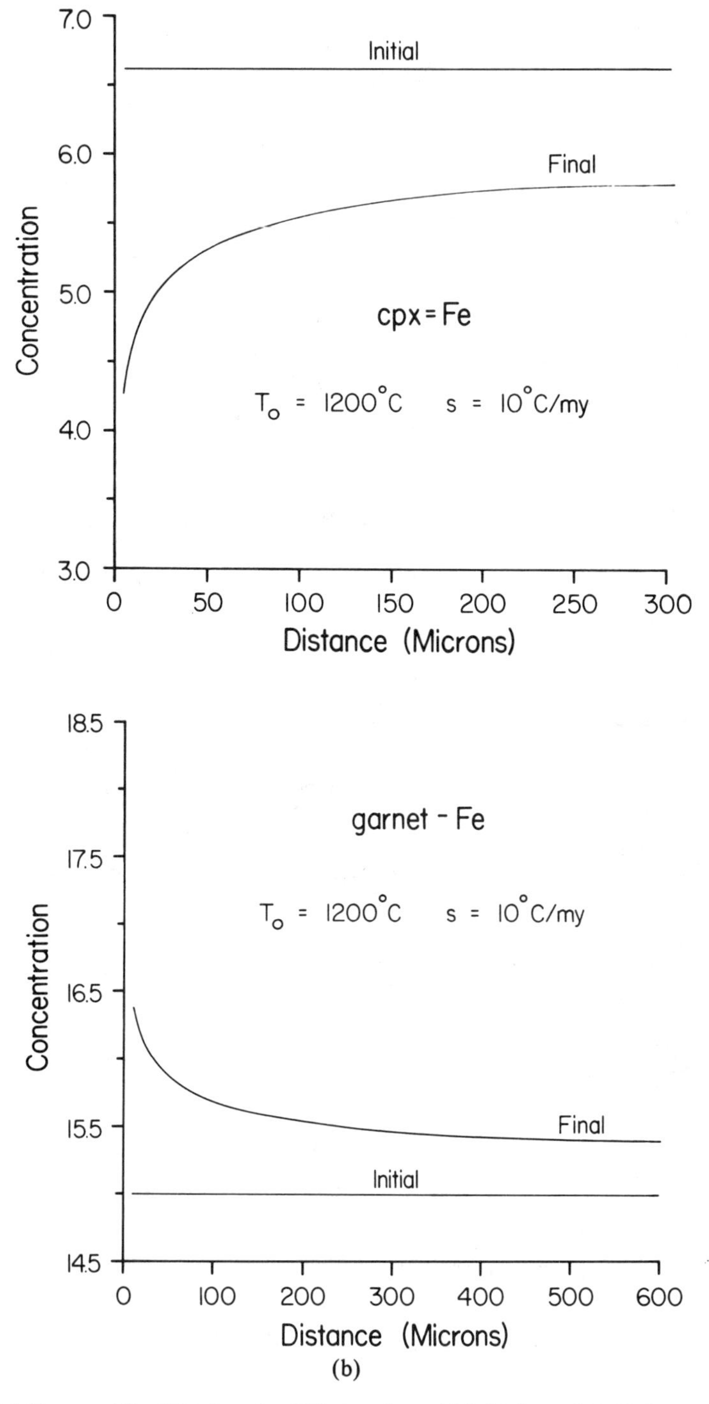

Fig. 8. (b) Same as Fig. 7(b) but the diffusion data of Elphick *et al.* (1981) are used for garnet.

Table 11. γ' values for clinopyroxene.[a]

$s = 10^\circ C/my$		$s = 100^\circ C/my$	
T_0 (°C)	γ'	T_0 (°C)	γ'
1200	0.009	1200	0.090
1100	0.089	1100	0.89
1000	1.24	1050	3.16
950	5.40	1000	12.4
900	26.60	950	54.0
850	150.6	900	266.0

$a = 300\ \mu m$.

[a]Diffusion data of Brady and McCallister (1983).

crystals could preserve the temperature. A simple calculation yields $T_0 = 930^\circ C$ if $\gamma' = 10$ and $s = 10^\circ C/my$. We can take the core regions of Fig. 7(a) to calculate K_D:

$$K_D = \frac{5.23/21.38}{15.84/21.16} = 0.3268,$$

which, using Eq. (46), yields

$$T = 948^\circ C$$

in very good agreement with T_0. Similarly $T_0 = 1010^\circ C$ if $\gamma' = 10$ and $s = 100^\circ C/my$. The K_D calculated from the cores of Fig. 7(b) yields $T = 1033^\circ C$.

Figure 8 uses different garnet diffusion data. According to our results, if garnet is the faster diffusing mineral, changing diffusion coefficients should *not* lead to major changes in the profiles. It is of interest to note that the core compositions in Figs. 8(a) and 8(b) yield corresponding temperatures of 965°C and 1051°C. Obviously, the similarity in the temperatures of Figs. 7 and 8 show that it is the slow diffusing mineral that controls the ultimate K_D obtained. In both cases, clinopyroxene is the mineral with the lowest D, although in the case of the Elphick *et al.* data the D's are beginning to approach each other (and there are some deviations).

These calculations show the great power of γ' and Figs. 4 and 5. They also quantify further the kinetic problem recently raised by Harte and Freer (1982).

Appendix A

Equations (7) and (8) are needed to satisfy conservation of mass at the boundary of the exchanging minerals. However, in most applications the concentration units are in mol % or mole fraction. While these units are perfectly acceptable in the K_D equation and in the diffusion equation if there are negligible changes in volume due to diffusion for each mineral, they must be corrected when comparing the fluxes as in (7) and (8). The reason for the correction stems from the need to match fluxes in units of moles/cm^2/sec. Let us assume c in Eqs. (7) and (8) is in units of mol %. Then to convert the diffusion flux to units of moles/cm^2/sec, it is necessary to use molar volumes (to change c to units of moles/cm^3). For example, suppose the minerals are orthopyroxene, $Mg_xFe_{1-x}SiO_3$, and garnet, $Mg_xFe_{3-x}Al_2Si_3O_{12}$. Then, utilizing oxides, the mol % of magnesium in opx and garnet is given by

$$\text{mol \% Mg} = \frac{x}{2} \cdot 100 \qquad \text{opx},$$

$$\text{mol \% Mg} = \frac{x}{7} \cdot 100 \qquad \text{garnet}.$$

Therefore the relation between mol % and moles/cm^3 is

$$\text{for opx moles Mg/cm}^3 = \frac{x}{\overline{V}_{\text{opx}}} = \frac{\text{mol \%}}{100} \frac{2}{\overline{V}_{\text{opx}}},$$

$$\text{for garnet moles Mg/cm}^3 = \frac{x}{\overline{V}_{\text{gar}}} = \frac{\text{mol \%}}{100} \frac{7}{\overline{V}_{\text{gar}}},$$

where $\overline{V}_i$ stands for the molar volume of i. Therefore the correct expression of the diffusion flux boundary condition is

$$-\frac{2}{100\overline{V}_{\text{opx}}} D_{\text{opx}} \left.\frac{\partial c_{\text{opx}}}{\partial x}\right|_{x=0} = -\frac{7}{100\overline{V}_{\text{gar}}} D_{\text{gar}} \left.\frac{\partial c_{\text{gar}}}{\partial x}\right|_{x=0}.$$

These equations reduce to Eqs. (7) and (8) only if $2/\overline{V}_{\text{opx}} \approx 7/\overline{V}_{\text{gar}}$. This is indeed reasonably true, i.e., using $\overline{V}_{\text{opx}} \approx \overline{V}_{\text{enstatite}} = 31.4$ cm^3/mol and $\overline{V}_{\text{gar}} \approx \overline{V}_{\text{pyrope}} = 113$ cm^3/mol, then $2/\overline{V}_{\text{opx}} = 0.064$ and $7/\overline{V}_{\text{gar}} = 0.062$! Analogous correction factors for olivine, $Mg_xFe_{2-x}SiO_4$, and clinopyroxene, $Mg_xFe_{1-x}CaSi_2O_6$, are $3/(100\overline{V}_{\text{oliv}})$ and $4/(100\overline{V}_{\text{cpx}})$, respectively. Again, using $\overline{V}_{\text{oliv}} \approx \overline{V}_{\text{forsterite}} = 43.79$ cm^3/mol and $\overline{V}_{\text{cpx}} \approx \overline{V}_{\text{diopside}} = 66.09$ cm^3/mol, the comparison is between $3/\overline{V}_{\text{oliv}} = 0.068$ and $4/\overline{V}_{\text{cpx}} = 0.061$. It should be clear that the effect of incorporating these volume factors on either side of Eq. (7) or (8) is such as to leave the equation unchanged (i.e., we multiply both sides by almost the same number!). In conclusion, Eqs. (7) and (8) are almost exact (and quite adequate for our purposes) as they stand.

Appendix B

Diffusion Data

We should make some remarks regarding the diffusion data in Table 3. There is still a *great* need to obtain more data on mineral diffusion, especially because of the complexities of mineral structures and because of the slow diffusion mechanisms. However, there is now a preliminary body of data that is tentatively growing. The data needed in applications such as those envisaged here are the interdiffusion coefficients of minerals rather than the tracer diffusion coefficients (see Anderson, 1981). The data in Table 3 list some of the recent numbers for minerals of importance to geothermometry. The Freer (1981) garnet data are for cation interdiffusion in an almandine–pyrope couple and were obtained at 1150–1330°C and 30 kbar. The garnet data of Elphick *et al.* (1981) were also for Fe–Mg interdiffusion in almandine–pyrope couples at 1200–1400°C. Their data yield

$$\tilde{D} = 0.023\, e^{-88.3/RT} \text{ to } 1.86\, e^{-118.5/RT}\ \text{cm}^2/\text{sec}.$$

We have chosen the expression with the lower activation energy because it conforms better with other results. The garnet data of Lasaga *et al.* (1977) were again for Fe–Mg interdiffusion in almandine–pyrope but were obtained for high-grade metamorphic conditions from field observations. The f_{O_2} dependence of garnet diffusion has not been studied in any of the works discussed above.

Pyroxene diffusion data seem to be still in a state of infancy. The data of Brady and McCallister (1983) were obtained from homogenization of fine-scale coherent (001) pigeonite lamellae in a subcalcic diopside and refer to Ca–Mg interdiffusion in the 1150–1250°C range and at 25 kbar. The equation under Sanford and Huebner (1979) was computed by us using their minimum interdiffusion coefficient of 4×10^{-11} cm²/sec at 1050°C and a "reasonable" activation energy of 80 kcal/mol. Likewise the equation under McCallister *et al.* (1978) uses their tracer ^{45}Ca diffusion data in diopside at 1300°C and an activation energy of 80 kcal/mol.

The olivine Mg–Fe interdiffusion data of Buening and Buseck (1973) are shown. These data were obtained at temperatures of 1000–1200°C. The f_{O_2} was controlled in this case at $f_{O_2} = 10^{-12}$ bar. The f_{O_2} dependence was studied and D was found to depend on $f_{O_2}^{1/6}$. Furthermore, the ln D vs. $1/T$ curves had a change in slope at 1125°C, suggesting a change in diffusion mechanism.

Finally, the spinel ($MgAl_2O_4$) data of Lindner and Akerstrom (1958) shown are for Mg tracer diffusion and were obtained at 1173–1673K.

It is clear that use of these data is tentative. In many cases, important variables such as f_{O_2} have been ignored. Most Fe-containing minerals will have diffusion coefficients that depend on f_{O_2} (see Lasaga, 1981). Furthermore, we will be using the data in temperature intervals that lie outside the experimental ones. Such extrapolations should be done cautiously because diffusion mechanisms may change (Lasaga, 1981).

Summary

The equations governing the kinetic response of ion exchange geothermometers to their thermal history can be solved and reduced to workable parameters. With the introduction of the t' transformation, the equations are simplified considerably. Further simplification is accomplished by the use of dimensionless variables,

t'_r and x'_r. The solution to the equations can then be characterized largely by just two dimensionless parameters, γ' and β/γ. These parameters depend on the diffusion coefficients of the minerals, the enthalpy change of the ion exchange reaction, the initial compositions and the thermal history. Therefore, a general set of curves was generated, which depicted the behavior for various typical values of γ' and β/γ. Some major conclusions obtained are that (a) the lack of zoning is not an absolute guarantee of equilibrium (i.e., low γ' values will also yield the same type of profile) and (b) the usefulness of a geothermometer is largely controlled by the mineral with the slowest diffusion process in the temperature range of interest.

The general curves can be used to analyze all the various common ion exchange geothermometers. In particular, the suitability of garnet geothermometers for most metamorphic conditions is quantitatively proven. Use of olivine, on the other hand, is strictly limited to rather fast cooling conditions ($s > 10°\text{C/year}$). The applicability of garnet–pyroxene geothermometry in upper mantle studies is problematic in that in some cases the thermometer will work while in others the temperatures obtained may be nonequilibrium.

References

Adams, G. E., and Bishop, F. C. (1982) Experimental investigation of Ca–Mg exchange between olivine, orthopyroxene, and clinopyroxene: potential for geobarometry, *Earth Planet. Sci. Lett.* **57**, 241–250.

Anderson, D. E. (1981) Diffusion in electrolyte mixtures, in *Kinetics of Geochemical Processes,* vol. 8, *Reviews in Mineralogy,* edited by A. C. Lasaga and R. J. Kirkpatrick, Ch. 6., pp. 211–260, Min. Soc. Amer., Washington, D.C.

Bishop, F. C. (1980) The distribution of Fe^{2+} and Mg between coexisting ilmenite and pyroxene with applications to geothermometry, *Amer. J. Sci.* **280**, 46–77.

Bottinga, I., and Allegre, C. (1976) Geophysical, petrological and geochemical models of the oceanic lithosphere. Tectonophysics, **32**, 9–59.

Brady, J. B., and McCallister, R. H. (1983) Diffusion data for clinopyroxenes from homogenization and self-diffusion experiments, *Amer. Mineral.,* **68,** 95–105.

Buening, D. K., and Buseck, P. R. (1973) Fe–Mg lattice diffusion in olivine, *J. Geophys. Res.* **78**, 6852–6861.

Carswell, D. A., and Gibb, F. G. F. (1980) The equilibration conditions and petrogenesis of European crustal garnet lherzolites, *Lithos* **13**, 19–29.

Dahl, P. S. (1979) Comparative geothermometry based on major-element and oxygen isotope distributions in Precambrian metamorphic rocks from southwestern Montana, *Amer. Mineral.* **64**, 1280–1293.

Dahl, P. S. (1980) The thermal-compositional dependence of Fe^{2+}–Mg distributions between coexisting garnet and pyroxene: Applications to geothermometry, *Amer. Mineral.* **65**, 854–866.

Dodson, M. H. (1973) Closure temperature in cooling geochronological and petrological systems, *Contrib. Mineral. Petrol.* **40**, 259–274.

Elphick, S. C., Ganguly, J., and Loomis, T. P. (1981) Experimental study of Fe–Mg interdiffusion in aluminumsilicate garnet, EOS (abstract), *Trans. Amer. Geophys. Union* **62**(17), 411.

Ferry, J. M. (1980) A comparative study of geothermometers and geobarometers in pelitic schists from south-central Maine, *Amer. Mineral.* **65**, 720–732.

Ferry, J. M., and Spear, F. S. (1978) Experimental calibration of the partitioning of Fe and Mg between biotite and garnet, *Contrib. Mineral. Perol.* **66**, 113–117.

Freer, R. (1981) Diffusion in silicate minerals and glasses: a data digest and guide to the literature, *Contrib. Mineral. Petrol.* **76**, 440–454.

Harte, B., and Freer, R. (1982) Diffusion data and their bearing on the interpretation of mantle nodules and the evolution of the mantle lithosphere, *Terra Cognita* **2**(3), 273–275.

Harte, B., Jackson, P. M., and Macintyre, R. M. (1981) Age of mineral equilibria in granulite facies nodules from kimberlites, *Nature* **291**, 147–148.

Lasaga, A. C. (1979) Multicomponent exchange and diffusion in silicates, *Geochim. Cosmochim. Acta* **43**, 455–469.

Lasaga, A. C. (1981) The atomistic basis of kinetics: defects in minerals, in *Kinetics of Geochemical Processes,* vol. 8, *Reviews in Mineralogy,* edited by A. C. Lasaga and R. J. Kirkpatrick, Ch. 7., pp. 261–320, Min. Soc. Amer., Washington, D.C.

Lasaga, A. C., Richardson, S. M., and Holland, H. D. (1977) The mathematics of cation diffusion and exchange between silicate minerals during retrograde metamorphism, in *Energetics of Geological Processes,* edited by S. K. Saxena and S. Bhattachanji, pp. 353–388, Springer-Verlag, New York.

Lindner, R., and Akerstrom, A. (1958) *Z. Phys. Chem.* **18**, 303.

McCallister, R. H., Brady, J. B., and Mysen, B. O. (1978) Self-diffusion of Ca in diopside, *Carnegie Institute Reports* **78**, 574–577.

Mori, T., and Green, D. H. (1978) Laboratory duplication of phase equilibria observed in natural garnet lherzolites, *J. Geol.* **86**, 83–97.

Navrotsky, A. (1976) Silicates and related minerals: Solid state chemistry and thermodynamics applied to geothermometry and geobarometry, *Prog. in Solid State Chem.* **11**, 203–264.

Newton, R. C. (1977) Thermochemistry of garnets and aluminous pyroxenes in the CMAS system, in *Thermodynamics in Geology,* edited by D. G. Fraser, pp. 29–55, Reidel Pub. Co., Boston.

Roheim, A., and Green, D. H. (1974) Experimental determination of the temperature and pressure dependence of the Fe–Mg partition coefficient for coexisting garnet and clinopyroxenes, *Contrib. Mineral. Petrol.* **48**, 179–203.

Sanford, R. F. and Huebner, J. S. (1979) Reexamination of diffusion processes in 77115 and 77215 (abstract), in *Lunar and Planetary Science,* Vol. X, pp. 1052–1054, Lunar and Planetary Science Institute, Houston, Texas.

Saxena, S. K. (1972) Retrieval of thermodynamic data from a study of inter-crystalline and intra-crystalline ion-exchange equilibrium, *Amer. Mineral.* **57**, 1782–1800.

Saxena, S. K. (1973) *Thermodynamics of Rock-Forming Crystalline Solutions,* Springer-Verlag, New York.

Smith, D., and Roden, M. F. (1981) Geothermometry and kinetics in a two-spinel peridotite nodule, Colorado Plateau, *Amer. Mineral.* **66**, 334–345.

Thompson, A. B. (1976) Mineral reactions in pelitic rocks: II. Calculation of some *P–T–X* (Fe–Mg) phase relations, *Amer. J. Sci.* **276**, 425–454.

Tracy, R. J., Robinson, P., and Thompson, A. B. (1976) Garnet composition and zoning in the determination of temperature and pressure of metamorphism, central Massachusetts, *Amer. Mineral.* **61**, 762–775.

Wood, B. J. (1977) Experimental determination of the mixing properties of solid solutions with particular reference to garnet and clinopyroxene solutions, in *Thermodynamics in Geology,* edited by D. G. Fraser, Ch. 2, pp. 11–28, Reidel Pub. Co., Boston.

Wood, B. J., and Fraser, D. G. (1978) *Elementary Thermodynamics for Geologists,* Oxford Univ. Press, New York/London.

Chapter 4
Mg–Fe Fractionation in Metamorphic Environments

H. P. Eugster and E. S. Ilton

Introduction

Metamorphic reactions generally proceed in and depend on the presence of an aqueous fluid, whether as a separate phase or as an aqueous grain boundary phase. Walther and Orville (1982) believe that

> Regionally metamorphosed rocks will have a discrete fluid phase only when devolatilization reactions are actually taking place. At other times only an absorbed surface monolayer of volatiles on the minerals will be present. (p. 252)

Whichever is the case, this fluid is the principal transport medium for chemical exchanges between reactants and product minerals. The composition of this fluid must be known in order to define concentration gradients and calculate mass balance conditions. In earlier contributions (Gunter and Eugster, 1980; Eugster, 1981, 1982; Eugster and Gunter, 1981) we have used mineral assemblages and their experimentally determined solubilities to estimate fluid compositions. This approach rests on the assumption that for most metamorphic reactions mineral assemblages buffer the composition of the surrounding fluid. Effects of solid solution on solid–fluid fractionation were also evaluated briefly.

In this chapter we attempt to build a bridge to the large amount of data available on solid–solid exchange reactions in natural and synthetic systems. For the time being, we restrict ourselves to Mg–Fe exchange. The importance of measuring distribution constants between two or more crystalline solutions was documented by Kretz (1959) and Mueller (1960). Since then, such information has been used extensively in geothermometry and geobarometry, as well as for the purpose of characterizing the solid solutions. Solubility data for silicates at elevated temperatures and pressures are confined to pure end members, but with the aid of exchange data calibrated in the laboratory, we can estimate solubilities of intermediate members. We can also use the extensive information available on natural assemblages in order to estimate fractionation effects between solids and fluids, which can be very pronounced.

Review of Cl–OH Systems

Systems dominated by Cl and OH have been discussed earlier (Eugster and Gunter, 1981) with respect to the major cationic solutes SiO_2, Na, K, Ca, Mg, and Fe. The few available data on carbonate-dominated fluids at elevated P and T were summarized by Eugster (1982). In some near-surface environments, SO_4 is an important solute, provided the conditions are sufficiently oxidizing. Anhydrite frequently occurs in the alteration products of basalt by sea water, while sulfate reduction has been invoked for the formation of large-scale sulfide deposits (e.g., Williams, 1978). Nonetheless, data on fluid inclusions point to the fact that most metamorphic fluids are dominated by chloride (see for instance Hollister and Crawford, 1981).

Mineral solubilities are strongly dependent upon chloride concentrations, with the lowest solute levels associated with the freshest waters. In the absence of chloride, OH becomes the dominant anion, representing the most alkaline conditions which can be reached by the chloride-free, limiting system. These relations are illustrated in Fig. 1 for fluids equilibrated with the assemblage albite + paragonite + quartz at 2000 bars pressure. Individual species abundance curves are based on the following exchange and dissociation reactions:

$$1.5\ NaAlSi_3O_8 + HCl \rightleftharpoons 0.5\ NaAl_3Si_3O_{10}(OH)_2 + 3\ SiO_2 + NaCl \quad (1)$$

$$H_2O \rightleftharpoons H^+ + OH^- \quad (2)$$

$$HCl \rightleftharpoons H^+ + Cl^- \quad (3)$$

$$NaCl \rightleftharpoons Na^+ + Cl^-. \quad (4)$$

The abundances of each of the seven species, H_2O, HCl, NaCl, H^+, OH^-, Na^+, and Cl^-, can be calculated from the four equations for a given temperature, making use of the two additional constraints of electrical neutrality and the partial pressure equation ($P_{gas} = \Sigma_{Pi}$). One of the solutes still remains to be specified. This is achieved most conveniently either by using pH as the master variable, or by plotting individual molalities against the sum of the molalities of all chloride species, the total chlorinity. Figure 1 illustrates the first choice and Fig. 2 the latter, both calculated for a range of temperatures.

As pointed out by Gunter and Eugster (1980), solubilities reach a minimum when chloride molality goes to zero and the molality of OH^- becomes equal to the molality of Na^+, the most abundant cation. For the mica–feldspar assemblages, this limiting pH is near 7, and it can be exceeded only if additional cations are present to balance OH^-. On the other hand, a decrease in pH, representing increasing chlorinity, leads to sharply increasing solute molalities. Slopes and position of the abundance curves can be calculated from the equilibrium constants K_1 to K_4 representing eqs. (1)–(4). The albite–paragonite–quartz assemblage buffers, for a given P and T, the NaCl/HCl ratio in the fluid:

$$\log K_1 = \log NaCl - \log HCl. \quad (5)$$

From Eqs. (3) and (4) we have

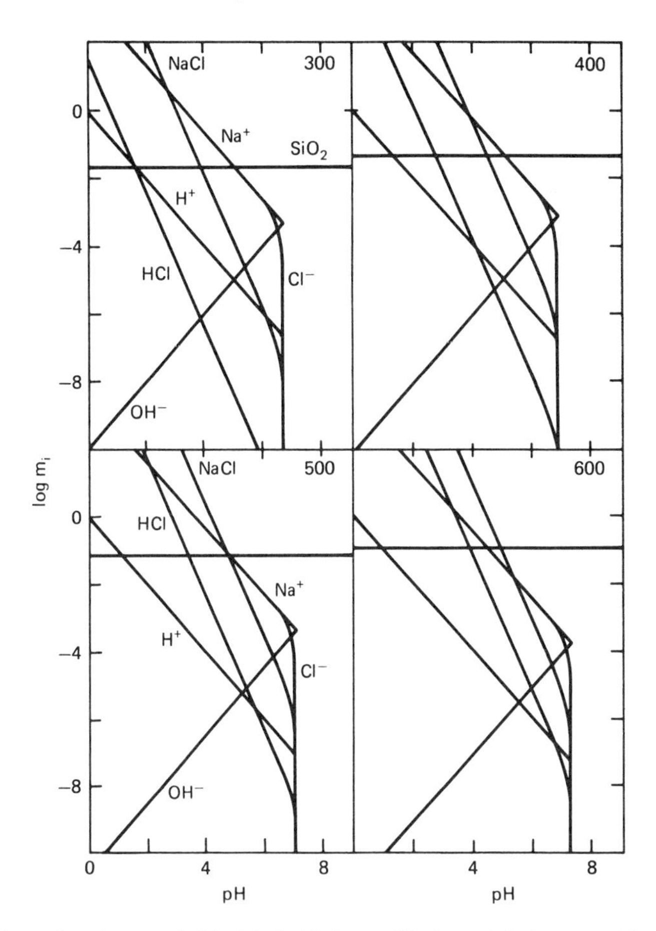

Fig. 1. Solute abundances of chloride fluids in equilibrium with the assemblage albite + paragonite + quartz as a function of pH at a water pressure of 2 kbars, and at temperatures from 300 to 600°C. At the lower pH values, the curves for Na^+ and Cl^- are superimposed, while at the limiting pH, Cl^- concentration goes to zero, because $OH^- = Na^+$. From Gunter and Eugster (1980, Fig. 9).

$$\log \mathrm{HCl} = \log \mathrm{Cl}^- - \log K_3 - \mathrm{pH} \tag{6}$$
$$\log \mathrm{NaCl} = \log \mathrm{Na}^+ + \log \mathrm{Cl}^- - \log K_4. \tag{7}$$

Combining Eqs. (5), (6), and (7) yields

$$\log \mathrm{Na} = (\log K_1 - \log K_3 + \log K_4) - \mathrm{pH}. \tag{8}$$

That is, sodium molality and pH vary inversely. Furthermore

$$m\mathrm{Na}^+ \gg m\mathrm{H}^+$$

and hence, because OH^- molality is small, except for the limiting, chloride-free system, we have

$$m\mathrm{Na}^+ = m\mathrm{Cl}^-. \tag{9}$$

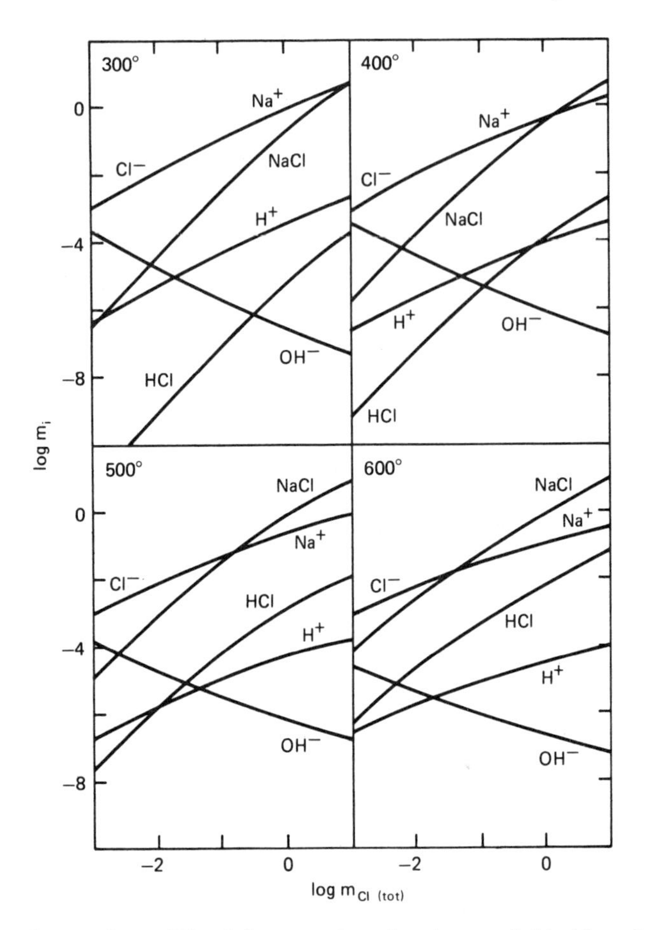

Fig. 2. Same data as those of Fig. 1, but now plotted against total chloride molality. From Eugster and Gunter (1981, Fig. 4).

From (5), (7), and (9) we get

$$\log \text{NaCl} = (2 \log K_1 - 2 \log K_3 + \log K_4) - 2\text{pH} \tag{10}$$

$$\log \text{HCl} = (\log K_1 - 2 \log K_3 + \log K_4) - 2\text{pH}. \tag{11}$$

Consequently, the curves for log NaCl and log HCl are inversely proportional to 2 pH. The value for SiO_2, on the other hand, is independent of pH and is controlled by the solubility of quartz. OH^- molality is defined by Eq. (2) and the intersection of H^+ and OH^-, the acid–base neutrality condition, varies as a function of P and T. Figure 1 applies to the pure system Na–Cl–O–H, buffered by albite + paragonite + quartz. Addition of other solutes will modify the species abundance curves through the electrical neutrality condition, Eq. (9).

Figure 2 is constructed from the same information as Fig. 1, but using the mass balance constraint of chlorinity

$$(m\text{Cl})\text{tot} = m\text{Cl}^- + m\text{NaCl} + m\text{HCl}, \tag{12}$$

assuming no other chloride-bearing solutes are present. Reasonable molalities are obtained for pH values from about 7 to 3 or 4, depending on T. Lower pH conditions lead to chlorinities of more than 10 moles for these assemblages. Figure 2 illustrates that increasing temperature and increasing chlorinities favor the associated species NaCl and HCl over the ionic solutes Na^+, Cl^-. Increases in water pressure favor dissociation, but the effect is much smaller (see Gunter and Eugster, 1980, Fig. 8). Diagrams similar to Fig. 1 and 2 have been constructed for Na–K–Cl–OH systems (Gunter and Eugster, 1980) as well as Mg–Cl–OH systems (Eugster and Gunter, 1981). They can be constructed for any solute system for which the appropriate information is available on mineral solubilities, dissociation constants and mineral buffers.

Speciation in Chloride Fluids at High *P* and *T*

Information on speciation of chloride solutes in hydrothermal fluids is incomplete, but some general remarks may be helpful. Frantz and Marshall (1982) presented information on the stepwise dissociation of $MgCl_2$ in supercritical aqueous fluids based on conductivity measurements, while Crerar *et al.* (1978) have used solubility data on the magnetite + pyrite + pyrrhotite buffer to extract dissociation data for $FeCl_2$ from 200 to 350°C along the vapor-saturated ($S + L + V$) curve. Figures 3 and 4 are Bjerrum diagrams (see, for instance, Stumm and Morgan, 1981, p. 145), showing the abundance of species for Mg, Fe and Na in chloride solutions. Total metal molalities, *mi*, are fixed arbitrarily (log *mi* is 0, −2, or −

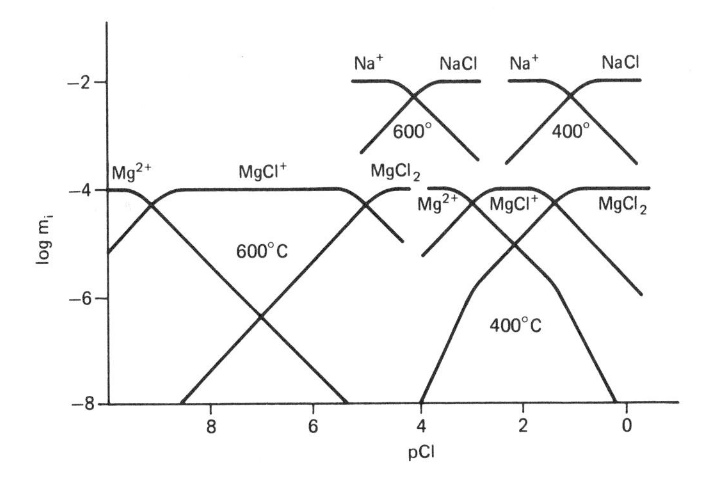

Fig. 3. Species abundance diagrams for $MgCl_2$ and NaCl and their dissociation products at 400° and 600°C and 1 kbar as a function of pCl, the negative logarithm of the total chloride molality. Total Mg molality is 10^{-4} moles and total Na molality is 10^{-2} moles. Data from Frantz and Marshall (1981) and Quist and Marshall (1968).

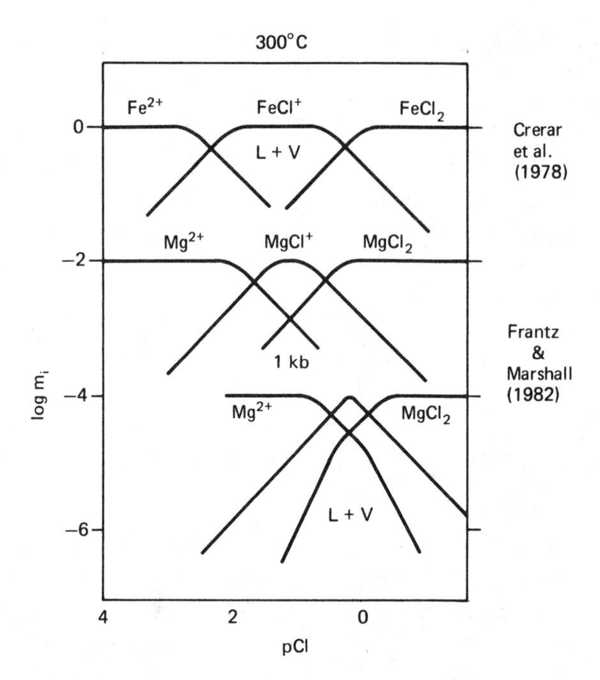

Fig. 4. Species abundance diagrams for $MgCl_2$ and $FeCl_2$ and their dissociation products at 300 °C as a function of $-\log(m\text{Cl})_{tot}$, pCl. The Mg diagram is drawn for two pressures, 1 kbar and the pressure of the saturated liquid ($L + V$, about 85 bars). For the former, magnesium molality was set at 10^{-2} and for the latter at 10^{-4} for ease of illustration. Data from Frantz and Marshall (1982). $FeCl_2$ dissociation from Crerar *et al.* (1978) is drawn for the $L + V$ pressure, and an iron molality of one.

4) to best portray the relationships. Because we are concerned with chloride species and not with hydrolysis, the master variable is pCl, the negative logarithm of the total chloride molality, rather than pH. Dilute solutions plot on the left, saline ones on the right.

Figure 3 is drawn for 1 kbar water pressure, at 400 and 600 °C and compares the behavior of Mg with that of Na (data from Quist and Marshall, 1968). The magnitude of the dissociation constants for the chloride complexes is indicated by the pCl value at the intersection of the participating metal species. Thus the first dissociation constant for $MgCl_2$ corresponds to the pCl value for the condition $m_{MgCl^-} = m_{MgCl_2}$, and so forth. Figure 3 documents a number of significant features:

1. At constant P and T, increasing chlorinity, $(m\text{Cl})_{tot}$, correlates with the transition from the most highly charged, chloride-poor cationic complex to the most chloride-rich anionic species:

$$\begin{aligned} &Mg^{2+} \rightarrow MgCl^+ \rightarrow MgCl_2 \rightarrow MgCl_3^- \rightarrow MgCl_4^{2-} \\ &Na^+ \rightarrow NaCl \rightarrow NaCl_2^-. \end{aligned} \tag{13}$$

Similar behavior has been documented for silver chloride complexes by Seward (1976) in his solubility study of chlorargyrite. With increasing chlorinity, Seward found the following sequence:

$$Ag^+ \rightarrow AgCl \rightarrow AgCl_2^- \rightarrow AgCl_3^{2-} \rightarrow AgCl_4^{3-}. \quad (14)$$

Seward (1981) has documented a similar series for lead chloride complexes.

2. Increasing temperature has an effect equivalent to increasing chlorinity, pCl. At constant pCl, an increase in temperature leads from chloride-poor to chloride-rich complexes, and from cations and anions to neutral species:

$$Mg^{2+} \rightarrow MgCl^+ \rightarrow MgCl_2 \leftarrow MgCl_3^-. \quad (15)$$

Again, this accords with the findings of Seward (1976, 1981) on silver and lead chloride complexes. Formation of chloride complexes appears to be attended by an overall decrease in solvation and hence is favored by higher salinity (lower a_{H_2O}) and higher temperature (decrease of the dielectric constant of the solvent):

$$Mg^{2+} \cdot k\ aq + Cl^- \cdot m\ aq \rightarrow MgCl^+ \cdot n\ aq, \qquad \text{where } n < (k + m). \quad (16)$$

For comparison, Fig. 3 also shows NaCl dissociation; the constants are not very different from the first dissociation constant for $MgCl_2$.

Figure 4 gives information on the effect of water pressure on dissociation as well as a comparison between Mg and Fe complexes for a constant temperature of 300°C. For Mg complexing, two pressures are shown: $L + V$ (about 85 bars) and 1000 bars. In this range, an increase in pressure also seems to favor the more highly chlorinated complexes for a given chlorinity. There appears to be little difference between the first dissociation constants for $MgCl_2$ and $FeCl_2$, whereas the second constants are quite different, with Fe more easily complexed. However, the data are too incomplete and uncertain to permit generalizations.

The data on chlorargyrite indicate that we must contemplate the existence of higher chloride complexes for Mg and Fe at higher temperatures and chlorinities. The data reported by Chou and Eugster (1977, table 1) on magnetite solubility in chloride solutions generally conform to the condition:

$$2m\text{Fe} = m\text{Cl}$$

except for temperature at and above 600°C, where clearly

$$2m\text{Fe} < m\text{Cl}$$

for solutions of moderate salinities. Perhaps this signals the presence of more highly chlorinated species, such as $FeCl_3^-$ and $FeCl_4^{2-}$. A further analysis of the data is in progress.

Figure 5 is a schematic diagram for a hypothetical bivalent metal and the effect of chloride complexing on its cation as a function of pCl. The upper portion is a Bjerrum distribution diagram equivalent to Fig. 3. The lower part is the corresponding solubility diagram for the solid $MeCl_2$ phase, with the solubility curve defined as the sum of all Me solute species. In metamorphic environments, temperatures and salinities may reach high enough values to stabilize chloride-rich

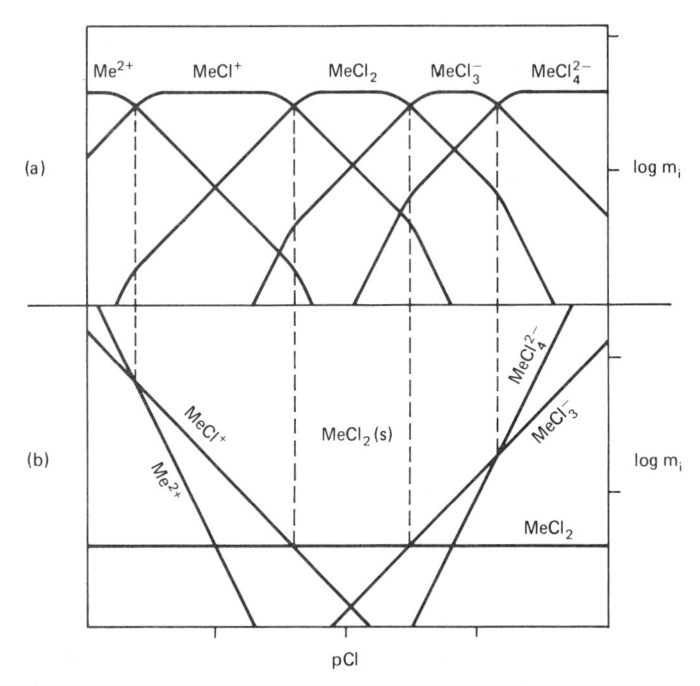

Fig. 5. Hypothetical diagram illustrating the chloride complexing behavior of a bivalent metal as a function of pCl. The upper diagram (a) is a distribution diagram for a fixed metal molality, *m*Me. The lower diagram (b) is the solubility diagram for the solid phase, $MeCl_2$. Corresponding equivalence points (dissociation constants) are indicated by dashed lines.

anionic complexes such as $FeCl_3^-$ and $MgCl_3^-$ or even $NaCl_2^-$ over the neutral, uncharged complexes $FeCl_2$, $MgCl_2$, and NaCl. If this is the case, mineral solubilities and transport in chloride solutions could be effected significantly at the higher grades of metamorphism. Appropriate solubility determinations are required to test this inference. In this chapter, for lack of information and for the sake of simplicity, we take $MgCl_2$ and $FeCl_2$ to represent all uncharged and anionic chloride complexes with $Cl \geq 2$.

Mixing of Waters

Mineral precipitation as a consequence of mixing of waters has been suggested for many geologic environments, including ore deposits and carbonate rocks, and it must also be considered as an important mechanism in metamorphic environments where conditions are appropriate for the flow of aqueous fluids. Because of the dependence of solute abundances and solute speciation on chlorinity, mixing effects can be pronounced. To be able to predict such effects quantitatively, we would have to know the chlorinities of the fluids before and after mixing, along with the appropriate solubility and dissociation constants. This seems a near impossible task, even if detailed fluid inclusion data are available.

Gunter and Eugster (1981) calculated activity gradients maintained in fluids by incompatible mineral assemblages. They documented that such incompatible assemblages impose upon the fluid gradients in either pH or pCl, which cannot be wiped out by mixing of waters or by diffusion without destroying at least one of the incompatible assemblages. Assuming that the proton diffuses most rapidly, it seems likely that gradients in chlorinity continued to exist between such incompatible assemblages and that such gradients might be reflected in the compositions of the fluid inclusions.

Mg–Fe Solubility and Exchange Reactions: Methodology

Solubility determinations for Fe–Mg solid solutions in supercritical chloride solutions are not available. In an attempt to estimate the relative abundances of $MgCl_2$ and $FeCl_2$ in metamorphic solutions, Eugster and Gunter (1981, Fig. 10) relied on the assemblages talc + quartz and magnetite. For a given chlorinity, $FeCl_2$ was found to be somewhat more abundant than $MgCl_2$. This conclusion depends of course on the nature of the assemblages chosen for the comparison.

Schulien (1973, 1975, 1980) measured the equilibrium constant, K_D, for the exchange reaction

$$\tfrac{1}{3}KFe_3AlSi_3O_{10}(OH)_2 + MgCl_2 \rightleftharpoons \tfrac{1}{3}KMg_3AlSi_3O_{10}(OH)_2 + FeCl_2 \quad \text{where } K_D = (Mg/Fe)\text{biotite}/(Mg/Fe)\text{fluid}, \tag{17}$$

in the temperature range 500–700 °C and at water pressures of 250 to 2000 bars, using a 2 molar chloride solution ($FeCl_2$ + $MgCl_2$ + KCl). In this temperature range, Mg is preferentially fractionated into biotite and Fe into the fluid. With increasing temperature, the effect increases, as documented in Fig. 6. At temperatures at and above 600 °C, K_D values are, within experimental uncertainty, independent of composition and the exchange equilibrium can be represented by a hyperbola in a Roozeboom diagram:

$$X_{Fe}^{fluid} + (K_D - 1)\, X_{Fe}^{fluid}\, X_{Fe}^{biotite} - K_D\, X_{Fe}^{biotite} = 0 \tag{18}$$

corresponding to the defintion of K_D,

$$K_D = X_{Fe}^{fluid}\,(1 - X_{Fe}^{biotite})/(1 - X_{Fe}^{fluid})\, X_{Fe}^{biotite}. \tag{19}$$

From this relationship we imply that both solution phases mix ideally at higher temperatures or that nonidealities cancel. At lower temperatures, for lack of more compelling information, we may assume that the fluid also mixes ideally regardless of the solid solution with which it is in equilibrium. It is important to remember that Schulien (1980) only considered synthetic biotites along the join

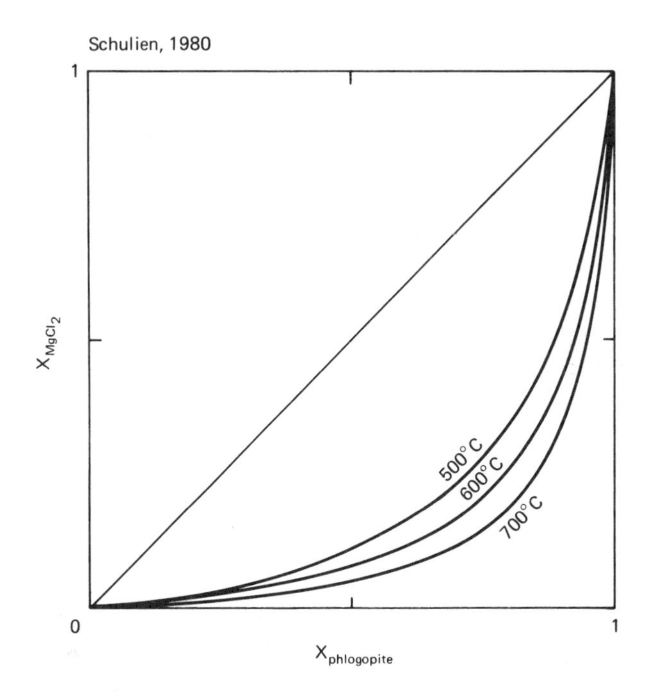

Fig. 6. Mg–Fe exchange equilibrium between biotite and a chloride fluid, as determined by Schulien (1980). Ideal exchange has been assumed throughout the compositional range, with the following constants calculated from the averages of the individual experimental determinations: 700°C (19.27), 600°C (11.1), 500°C (7.78), where K_D = (Mg/Fe) biotite/(Mg/Fe) fluid. For 700 and 600°C the predominant pressure was 1 kbar, for 500° it was 0.3 kbar.

phlogopite–annite. Additional components can exert significant influence. On the other hand, as much as 10–15% ferric iron was found in a few samples, apparently with no effect on the K_D values.

Schulien *et al.* (1970) measured K_D values for the exchange between olivines and Fe–Mg fluids. Olivine fractionates less effectively than biotite (see Fig. 7). At 600°C, K_D olivine/fluid is 8.33, whereas K_D biotite/fluid is 11.11. Having obtained the biotite–fluid and olivine–fluid calibrations at 600°C and 1 kbar, we can use data on biotite–(mineral 1) and olivine–(mineral 2) pairs in order to evaluate the (mineral 1)–fluid and (mineral 2)–fluid exchange equilibria. Treating the chloride fluid as a monitoring solution allows us to compare the behavior of any Mg–Fe solid solution with respect to that standard. Because of the narrow temperature range for the calibrations, we will stick to 600°C for this initial summary. In this manner we can correlate quantitatively a wide range of mineral solutions.

We propose to use the following bridges: fluid–biotite–garnet–clinopyroxene–orthopyroxene (and hornblende); fluid–garnet–cordierite; fluid–olivine–orthopyroxene and fluid–olivine–spinel in order to cover the data available.

Garnet–Fluid Exchange

Ferry and Spear (1978) carried out experiments on the partitioning of Fe and Mg between garnet and biotite in the temperature range from 550 to 800°C and a pressure of 2.07 kbar, using synthetic biotites on the join phlogopite–annite and synthetic garnets along the join almandine–pyrope. The experimental results best fit the expression

$$\text{Log } K_D = -916/T(K) + 0.340,$$

where $K_D = (\text{Mg/Fe})\text{garnet}/(\text{Mg/Fe})\text{biotite}$.

Adding a pressure correction term, this expression becomes

$$52.107\ KJ - 19.506\ T(K) + 0.238\ P(\text{bars}) + 3\ RT \ln K_D = 0. \quad (20)$$

For 600°C and 1 kbar we obtain a K_D value of 0.198. Next we combine the biotite–fluid exchange reaction with the garnet–biotite information to derive the garnet–fluid exchange:

$$\text{Fe-biotite} + MgCl_2 \rightleftharpoons \text{Mg-biotite} + FeCl_2, \quad K_D = 11.11 \quad (21)$$

$$\text{Fe-garnet} + \text{Mg-biotite} \rightleftharpoons \text{Mg-garnet} + \text{Fe-biotite}, \quad K_D = 0.198 \quad (22)$$

$$\text{Fe-garnet} + MgCl_2 \rightleftharpoons \text{Mg-garnet} + FeCl_2, \quad K_D = 2.2. \quad (23)$$

The first two reactions are added so that biotite cancels and the coefficients are multiplied. Since K_D for garnet/fluid is 2.2, it is clear that garnet fractionates Mg

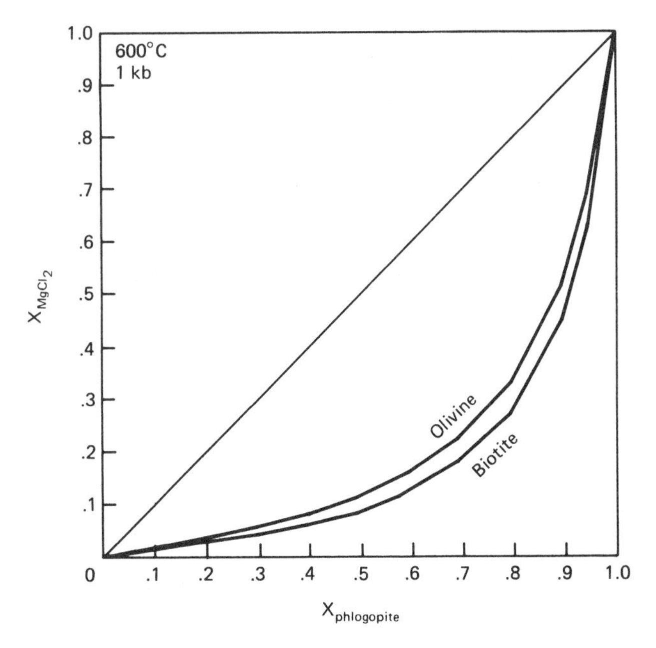

Fig. 7. Comparison of the Mg–Fe fractionation between biotite–fluid and olivine–fluid at 600°C, 1 kbar. Data from Schulien (1980) and Schulien *et al.* (1970).

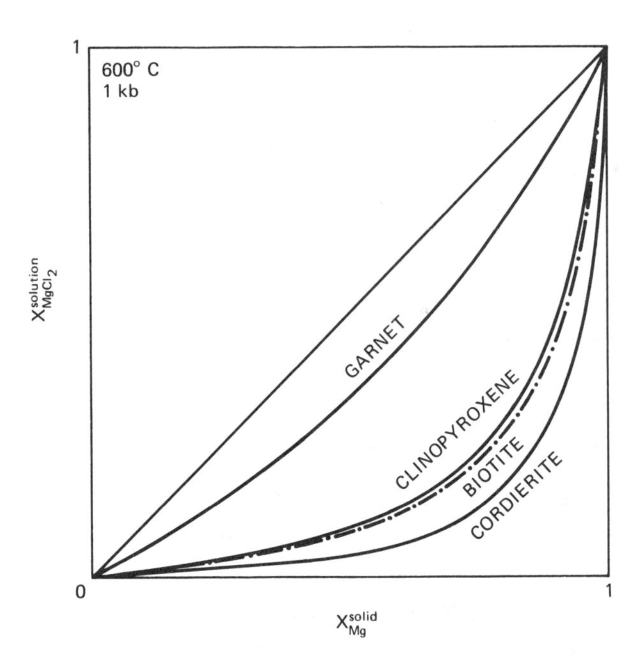

Fig. 8. Comparison of the Mg–Fe fractionation at 600 °C, 1 kbar between a chloride fluid and the minerals garnet, clinopyroxene and cordierite. Also shown, as a dashed curve, is the biotite–fluid curve of Fig. 7. For sources of data see text.

very much less strongly than does biotite. In Fig. 8 the two curves are contrasted, along with comparative curves for clinopyroxene and cordierite to be discussed later.

The calibration of Ferry and Spear (1978) was carried out with pure synthetic minerals, but it was tested only for the almandine-rich portion of the garnet solid solution. Also, because of the lack of reactivity of the garnets, very large garnet/biotite ratios were used, so as not to have to rely on a change of garnet composition. We have extrapolated the ideal exchange. As garnet is a crucial link in the chain, this extrapolation must be kept in mind for all subsequent results based on the garnet–biotite bridge. The calibration of Ferry and Spear (1978) agrees with the results of Thompson (1975) for the temperature range 400–600 °C; however it lies about 50 °C higher than the data of Goldman (1977) for the same range.

Since K_D (biotite–fluid) and K_D (garnet–biotite) both decrease with temperature, K_D (garnet–fluid) must also decrease with decreasing temperature. Furthermore, the fractionation between garnet and fluid is weak and it is quite possible that at lower temperatures a reversal will be encountered, with garnet favoring Fe over Mg, whereas the fluid will be enriched in Mg. Such behavior is illustrated by Fe–Mg carbonates in chloride solutions, with the reversal located at about 350 °C (Eugster, 1982).

When comparing data from natural mineral assemblages, it must be remembered that Ferry and Spear (1978) did not evaluate the influence of Ca and Mn on the Mg–Fe exchange behavior. O'Neill and Wood (1979), in their work on garnet–olivine, found that partitioning between 900° and 1400°C at 30 kbar is strongly dependent on the Mg/Fe ratio and the calcium content of the garnet. They interpreted their data in terms of a regular solution model. We did not use the O'Neill and Wood (1979) data, because of the large temperature and pressure gap.

Clinopyroxene–Fluid Exchange

Two experimental calibrations are available for the garnet–clinopyroxene Fe–Mg exchange reaction. Raheim and Green (1974a) studied this mineral pair in eclogites and their work was carried out in the temperature range 600–1500°C and the pressure range 20–40 kbars. They concluded that, for $6.2 < 100$ Mg/(Mg + Fe) < 85, the bulk chemical composition does not perceptibly affect the K_D values. This is a much wider compositional range than normally encountered in basalts. Furthermore, the authors claim that, within basaltic compositions, minor chemical variations have no appreciable effect on the K_D values; but they do caution the application of this calibration to rocks of vastly different chemical composition.

Ellis and Green (1979) attempted a similar calibration. They also used basaltic compositions and compositions within the simple system CaO–MgO–FeO–Al_2O_3–SiO_2. Garnet and clinopyroxene were crystallized at temperatures ranging from 750 to 1300°C and at pressures from 24 to 30 kbars. Unlike Raheim and Green (1974), Ellis and Green (1979) also attempted compositional calibrations for components other than MgO and FeO. Apparently, up to 30% jadeite in clinopyroxene did not significantly affect the K_D values for a fixed P and T. On the other hand, for a given P and T, the spread in K_D values could be correlated with the CaO content of garnet. The authors claimed that, using their calibration, they were able to reconcile K_D values from different bulk compositions.

Both experiments are consistent with respect to the trend of partitioning with changing temperatures and pressures. $K_D = (Mg/Fe^{2+})cpx/(Mg/Fe^{2+})gr$ decreases with increasing temperatures and with decreasing pressure. However, Raheim and Green (1974) obtained a rather large pressure coefficient, while Ellis and Green (1979) argued that this pressure coefficient was erroneously high and was caused by variable CaO content in garnet. In any case, our pressure extrapolation is excessively large and the results perhaps are semiquantitative at best.

For 600°C and 1 kbar, the data of Raheim and Green (1974) give a K_D value of 6.86. At 600°C, 1 kbar and extrapolated to 0% CaO in garnet, Ellis and Green (1979) give a K_D of 4.85. Both results were combined with K_D (garnet–fluid) to give K_D (cpx–fluid) = 10.67 and 15.09, respectively, where K_D (cpx–fluid) = $(Mg/Fe^{2+})cpx/(Mg/Fe^{2+})$ fluid. The two distribution curves are shown in Fig. 9 together with the biotite curve. Apparently, clinopyroxene partitions Mg as effectively as biotite does, although large extrapolations in both P and T are involved in this conclusion.

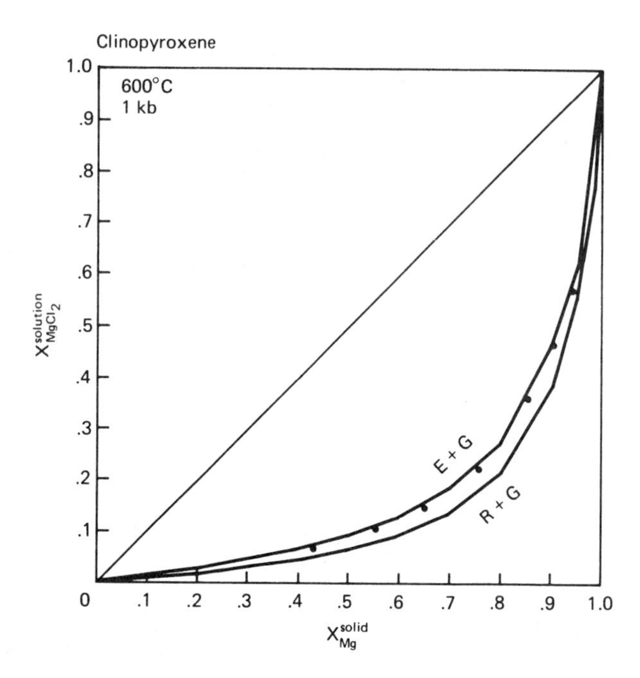

Fig. 9. Mg–Fe fractionation between clinopyroxene and a chloride fluid at 600°C, 1 kbar. Data shown are those of Ellis and Green (1979), E + G; and Raheim and Green (1974), R + G. The dots represent the biotite curve from Fig. 7.

Cordierite–Fluid Exchange

Holdaway and Lee (1977) attempted to calibrate the garnet–cordierite and biotite–cordierite Mg–Fe exchange reactions by using natural occurrences of pelitic rocks, as a function of metamorphic grade. Garnet with more than 13 mol % (spessartite + grossularite) was excluded. The analyses they used, for the most part, did not distinguish between Fe^{2+} and Fe^{3+} and K_D values were determined using total Fe. Although we do have some constraints on garnet compositions, it is clear that the method does not take into account the compositional dependence of the K_D values. In particular, we do not know how K_D might vary with the Mg/(Mg + Fe) ratio.

We combined K_D (garnet–cordierite) and K_D (biotite–cordierite), at 600°C and 1 kbar with the garnet–fluid and biotite–fluid constants to obtain two separate estimates for the cordierite-fluid exchange:

$$K_D\,(\text{garnet–cordierite}) * K_D\,(\text{garnet–fluid}) = 20 = K_D\,(\text{cordierite–fluid})$$
$$K_D\,(\text{biotite–cordierite}) * K_D\,(\text{biotite–fluid}) = 20.8 = K_D\,(\text{cordierite–fluid})$$

where

$$K_D\,(\text{cordierite–fluid}) = (\text{Mg/Fe})\ \text{cordierite}/(\text{Mg/Fe})\ \text{fluid}.$$

As it involves fewer combinations, the second result should be more trustworthy, but the agreement is more than satisfactory. The distribution curve for cordierite–fluid Mg–Fe exchange, shown in Fig. 8, was drawn assuming ideal solution over the entire Mg–Fe compositional range. Cordierite fractionates Mg into its structure more strongly than any other mineral we have looked at thus far.

A controversy has surrounded the garnet–cordierite exchange equilibria. The experiments of Currie (1971) indicate that K_D (garnet–cordierite) increases with temperature. This is in disagreement with Holdaway and Lee's (1977) calibrations, experimental studies (Henson and Green, 1971, 1972; Holdaway, 1976) and other work based on natural occurrences (Thompson, 1976; Perchuk, 1977). Furthermore, Lonker (1981) shows that the temperature calibrations, at 4 kbar, of Thompson (1976) and Perchuk (1977) are very close to that of Holdaway and Lee (1977). All three calibrations are particularly close at 600 °C (Lonker, 1981).

Wood (1973) tried to reconcile Currie's (1971) "wet" experiments with Henson and Green's (1971, 1972) "dry" experiments by involving water in the exchange reactions (i.e., Mg–crd and Fe–crd were given different XH_2O). This would make the exchange equilibria dependent on P_{H_2O} with the dependence governed by the difference in water content of the two end member cordierites. Lonker (1981), however, cited previous work to indicate that Fe–cordierite and Mg–cordierite contain similar amounts of water. Also, Holdaway and Lee (1977) demonstrated that neither K_D (garnet–cordierite) nor K_D (biotite–cordierite) varied significantly with pressure. This constitutes further evidence that large amounts of water are not involved in the exchange reaction. Holdaway and Lee (1977) implied that Currie's results may not have represented equilibrium since run times were short.

We did not use the experimental results of Henson and Green (1971, 1972) because their data do not reach below 800 °C and their K_D values show considerable scatter. It should be noted that Labotka *et al.* (1981) describe a compositional dependence of K_D (cordierite–biotite) for metamorphosed argillites ($T \cong$ 500–600 °C, $P \cong 1.5$ kbar). Apparently K_D = (Mg/Fe) cordierite/(Mg/Fe) biotite decreases with decreasing Al(IV) in biotite and eventually a reversal occurs. At relatively high Al(IV) in biotite we obtain a K_D (cordierite–fluid) of 19.8.

One further check is possible. Engi (1978) presents data for 700 °C and 800 °C at 1 kbar for the exchange reaction olivine–cordierite. At 700 °C $K_D = 1.89$ and at 800 °C $K_D = 1.64$, where K_D = (Mg/Fe) cordierite/(Mg/Fe) olivine. We assumed a linear relationship between log K_D and $1/T$ and extrapolated to 600 °C, where $K_D = 2.26$. Using Schulien's value for olivine–fluid of 8.33, we obtain a value of 18.82 for K_D (cordierite–fluid) at 600 °C, 1 kbar in surprisingly good agreement with the numbers derived from the garnet–cordierite and biotite–cordierite data of Holdaway and Lee (1977).

Caution should be exercised with Engi's data, however, since cordierite was a byproduct phase. The initial cordierite composition was unknown and hence reversibility was not demonstrated. Furthermore, these K_D values were good only for low values of Fe/(Mg + Fe) and Engi (1978) detected a deviation in K_D at higher values of Fe/(Mg + Fe). Nonetheless, the agreement lends credence to the cordierite–fluid constant extracted here.

Orthopyroxene–Fluid Exchange

Kretz (1963) fashioned a tentative geothermometer from the distribution of Mg and Fe between calcic pyroxene and orthopyroxene. Mineral pairs from natural occurrences were grouped according to metamorphic grade. The K_D values, for three different metamorphic grades, were averaged and temperature estimates were assigned to these groupings. The three averages spanned temperatures from 1150°C to 670°C and showed a linear relationship when plotted on a log K_D vs. $1/T$ graph. K_D (orthopyroxene–clinopyroxene) increased with increasing temperature, where K_D = (Mg/Fe) orthopyroxene/(Mg/Fe) clinopyroxene. No compositional calibration was given.

We extrapolated to 600°C to obtain an average K_D (orthopyroxene–clinopyroxene) of 0.50. Next, we combined this K_D with K_D (clinopyroxene–fluid) obtained from Ellis and Green (1979) to derive K_D (orthopyroxene–fluid) = 5.4, at 600°C and ~1 kbar, where K_D (orthopyroxene-fluid) = (Mg/Fe) orthopyroxene/(Mg/Fe) fluid. The distribution curve of Fig. 10 was drawn assuming ideal solution over the entire Mg–Fe range.

Kretz (1981) found a slight compositional dependence of K_D (orthopyroxene–clinopyroxene) for granulite facies rocks. K_D = (Fe/Mg)opx/(Fe/Mg)cpx decreases with increasing X_{Fe} in opx such that $K_D = 2.0 - 0.45\ X_{Fe}$(opx). How-

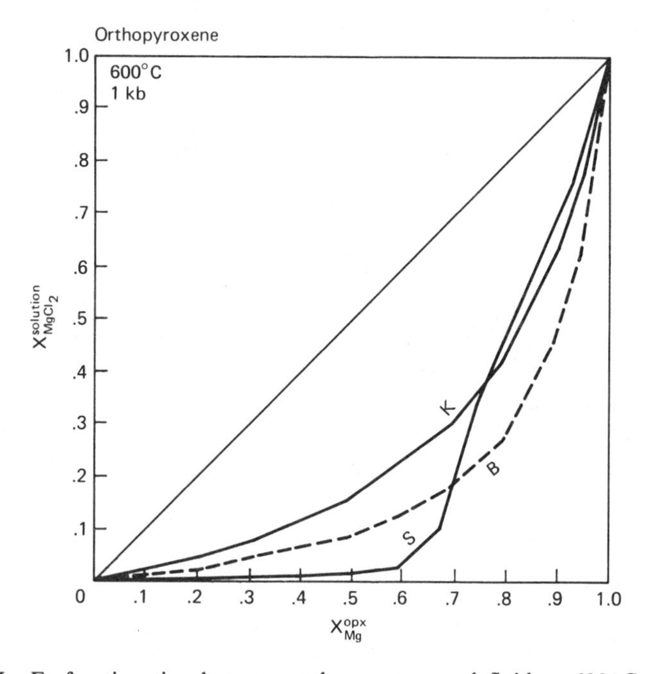

Fig. 10. Mg–Fe fractionation between orthopyroxene and fluid at 600°C, 1 kbar. K refers to the data of Kretz (1963), S to those of Saxena (1968). The biotite curve, B (dashed), is from Fig. 7 and is shown for comparison.

ever, Maxey and Vogel (1974) describe the opposite trend for amphibolite facies rocks with K_D (opx–cpx) increasing with increasing Fe content in the pyroxenes. K_D values ranged from 1.4 to 1.9.

Saxena (1968) attempted a calibration of the olivine–orthopyroxene Mg–Fe exchange reaction. His data came from coexisting olivine and orthopyroxene in rocks and from high-temperature experiments. The model developed implies a remarkable dependence of K_D on the Mg/(Mg + Fe) ratio. We combined the various values for K_D (orthopyroxene–olivine) at 600°C with K_D (olivine–fluid) to derive K_D (orthopyroxene–fluid) values. The distribution curve obtained for orthopyroxene–fluid in this manner is also shown in Fig. 10 and compared with the curve based on Kretz (1963). The Saxena curve is probably not trustworthy, in part because its 600°C isotherm is based entirely on optical data.

Engi (1978) also has presented data for the olivine–orthopyroxene exchange reaction. He used a simple solution model for both olivine and orthopyroxene at 700, 800, and 900°C. The shapes of the curves are similar to those of Saxena (1968), but not as extreme. However, the experiments covered only a very small compositional range (X_{Fe} from 0 to 0.15) and reversibility for orthopyroxene was not demonstrated.

Medaris (1969) also measured the olivine–orthopyroxene exchange reaction at 700, 800, and 900°C and 0.5 kbar. Reversibility was demonstrated only at 800 and 900°C. The reversibility brackets, however, were so large that his results are open to interpretation.

Hornblende–Fluid Exchange

Kretz and Jen (1978) attempted a calibration of the clinopyroxene–hornblende Fe–Mg exchange reaction. They used natural coexisting mineral pairs from amphibolite and granulite facies rocks. For higher temperature estimates, experimental data were incorporated. Sodium-rich minerals and minerals with unknown Fe^{2+} contents were excluded and the pressure effect was considered to be small. Kretz and Jen (1978) believed the distribution coefficient to be independent of the Mg/(Fe + Mg) ratio. The variation in K_D values, for a given metamorphic grade, was thought to arise, in part, from the dependence of K_D on the Al (IV) content in Ca–amphibole. Apparently, K_D decreases with increasing Al (IV) content in Ca–amphibole, where K_D (hornblende–clinopyroxene) = (Mg/Fe) hornblende/(Mg/Fe) clinopyroxene.

Extrapolating the curve of Kretz and Jen (1978) from 650 to 600°C along with the variance of their data and combining the result with the clinopyroxene–fluid data based on Raheim and Green (1974) and Ellis and Green (1979), yields K_D values for hornblende–fluid, K_D = (Mg/Fe) hornblende/(Mg/Fe) fluid of 4.96 to 9.36. These two sets of curves are shown in Fig. 11, but the results are very preliminary and uncertain. Nevertheless, the Mg–Fe fractionation between hornblende and fluid appears to be less pronounced than that between biotite and fluid.

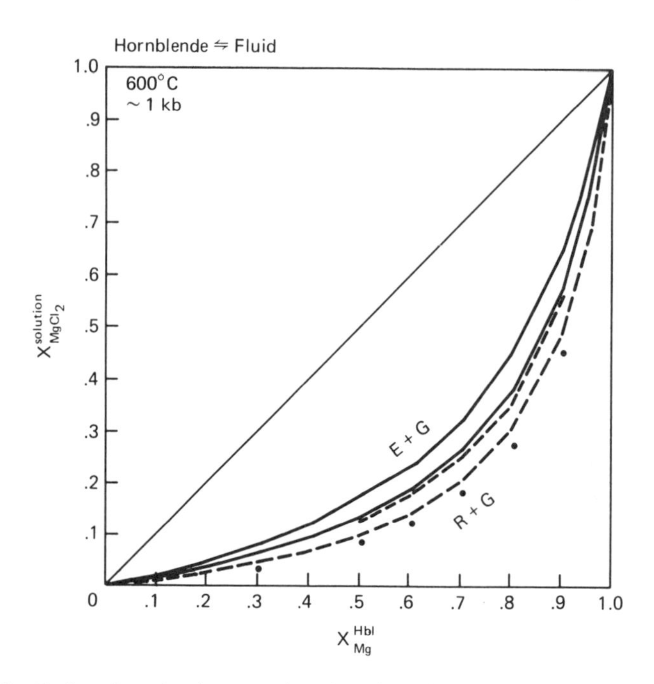

Fig. 11. Mg–Fe fractionation between hornblende and a chloride fluid at 600 °C, 1 kbar. The solid curves represent the upper and lower limits based on Ellis and Green (1979), while the lower two curves are based on Raheim and Green (1974). The dots represent the biotite curve of Fig. 7.

Exchange Between the M_1 and M_2 Sites of Orthopyroxenes and Fluid

The M_1 and M_2 sites in orthopyroxenes should behave quite differently during exchange reactions, with Fe strongly fractionated into the larger M_2 site (see Saxena and Ghose, 1971). Sufficient data has been published for 600 °C to derive M_1–fluid and M_2–fluid exchange curves for a direct comparison of the effectiveness of the two sites in Mg–Fe fractionation. From Saxena (1973) we have information on Fe–Mg exchange in M_1 sites between orthopyroxene and clinopyroxene. We combine this with the clinopyroxene–fluid exchange data derived earlier and based on Ellis and Green (1979). If we make the assumption that all M_2 sites in clinopyroxene are occupied by Ca, adding these two reactions yields information on the Fe–Mg exchange between M_1 in orthopyroxene and fluid

$$\mathrm{Fe}^{\mathrm{opx}}_{(M1)} + \mathrm{Mg}^{\mathrm{cpx}}_{(M1)} = \mathrm{Mg}^{\mathrm{opx}}_{(M1)} + \mathrm{Fe}^{\mathrm{cpx}}_{(M1)} \tag{24}$$

$$\mathrm{Fe}^{\mathrm{cpx}}_{(M1)} + \mathrm{MgCl_2} = \mathrm{Mg}^{\mathrm{cpx}}_{(M1)} + \mathrm{FeCl_2} \tag{25}$$

$$\mathrm{Fe}^{\mathrm{opx}}_{(M1)} + \mathrm{MgCl_2} = \mathrm{Mg}^{\mathrm{opx}}_{(M1)} + \mathrm{FeCl_2}. \tag{26}$$

Next we combine this information with the data of Saxena and Ghose (1971) on the reaction

$$\mathrm{Mg}^{\mathrm{opx}}_{(M2)} + \mathrm{Fe}^{\mathrm{opx}}_{(M1)} = \mathrm{Mg}^{\mathrm{opx}}_{(M1)} + \mathrm{Fe}^{\mathrm{opx}}_{(M2)} \quad (27)$$

to define

$$\mathrm{Fe}^{\mathrm{opx}}_{(M2)} + \mathrm{MgCl}_2 = \mathrm{Mg}^{\mathrm{opx}}_{(M2)} + \mathrm{FeCl}_2. \quad (28)$$

For the M_1–M_2 exchange in orthopyroxene we have the following relations

$$X^{\mathrm{opx}}_{\mathrm{Fe}} = \frac{\mathrm{Fe}(M_1) + \mathrm{Fe}(M_2)}{\mathrm{Fe}(M_1) + \mathrm{Mg}(M_1) + \mathrm{Fe}(M_2) + \mathrm{Mg}(M_2)} \quad (29)$$

$$\mathrm{Fe}(M_1) + \mathrm{Mg}(M_1) = \mathrm{Fe}(M_2) + \mathrm{Mg}(M_2) \quad (30)$$

and hence

$$2X_{\mathrm{Fe}} = X_{\mathrm{Fe}(M1)} + X_{\mathrm{Fe}(M2)} \quad (31)$$

$$K_D(27) = \frac{X_{\mathrm{Fe}(M2)}X_{\mathrm{Mg}(M1)}}{X_{\mathrm{Fe}(M1)}X_{\mathrm{Mg}(M2)}} = \frac{X_{\mathrm{Fe}(M2)}\,[1 - X_{\mathrm{Fe}(M1)}]}{[1 - X_{\mathrm{Fe}(M2)}]X_{\mathrm{Fe}(M1)}} \quad (32)$$

$$K_D(27) = \frac{[2X_{\mathrm{Fe}} - X_{\mathrm{Fe}(M1)}]\,[1 - X_{\mathrm{Fe}(M1)}]}{[1 - 2X_{\mathrm{Fe}} + X_{\mathrm{Fe}(M1)}]X_{\mathrm{Fe}(M1)}}. \quad (33)$$

Both K_D (27) and K_D (24) are composition dependent. Consequently, we must solve the problem by a stepwise calculation. We assume a value for the mole fraction of Fe in orthopyroxene, $X^{\mathrm{opx}}_{\mathrm{Fe}}$. From Saxena and Ghose (1971, Fig. 3) we

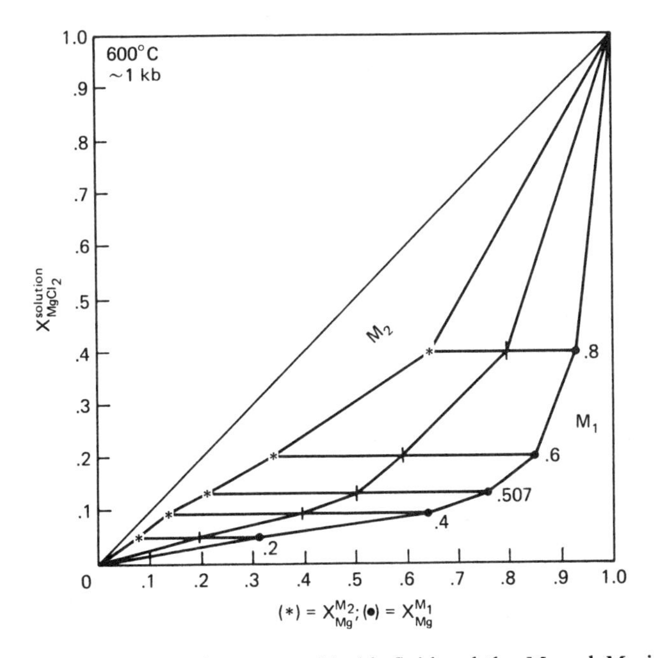

Fig. 12. Mg–Fe fractionation between a chloride fluid and the M_1 and M_2 sites of orthopyroxene at 600 °C, 1 kbar. Also shown is the average curve for orthopyroxene–fluid fractionation. For data sources, see text.

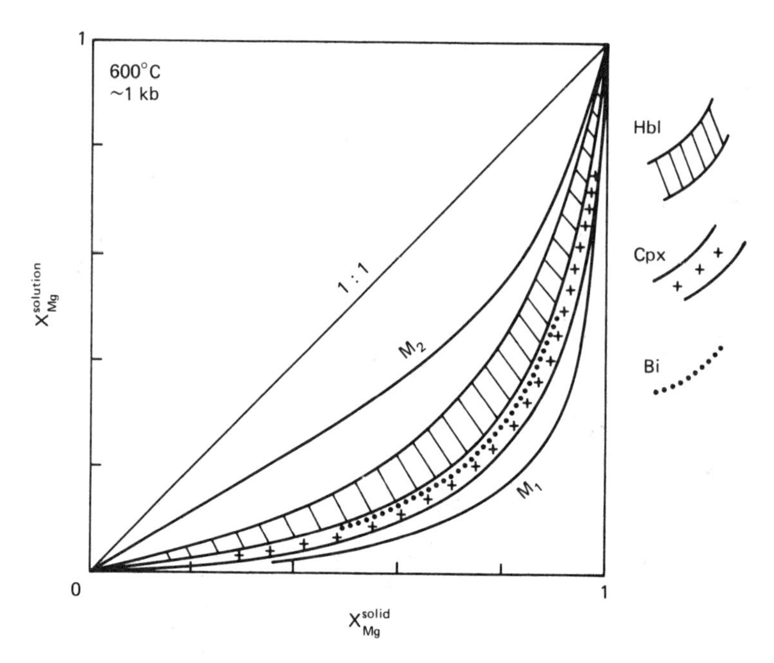

Fig. 13. Comparison of the Mg–Fe fractionation between silicates and a chloride fluid at 600°C, 1 kbar as a function of the M_2/M_1 site ratios of the silicate. M_1 and M_2 are the orthopyroxene curves of Fig. 12, crosses indicate the clinopyroxene range of Fig. 9. The hornblende curves of Fig. 11 are shown by the hatchured area, while the biotite curve of Fig. 7 is shown by the dotted curve. The orthopyroxene curve would lie just below the uppermost hornblende curve. For the corresponding M_2/M_1 ratios, see text.

obtain K_D (27) and from Eqs. (33) a value for $X_{Fe(M1)}$ for that bulk composition. Next, using K_D (26), we solve for X_{FeCl_2}, and from Eq. (31) obtain the $X_{Fe(M2)}$ value associated with this fluid composition. As a check, we can use K_D (28) to calculate X_{FeCl_2} for the same value of $X_{Fe(M2)}$. The results must agree. This procedure is then repeated for the whole compositional range. The results are shown in Fig. 12. Obviously, for a given bulk composition, the M_1 sites fractionate Mg much more effectively than do the M_2 sites.

The data base on which this extrapolation was carried out is very limited and the results are intended mainly to demonstrate how useful the fluid can be as a common medium for comparison.

The difference in the fractionation behavior of Mg and Fe between M_1 sites and fluid and between M_2 sites and fluid obviously must have a structural basis. The M_1 site is smaller than the M_2 site and shows a pronounced preference for Mg, as is illustrated in Fig. 12. Consequently, minerals with a high proportion of M_1-like sites should exhibit the strongest Mg fractionation. This relationship is born out in Fig. 13, where we compare biotite ($M_2/M_1 = 0$), clinopyroxene ($M_2/M_1 = 0$–0.2), hornblende ($M_2/M_1 = 0.4$) with orthopyroxene ($M_2/M_1 = 1$).

Spinel-Fluid Exchange

Engi (personal communication) has calibrated the spinel–olivine geothermometer with experiments at 1 kbar pressure and temperatures from 650 to 902°C. The structural formula of the spinel used is

$$(Mg,Fe^{2+})\ (Al,Cr,Fe^{3+})_2O_4.$$

Engi (1978) reported nonideal mixing of Fe and Mg in olivine, in contrast to the finding of Schulien *et al.* (1970). To calculate his distribution curves, Engi (1978, in preparation) used an ideal distribution of species solution model for spinels, and an asymmetric, subregular solution model for olivines. Ferric iron was low in all run products. Engi constructed isotherms for bulk mole fractions of Mg of 0.88 and 0.64. From the isotherms we have derived K_D (spinel–olivine) constants for spinels with a range of chromium contents. These values of K_D were combined with the K_D (olivine–fluid) data of Schulien *et al.* (1970) and Schulien (1980) to derive K_D (spinel–fluid) constants. The results are plotted in Fig. 14 and the agreement for the two sets of data, based on different bulk compositions, is good. The dependence of K_D on the chromium content is a strong reminder of the possible compositional dependence of distribution coefficients in general.

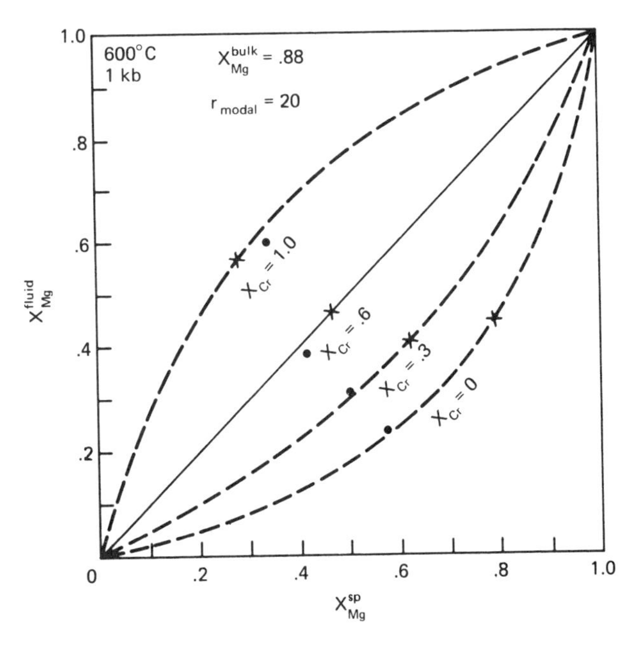

Fig. 14. Mg–Fe fractionation between spinel and a chloride fluid at 600°C, 1 kbar. Data from Engi (personal communication). Individual curves are drawn for fixed chromium content of the spinel, X_{Cr}. Crosses are for a constant magnesium mol fraction (bulk) of 0.88, circles for X_{Mg} = 0.64. r is the model ratio of olivin to spinel.

Discussion and Summary

Judging from fluid inclusion data, many metamorphic fluids are dominated by chloride and chloride complexes. Mineral solubilities are strongly dependent upon total chloride molalities, with lowest solubilities associated with the limiting, chloride-free OH system, but they also depend on speciation in the fluid. With increasing total chloride molality (decreasing pCl), and with increasing temperature, a series of association–dissociation reactions take place, leading from the chloride-free cation to chloride-rich anionic complexes. We suggest, for instance, the sequence

$$Fe^{2+} \rightarrow FeCl^{+} \rightarrow FeCl_2 \rightarrow FeCl_3^{-} \rightarrow FeCl_4^{2-}.$$

Solubility determinations on rock-forming minerals as a function of P_{H_2O}, T, and pCl are still so scarce, that only a crude estimate can be made of the solute loads to be expected in metamorphic fluids. Data are available, however, to evaluate the effects of exchange reactions on mineral–fluid equilibria, where the mineral is a solid solution. Information is most abundant on Mg–Fe exchange.

By combining the Mg–Fe fractionation data for biotite–fluid and olivine–fluid of Schulien (1980) and Schulien *et al.* (1970) with a wide range of solid–solid exchange data available from synthetic and natural assemblages, we have constructed a set of mineral–fluid exchange diagrams at 600°C and 1 kbar, where the fluid is a chloride-rich aqueous solution. The minerals include garnet, cordierite, clinopyroxene, orthopyroxene, hornblende and spinel. The intensity of the Mg preference of the silicates decreases as follows (numbers in parentheses are preferred K_D values for 600°C):

Cordierite (20.8) $>$ biotite (11.1) $>$ clinopyroxene (10.7) $>$ olivine (8.3) $>$ hornblende (6.6-5.0) $\sim$ orthopyroxene (5.4) $>$ garnet (2.2).

In Fig. 15 we have contrasted the two extreme cases, cordierite and garnet, and have illustrated the effect of these solid solutions on the abundances of $MgCl_2$ and $FeCl_2$ in the fluid at 600°C and 1 kbar. Obviously, for cordierites even a very Mg-rich cordierite is in equilibrium with a fluid in which $FeCl_2$ predominates over $MgCl_2$. A similar diagram, for biotite–fluid equilibria, is given in Eugster and Gunter (1981, Fig. 15). If chloride fluids come in contact with Fe–Mg silicates such as cordierite, biotite, pyroxene at elevated temperature, such fluids will become moderately to greatly enriched in $FeCl_2$ relative to $MgCl_2$. For those solid solutions, for which information is available, Mg preference by the solid phase decreases with decreasing temperature. It is quite possible that silicates exhibit the same kind of trend observed in the carbonates, where at lower temperatures magnesium is fractionated preferentially into the fluid (see Eugster, 1982). We have calibrated the very strong preference of Mg for the M_1 position of orthopyroxenes. Using the M_1; M_2 (opx)–fluid exchange reactions, we have illustrated that there is an underlying structural basis for the observed partitioning behavior of Fe and Mg.

The results allow us to compare the Mg–Fe exchange behavior of a wide range

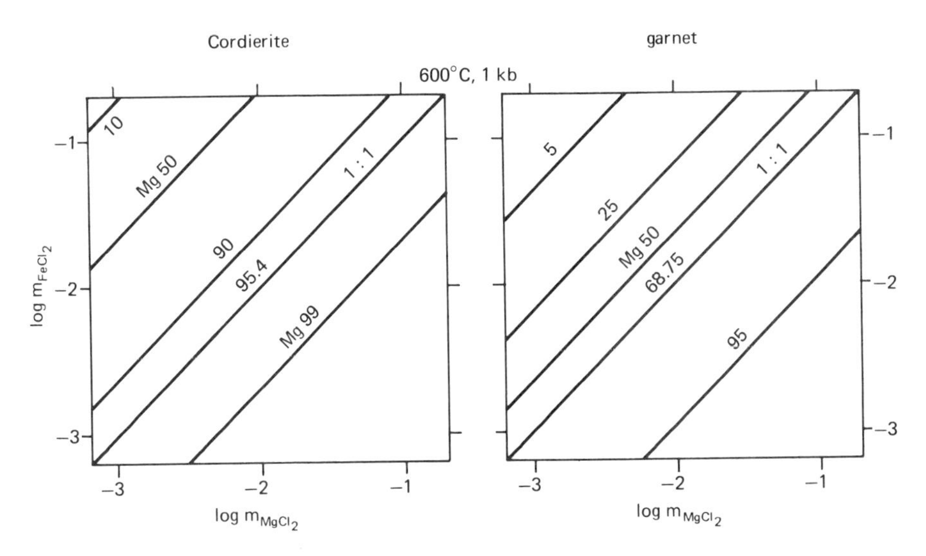

Fig. 15. Effect of Mg–Fe fractionation between solid solutions and a chloride fluid at 600 °C, 1 kbar. The molalities of $MgCl_2$ and $FeCl_2$ in the fluid are plotted against each other as a function of the compositions varying from 0 to 100. For cordierite, $MgCl_2$ equals $FeCl_2$ at Mg-cordierite 94.5, while the corresponding condition applies to Mg-garnet of 68.75.

of solid solutions to each other. However, the constants obtained must be used with great caution. Some of the extrapolations are considerable, and some of the bridges are quite uncertain, with errors piling up. Garnet, for instance, is a crucial link in many of our derivations, and yet the calibration of Ferry and Spear (1978) is restricted to a narrow compositional range. Temperatures and pressures have been extrapolated to 600 °C and 1 kbar, often with little supporting information.

In all cases, we have assumed an ideal Mg–Fe exchange between solids and fluids. This appears to be supported by the data of Schulien *et al.* (1970) and Schulien (1980) for olivine and biotite at 600 °C, but seems hardly likely to be the case for the other silicates. In fact, Wood and Kleppa (1981) documented small departures from ideal mixing for forsterite–fayalite solid solutions at temperatures as low as 970 K.

Perhaps the principal value of this study lies in pointing out how little quantitative information we have on metamorphic fluids and how much work remains to be done. Eventually, we should be able to monitor structural changes in the solids by compositional effects in the associated fluids, but at the present time, our chain of extrapolations is too long and fluid composition calibrations are too imperfect to realize this hope.

Acknowledgments

We thank Martin Engi for access to an unpublished manuscript and for a careful review of an earlier version of the paper, and we are grateful to David Veblen for advice in crystallographic matters. Work supported by NSF Grant EAR-8206177.

References

Chou, I-Ming, and Eugster, H. P. (1977) Solubility of magnetite in supercritical chloride solutions, *Am. J. Sci.* **277**, 1296–1314.

Crerar, D. A., Susak, N. J., Borcsik, M., and Schwartz, S. (1978) Solubility of the buffer assemblage pyrite + pyrrhotite + magnetite in NaCl solutions from 200 to 350°C, *Geochim. Cosmochim. Acta* **42**, 1427–1437.

Currie, K. L. (1971) The reaction 3 cordierite = 2 garnet + 4 silliminite + 5 quartz as a geological thermometer in the Opicon Lake region, Ontario, *Contrib. Mineral. Petrol.* **33**, 215–226.

Ellis, D. J., and Green, D. H. (1979) An experimental study of the effect of Ca upon the garnet-clinopyroxene Fe–Mg exchange equilibria, *Contrib. Mineral Petrol.* **71**, 13–22.

Engi, M. (1978) Mg–Fe exchange equilibria among Al–Cr spinel, olivine, orthopyroxene, and cordierite, Diss. ETH 6256, Zurich.

Engi, M. (in press) Equilibria involving Al–Cr spinel: I. Mg–Fe exchange with olivine. Experiments, thermodynamic analysis and consequences for geothermometry. *Am. J. Sci.*

Eugster, H. P. (1981) Metamorphic solutions and reactions, in *Physics and Chemistry of the Earth,* Vols. 13 and 14, *Chemistry and Geochemistry at High Temperatures and Pressures,* edited by D. T. Richards and F. E. Wickman, pp. 461–507, Pergamon Press, Oxford.

Eugster, H. P. (1982) Rock-fluid equilibrium systems, in *High-Pressure Researches in Geoscience,* edited by W. Schreyer, pp. 501–518, Schweitzerbart, Stuttgart.

Eugster, H. P., and Gunter, W. D., (1981) The compositions of supercritical metamorphic solutions, *Bull. Mineral.* **104**, 817–826.

Ferry, J. M., and Spear, F. S., (1978) Experimental calibration of the partitioning of Fe and Mg between biotite and garnet, *Contrib. Mineral. Petrol.* **66**, 113–117.

Frantz, J. D., and Marshall, W. L. (1982) Electrical conductances and ionization constants of calcium chloride and magnesium chloride in aqueous solutions at temperatures to 600°C and pressures to 4000 bars, *Amer. J. Sci.* **282**, 1666–1693.

Goldman, D. S., and Albee, A. L. (1977) Correlation of Mg/Fe partitioning between garnet and biotite with 0(18)/0(16) partitioning between quartz and magnetite, *Am. J. Sci.* **277**, 750–761.

Gunter, W. D., and Eugster, H. P. (1978) Wollastonite solubility and free energy of supercritical aqueous $CaCl_2$, *Contrib. Mineral. Petrol.,* **66**, 271–281.

Gunter, W. D., and Eugster, H. P. (1980) Mica–feldspar equilibria in supercritical alkali chloride solutions, *Contrib. Mineral. Petrol.* **75**, 235–250.

Hensen, B. J., and Green, D. H. (1971) Experimental study of the stability of cordierite and garnet in pelitic compositions at high pressures and temperatures. I. Compositions with excess alumino silicate, *Contrib. Mineral. Petrol.* **33**, 309–330.

Henson, B. J., and Green, D. H. (1972) Experimental study of the stability of cordierite and garnet in pelitic compositions at high pressures and temperatures. II. Compositions without excess alumino silicate, *Contrib. Mineral. Petrol.* **35**, 331–354.

Holdaway, M. J. (1976) Mutual compatibility relations of the Fe^{+2}–Mg–Al silicates at 800°C and 3kb., *Am. J. Sci.* **276**, 285–308.

Holdaway, M. J., and Lee, S. M. (1977) Fe–Mg cordierite stability in high-grade pelitic rocks based on theoretical and natural observation, *Contrib. Mineral. Petrol.* **63**, 175–198.

Hollister, L. S., and Crawford, M. L. (1981) Fluid inclusions: Applications to Petrology, *Mineral. Assoc. Canada, Short Course Handbook* **6**, 304pp.

Kretz, R. (1959) Chemical study of garnet, biotite, and hornblende from gneisses of south-western Quebec, with emphasis on distribution of elements in coexisting minerals, *J. Geol.* **67**, 371–402.

Kretz, R. (1963) Distribution of Mg and Fe between orthopyroxene and calcic pyroxene in natural mineral assemblages, *J. Geol.* **71**, 773–785.

Kretz, R. (1981) Site-occupancy interpretation of the distribution of Mg and Fe between orthopyroxene and clinopyroxene in metamorphic rocks, *Can. Mineral.* **19**, 493–500.

Kretz, R., and Jen, L. S. (1978) Effect of temperature on the distribution of Mg and Fe between calcic pyroxene and hornblende, *Can. Mineral.* **16**, 533–537.

Labotka, T. C., Papike, J. J., Vaniman, D. T., and Morey, G. B. (1981) Petrology of contact metamorphosed argillite from the Rove Formation, Gunflint Trail, Minnesota, *Amer. Mineral.* **66**, 70–86.

Lonker, S. W. (1981) The *P-T-X* relation of the cordierite–garnet–sillimanite–quartz equilibrium, *Am. J. Sci.* **281**, 1056–1090.

Maxey, L. R., and Vogel, T. A. (1974) Compositional dependence of the coexisting pyroxene iron-magnesium distribution coefficient, *Contrib. Mineral. Petrol.* **43**, 295–306.

Medaris, L. G., Jr. (1969) Partitioning of Fe^{++} and Mg between synthetic olivine and orthopyroxene, *Am. J. Sci.* **267**, 945–968.

Mueller, R. F. (1960) Compositional characteristics and equilibrium relations in mineral assemblages of a metamorphosed iron formation, *Am. J. Sci.* **258**, 449–497.

O'Neill, H. St.C., and Wood, B. J. (1979) An experimental study of Fe–Mg partitioning between garnet and olivine and its calibration as a geothermometer, *Contrib. Mineral. Petrol.* **70**, 59–70.

Perchuk, L. L. (1977) Thermodynamic control of metamorphic processes, in *Energetics of Geological Processes,* edited by S. K. Saxena and S. Bhattadrarji, Springer-Verlag, New York.

Quist, A. S., and Marshall, W. L. (1968) Electrical conductances of aqueous sodium chloride solutions from 0° to 800° and at pressures to 4000 bars, *J. Phys. Chem.* **72**, 684–703.

Raheim, A., and Green, D. H. (1974) Experimental determination of the temperature and pressure dependence of the Fe–Mg partition coefficient for coexisting garnet and clinopyroxene, *Contrib. Mineral. Petrol.* **48**, 170–203.

Saxena, S. K. (1968) Silicate solid solutions and geothermometry. 2. Distribution of Fe^{++} and Mg between coexisting olivine and pyroxene, *Contrib. Mineral. Petrol.* **22**, 147–156.

Saxena, S. K., (1973) Thermodynamics of Rock-Forming Crystalline Solutions. Springer-Verlag, New York, 140pp.

Saxena, S. K., and Ghose, S. (1971) Mg^{++}–Fe^{++} order–disorder and the thermodynamics of the orthopyroxene crystalline solution, *Amer. Mineral.* **56**, 532–559.

Schulien, S. (1973) Das Mischkristallverhalten des Biotites im Temperaturbereich 500–700°C bei Wasserdampfdrucken von 250–2000 bar. Diss. Marburg/Lahn, 133pp.

Schulien, S. (1975) Determination of the equilibrium constant and the enthalpy of reaction for the Mg^{2+}–Fe^{2+} exchange between biotite and a salt solution, *Fortschr. Mineral.* **52**, 133–139.

Schulien, S. (1980) Mg–Fe partitioning between biotite and a supercritical chloride solution, *Contrib. Mineral. Petrol.* **74**, 85–93.

Schulien, S., Friedrichsen, H., and Hellner, E. (1970) Das Mischkristallverhalten des Olivins zwischen 450°C und 650°C bei 1 kb Druck, *Neues Jahrb. Mineral., Monatsh,* 141–147.

Seward, T. M. (1976) The stability of chloride complexes of silver in hydrothermal solutions up to 350°C, *Geochim. Cosmochim. Acta* **40**, 1329–1341.

Seward, T. M. (1981) Metal complex formation in aqueous solutions at elevated temperatures and pressures, in *Physics and Chemistry of the Earth,* Vols. 13 and 14, *Chemistry and Geochemistry at High Temperatures and Pressures,* edited D. T. Richards and F. E. Wickman, pp. 113–132, Pergamon Press, Oxford.

Stumm, W., and Morgan, J. J. (1981) *Aquatic Chemistry,* 2nd ed., John Wiley, New York.

Thompson, A. B. (1975) Mineral reactions in pelitic rocks: II. Calculations of some *P–T–X* (Fe–Mg), phase relations, *Am. J. Sci.* **276**, 425–454.

Walther, J. V., and Orville, P. M. (1982) Volatile production and transport in regional metamorphism, *Contrib. Mineral. Petrol.* **79**, 252–257.

Williams, N. (1978) Studies of the base metal sulfide deposits at McArthur River, Northern Territory, Australia, *Econ. Geol.* **73**, 1005–1056.

Wood, B. J. (1973) Fe^{++}–Mg^{++} partitioning between coexisting cordierite and garnet—A discussion of the experimental data, *Contrib. Mineral. Petrol.* **40**, 253–258.

Wood, B. J., and O. J. Kleppa (1981) Thermochemistry of forsterite–fayalite olivine solutions, *Geochim. Cosmochim. Acta* **45**, 529–534.

Chapter 5
Geobarometry in Granulites

Steven R. Bohlen, Victor J. Wall, and A. L. Boettcher

Introduction

It is apparent from the initial attempts at quantitative tectonic modeling of regional metamorphism (England and Richardson, 1977; Wells, 1980) that a detailed knowledge of metamorphic pressure is essential to the accurate formulation of any tectonic model attempting to incorporate metamorphic pressure—temperature—time relationships. There is, therefore, a need for pressure data of greater accuracy and precision than has been available. Of greatest importance is knowledge not only of the so-called peak metamorphic pressures but also regional variations in the inferred maximum pressures, as well as the prograde and retrograde pressure path of rocks, especially of those that were once buried to the deepest levels of the crust. Unfortunately, at least some of the information necessary to infer the prograde metamorphic path is destroyed by the metamorphic process itself, although in a few terranes, limited data on the prograde pressure–temperature path can be inferred (e.g., Tracy *et al.*, 1976; Phillips and Wall, 1981). Ability to infer such data depends in part on availability of numerous well-calibrated geobarometers applicable in a variety of bulk compositions.

The extremes in estimates of peak pressures in regional granulite terranes range from 15–18.5 kbar for the Scourie gneisses (O'Hara 1975, 1977) to 2–4 kbar (Saxena, 1977), but most estimates of pressure range between 5 and 10 kbar. Recently, Newton and Perkins (1982) noted a clustering of pressures (8.9 ± 1.5 kbar) for granulite massif areas based on results of their calculated garnet barometers. If substantiated by results from other well-calibrated barometers, the clustering of pressures has important consequences for the modeling of tectonic regimes that generate granulites. Newton and Perkins (1982) proposed that continental overthrusting, such as that in progress under the Tibetan Plateau, was the primary tectonic mechanism that could generate regional granulite metamorphism and explain the recurrent 8.9 ± 1.5 kbar pressures. However, appreciable modifications to such a model, if not alternate models, might be necessary to explain the extreme pressure determinations if they too withstand further scrutiny.

Because garnet is stable over a broad range of pressure and temperature as a result of its varied chemistry, its formation and stability relationships can provide a wealth of information regarding the thermal and barometric histories of rocks. Garnet-bearing reaction equilibria frequently have relatively large volume changes, making them appropriate for geobarometry (Essene, 1982). Therefore, a variety of "garnet" geobarometers have been proposed for both metaluminous and peraluminous bulk compositions, and these have been applied in granulites with widely varying degrees of success.

Barometry in Metaluminous Rocks

In metaluminous bulk compositions, greatest attention has been given to assemblages of coexisting orthopyroxene–garnet, olivine–plagioclase–garnet, orthopyroxene and/or clinopyroxene–plagioclase–garnet–quartz, from which pressure can be determined if equilibration temperature, mineral compositions, and appropriate activity data are known. Stability of these assemblages have been investigated by experiments on multicomponent natural rock compositions and on three- or four-component synthetic systems. Investigations of multicomponent systems (Ringwood and Green, 1966; Green and Ringwood, 1967; Green, 1970; Ito and Kennedy, 1971) cannot be quantitatively applied to granulites because no compositional data of appropriate phases were obtained. However, these studies do serve to broadly outline granulite and garnet–granulite facies relationships. The experiments also show that garnet is stable at lower pressures in olivine normative compositions than in quartz normative rocks. This reflects the growth of garnet by different reactions:

olivine + plagioclase ⇄ garnet + Na-enriched plagioclase (olivine normative); (1)

orthopyroxene + plagioclase ⇄ garnet + quartz + Na-enriched plagioclase (quartz-normative). (2)

Similar phase relations have been inferred from field observations (e.g., Buddington, 1965, 1966; DeWaard, 1965, 1967; Wood, 1975). Experiments in the synthetic systems $MgO–Al_2O_3–SiO_2$ (MAS) and $CaO–MgO–Al_2O_3–SiO_2$ (CMAS) provided the data necessary for formulation of barometers based on the equilibrium relations:

enstatite + Mg–tschermak's pyroxene ⇄ garnet (Py[1]); (3)

forsterite + anorthite ⇄ orthopyroxene + clinopyroxene + spinel; (4)

orthopyroxene + clinopyroxene + spinel + anorthite ⇄ garnet (Gr_1Py_2); (5)

orthopyroxene + clinopyroxene + spinel ⇄ garnet (Gr_1Py_2) + forsterite; (6)

enstatite + anorthite ⇄ garnet (Gr_1Py_2) + quartz; (7)

forsterite + anorthite ⇄ garnet (Gr_1Py_2). (8)

With the exception of reaction (7), barometers based on the above phase relations were formulated for use in garnet and spinel lherzolites, but in every case, their

[1]Gr, Py, Alm, and Sp refer to grossular, pyrope, almandine, and spessartine components, respectively, in garnet.

use in relatively iron-rich, crustal rocks has been attempted. Reaction (3), originally calibrated by Wood and Banno (1973), has been recalibrated by Harley and Green (1983) on the basis of their recent experiments in the system FeO–MgO–Al_2O_3–SiO_2. Harley and Green have applied their barometer in a variety of rock types but greater accuracy and precision is obtained in magnesian rocks of mantle origin. Johnson and Essene (1982) attempted to calibrate barometers for crustal rocks based on reactions (4), (5), (6), and (8) and found that the barometer based on (8) gave the most reasonable pressures in the Adirondacks. Reaction (7) forms the basis of barometers calibrated by Wood (1975), Perkins (1979), and Wells (1979). In comparison with limited data from other well-calibrated barometers all three of these formulations yield pressures that are erratic and usually too high. Despite limited success in application a few of the various barometers (e.g., Johnson and Essene, 1982), these barometers are subject to large uncertainties of, perhaps, several kilobars because they are based on experimentally determined equilibria that are not tightly reversed and that require large and uncertain extrapolations to lower temperatures appropriate for granulites. In order to avoid such problems, Newton and Perkins have calibrated two different garnet barometers based on reaction (7) and the equilibrium relation:

$$\text{diopside} + \text{anorthite} \rightleftarrows \text{garnet}\ (Py_1Gr_2) + \text{quartz} \qquad (9)$$

from available thermochemical data. The Newton and Perkins barometers appear to yield lower, more reasonable pressures than those calculated from Wood, Perkins, or Wells for crustal rocks. However, comparison of results from the Newton and Perkins orthopyroxene barometer [based on (7)] with other well-calibrated barometers suggests that it too yields pressures that are slightly high. This may be the result of uncertainties in the thermochemical data base or in solution models for constituent minerals.

Uncertainty in experimental calibrations and in thermochemical data as well as the lack of tight experimental reversals notwithstanding, the inference of accurate and precise pressure data using the above barometers is severely hampered by the lack of unequivocally well-determined activity–composition data for garnets and pyroxenes. The problem is compounded for those barometers based on the CMAS or MAS systems because garnets and pyroxenes found in the majority or granulite terranes are greater than 50 mol% almandine and ferrosilite components, respectively. To accurately infer pressures from barometers based on Fe-deficient systems, accurate activity data are required to properly account for dilute concentrations of grossular and pyrope components in almandine-rich garnets and enstatite in ferrosilite-rich orthopyroxenes. As a consequence, pressures in granulites determined from Fe-silicate reactions should be more accurate and precise at the present time.

Barometry in Peraluminous Rocks

Historically relative pressures have been inferred in peraluminous rocks using Al_2SiO_5 index minerals. In terranes where one of three possible Al_2SiO_5 isograds can be mapped, quantitative assessment of pressure is possible if metamorphic temperatures are well known. Potential complications in using this system for barometry involve the metastable persistence of one Al_2SiO_5 polymorph into the stability field of another or uncertainty regarding textural relations especially for fibrolitic sillimanite. Alternatively pressure and temperature have been inferred simultaneously from assemblages containing two Al_2SiO_5 minerals as well as the products and reactants of a calibrated, pressure-insensitive dehydration reaction (Carmichael, 1978). Potential problems of application include the lack of numerous assemblages of low variance and effects of variable activity of H_2O at high metamorphic grade.

Other mineralogic barometers have been formulated in recent years. One such barometer is based on assemblages of plagioclase–garnet–Al_2SiO_5–quartz (Kretz, 1964; Ghent, 1976; Newton and Haselton, 1981) wherein the grossular component of the garnet is buffered by the equilibrium relation:

$$3 \text{ anorthite (in plag)} \rightleftarrows \text{grossular (in garnet)} + 2Al_2SiO_5 + 1 \text{ quartz}. \quad (10)$$

This assemblage is multivariant as the result of solid solution in plagioclase and garnet and, therefore, is stable over a wide range of pressure and temperature. The primary difficulties in application of the barometer include: (1) the substantial temperature dependence of reaction (10) requiring accurate knowledge of temperature for inference of reliable pressures; (2) significant uncertainty in the $\Delta\overline{V}$ required for calculation of pressures from the experimentally determined phase relation (10) containing end member grossular that occurs at pressures approximately 10–15 kbar above those of granulite terranes in which the barometer applies; (3) the extreme dilution of grossular component in garnet coexisting with plagioclase, Al_2SiO_5 and quartz in granulites (typically garnets contain < 5 mol% grossular and commonly < 2 mol% grossular) and the relatively large errors that accompany analysis of minor components; (4) uncertainty in activity data for grossular at extreme dilutions. Yet despite these difficulties the barometer does appear to yield reasonable pressures in a number of terranes (Schmid and Wood, 1976; Ghent *et al.*, 1979; Newton and Haselton, 1981; Perkins *et al.*, 1982). However, the precision of the barometer has never been tested in a terrane wherein rocks containing the assemblage have widely varying plagioclase and garnet compositions. Another potentially widely applicable barometer is based on equilibria between cordierite and garnet–sillimanite–quartz and has considerable potential in cordierite granulites. Experimental, theoretical and/or semiempirical calibrations of the barometer have been attempted by Currie (1971), Hensen and Green (1971), Weisbrod (1973), Thompson (1976), Holdaway and Lee (1977), Newton and Wood (1979), Lonker (1981), Martignole and Sisi (1981) and Perchuk *et al.* (1981). The calibrations are conflicting, and there is little agreement as to the P–T location and slope of the end member reactions, the role of P_{H_2O} vs. P_{tot} and the activity of anhydrous cordierite in H_2O-bearing cordierite. An additional problem, yet to be addressed, is the potential complication arising from attempts to apply results obtained with disordered (?) experimental cordierites to the natural assemblages containing ordered cordierite. As a consequence of the above problems, pressures inferred from assemblages of cordierite–garnet–Al_2SiO_5–quartz have large uncertainties. Similar problems of calibration and effect of H_2O, etc., also plague the barometer based on the equilibrium relation:

$$\text{cordierite} + \text{orthopyroxene} \rightleftarrows \text{garnet} + \text{quartz} \qquad (11)$$

(Hensen and Green, 1972). Assemblages of cordierite–orthopyroxene–garnet–quartz are found in only a few granulite terranes (e.g., Reinhardt, 1968; Perkins *et al.,* 1982).

It is apparent from the preceding discussion that the quality of geobarometric information obtained from available thermometers is highly variable and generally imprecise. No doubt geologic intuition plays a greater role than it should in deciphering the conflicting results of various barometers and in establishing the general pressure framework of high grade metamorphic terranes. Better calibration of existing barometers as well as proper calibration of additional pressure sensitive systems is necessary.

In metaluminous rocks the reaction separating granulite and garnet–granulite facies and the reaction of greatest potential for barometry is the Fe-analog of reaction (8):

$$\text{ferrosilite} + \text{anorthite} \rightleftarrows \text{garnet } (Gr_1Alm_2) + \text{quartz}. \qquad (12)$$

This reaction is metastable as the result of instability of ferrosilite at low pressures. However, the pressure–temperature locus of (12) can be accurately calculated from:

$$\text{fayalite} + \text{anorthite} \rightleftarrows \text{garnet } (Gr_1Alm_2), \qquad (13)$$

$$\text{ferrosilite} \rightleftarrows \text{fayalite} + \text{quartz}. \qquad (14)$$

Both (13) and (14) are themselves useful barometers. In peraluminous rocks an equilibrium of great potential for geobarometry is:

$$3 \text{ ilmenite} + Al_2SiO_5 + 2 \text{ quartz} \rightleftarrows \text{almandine} + 3 \text{ rutile}. \qquad (15)$$

Experiments calibrating barometers based on (12), (13), (14), and (15) have recently become available (Bohlen *et al.,* 1980a, 1980b, 1983a, 1983b). Application of these barometers in granulites and the significance of the pressure information is discussed below.

Experimental Methods

Details of the experiments are given in Bohlen *et al.* (1980a,b; 1983a,b) and Bohlen and Boettcher (1982). Important aspects and general procedures are outlined briefly.

Experiments were conducted in piston-cylinder aparatus with 2.54-cm diameter furnace assemblies. The furnace assemblies are similar to those described by Johannes (1978) and Boettcher *et al.* (1981). They consist of NaCl and graphite and differ from those used by Johannes in the following ways: (1) The bottom (piston end) graphite plug extends 3 mm into the internal portion of the assembly; (2) the sample is placed horizontally in the notched top surface of a cylinder of

NaCl and surrounded by powdered BN; and (3) the thermocouple ceramic is sheathed by 3-mm-diameter MgO tubing that in turn is surrounded by a cylinder of NaCl. Temperature was measured with Pt_{100}–$Pt_{90}Rh_{10}$ thermocouples with no correction for the effects of pressure on emf. We used the piston-in technique by bringing the pressure to 10–20% below that of the final value, increasing the temperature to that desired for the experiment and then increasing pressure to the final value. During the period over which the temperature was increased, there was a concomitant increase in pressure as a result of the thermal expansion of NaCl. However, the piston was always advanced to the final value. Pressures (temperatures) were controlled to within 0.1 kbar (3 °C).

The furnace assemblies have been calibrated over a wide range of pressure and temperature using the melting points of CsCl, LiCl, and NaCl (Clark 1959; 5-kbar data revised by Bohlen and Boettcher, 1982) and the reaction albite $\rightleftarrows$ jadeite + quartz at 600 °C (Hays and Bell, 1973). The furnace assemblies require no pressure correction even at pressures as low as 5 kbar and temperatures 300–500 °C below the solidus of NaCl. Details of the calibration are in Bohlen and Boettcher (1982).

With the exception of coarse-grained natural sillimanite and quartz, all starting materials were synthetic crystalline phases synthesized from gels, oxide mixes or glasses at appropriate P, T, and f_{O_2}. Starting materials and experimental products were analyzed optically and with X-ray, electron microprobe, and Mössbauer techniques before and after the experiments to ensure that the phases were homogeneous and stoichiometric. Experimental charges consisted of proportions of reactants and products consistent with the stoichiometry of the reaction of interest, and were contained in $Ag_{80}Pd_{20}$ capsules. These capsules were sealed in 5-mm Ag_{100} or Pt_{100} capsules along with 300–500 mg Fe° and sufficient H_2O to react approximately 75% of the Fe° to $Fe_{1-x}O$.

Experiments to determine stability of ferrosilite (reaction (14)) were originally conducted in talc furnace assemblies to which a −6% pressure correction was applied. These experiments have been repeated at 800 and 1000 °C using the NaCl furnace assemblies and procedures outlined above. The results agree precisely with the original experiments.

Results

Definitive experiments that tightly constrain equilibria (13), (14), and (15) are in Figs. 1, 3, and 2, respectively. These data are the bases for four useful, well-calibrated barometers shown in Figs. 4–6, 7 and 9. The barometer based on equilibrium (12) (Fig. 7) must be calculated from the data in Fig. 1 and the end member ferrosilite equilibrium in Fig. 3 using the scheme: $\Delta G_{(12)} = \Delta G_{(13)} + \Delta G_{(14)}$. The precise determinations of equilibria (13) and (14) allow precise calculation of the pressure–temperature locus of equilibrium (12).

Stability relationships between ferrosilite-rich orthopyroxenes and fayalite-rich olivines and quartz are themselves useful for barometry in granulites. Unfortu-

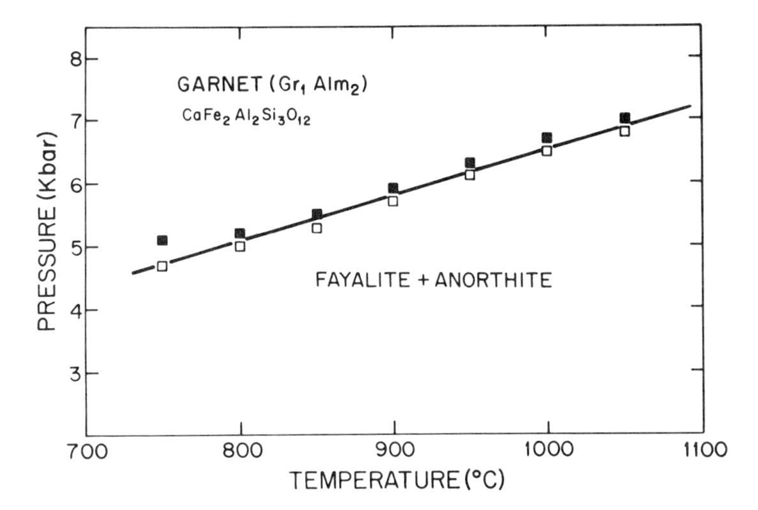

Fig. 1. P–T projection showing relative stabilities of anorthite + fayalite and garnet. □ indicates garnet reacted to fayalite + anorthite. ■ indicates that fayalite + anorthite reacted to garnet.

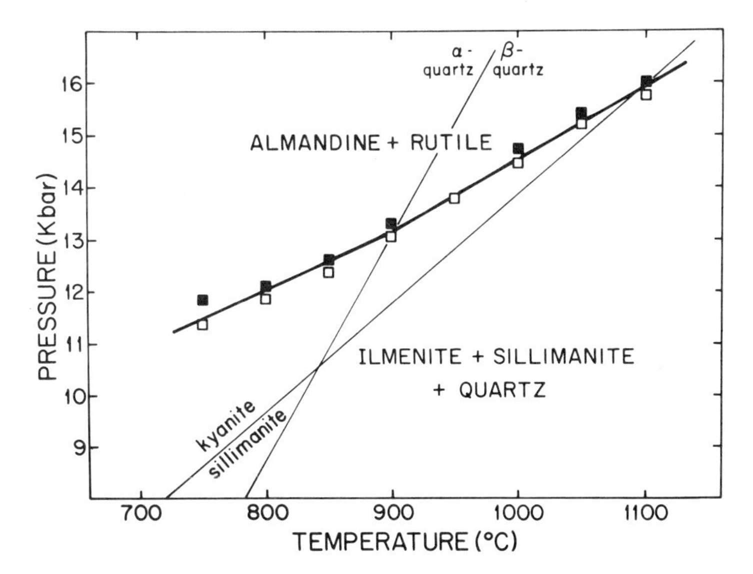

Fig. 2. P–T projection showing relative stabilities of almandine + rutile and ilmenite + sillimanite + quartz. □ indicates that almandine + rutile reacted to ilmenite + sillimanite + quartz. ■ indicates that ilmenite + sillimanite + quartz reacted to almandine + rutile. Kyanite-sillimanite and α-β quartz equilibria from Richardson *et al.* (1968) and Cohen and Klement (1967), respectively.

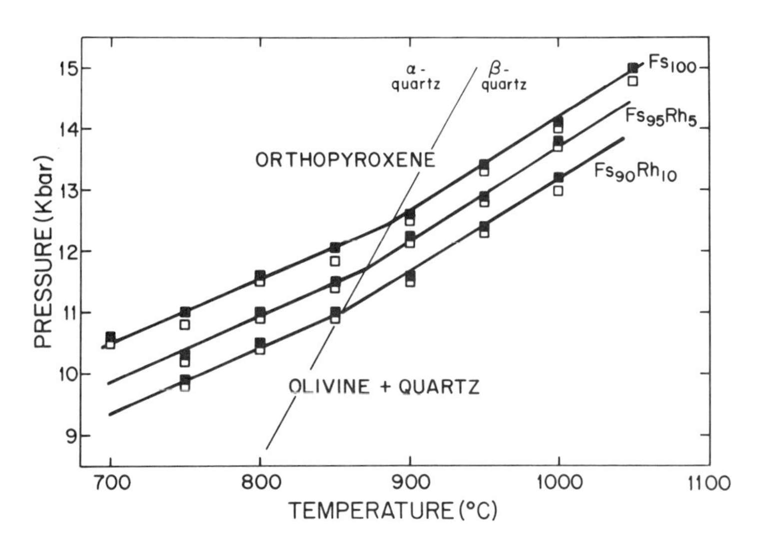

Fig. 3. P–T projection showing relative stabilities of ferrosilite and fayalite + quartz and the effects of 5 and 10 mol. % rhodonite solution in ferrosilite-rich orthopyroxene. □ indicates that orthopyroxene reacted to olivine + quartz. ■ indicates that olivine + quartz reacted to orthopyroxene. α-β quartz from Cohen and Klement (1967).

nately, the barometer is not generally applicable because assemblages of coexisting orthopyroxene–olivine–quartz are uncommon. Figures 3, 4, and 5 show definitive experiments that determine the effects of manganese (Fig. 3) and magnesium on the stability of ferrosilite-rich orthopyroxene (Fig. 4) and the assemblage fayalite-rich olivine + quartz (Fig. 5) (Bohlen *et al.,* 1980a,b; Bohlen and Boettcher, 1982). Also shown in Figs. 4 and 5 are the calculated lower stability of $Fs_{75}En_{25}$[2] and the calculated upper stability of $Fa_{85}Fo_{15}$[2] + quartz, respectively. The calculations are similar to those described in Bohlen *et al.* (1980a). In the calculations we used an ideal two-site mixing model for orthopyroxene (Wood and Banno, 1973) and the activity data of Engi (1980) for olivines as well as available molar volume data (Turnock *et al.* 1973; Akimoto *et al.,* 1976), thermal expansion (Smyth, 1975; Sueno *et al.,* 1976) and compressibility data (Birch, 1966). We used the experimentally determined pressure–temperature locus of the lower stability of $Fs_{80}En_{20}$ and upper stability of $Fa_{90}Fo_{10}$ + quartz as starting points for the calculation of the $Fs_{75}En_{25}$ and $Fa_{85}Fo_{15}$ + quartz curves, respectively. Uncertainty in the location of the calculated curves is almost certainly less than 300–400 bars. Since Mg and Mn are the most significant diluents in ferrosilite-rich orthopyroxenes, Figs. 3, 4, and 5 can be used to infer pressure information from Fe-rich orthopyroxene and/or olivine + quartz assemblages. Terranes in which this barometer is useful are listed in Table 1. The barometer is imprecise to the extent that minor proportions of other components, notably Ca, Fe^{3+}, and Al, must be estimated by using ideal solution modeling since appropriate activity data

[2]$Fs_wEn_xRh_yWo_z$ refer to molar proportions of ferrosilite, enstatite, rhodonite and wollastonite components, respectively, in orthopyroxenes; $Fa_xFo_yTe_z$ refer to molar proportions of fayalite, forsterite, and tephroite components, respectively, in olivine.

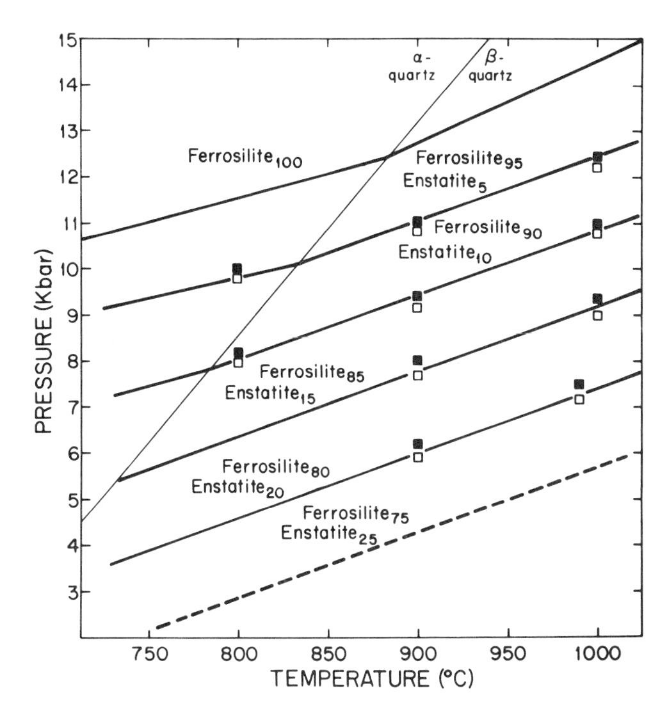

Fig. 4. P–T projection showing the effects of 5, 10, 15, 20, and 25 mol. % enstatite solution on ferrosilite-rich orthopyroxene stability. □ indicates the orthopyroxene reacted to an assemblage of orthopyroxene + olivine + quartz. ■ indicates olivine + orthopyroxene + quartz reacted completely to orthopyroxene. The dashed curve is calculated. α-β quartz from Cohen and Klement.

are not available. These elements typically comprise less than 5 mol % (combined) of most Fe-rich orthopyroxenes and, therefore, any errors in the estimated effects of uncalibrated components will not affect the inferred pressures significantly.

Apart from the barometry, phase relations in the Fe-rich portion of the pyroxene quadrilateral can also restrict maximum metamorphic temperatures. In a number of terranes (e.g., Adirondacks, Hobie Province, Lofoten) the Fe-rich orthopyroxenes (Table 1) coexist with hedenbergite-rich clinopyroxenes. The absence of metamorphic pigeonite in these assemblages restricts metamorphic temperatures to less than 825 °C (Podpora and Lindsley, 1979). If the pyroxenes are manganiferous (Lofoten), the maximum is reduced to about 775 °C for 10 mol % rhodonite solution in orthopyroxene (Bostwick, 1976).

Garnet Barometers

The experimental results shown in Fig. 1 have been used to calibrate barometers for olivine–plagioclase–garnet assemblages (Fig. 6) and for orthopyroxene–plagioclase–garnet assemblages (Fig. 7). The equilibrium relations in Fig. 1 were

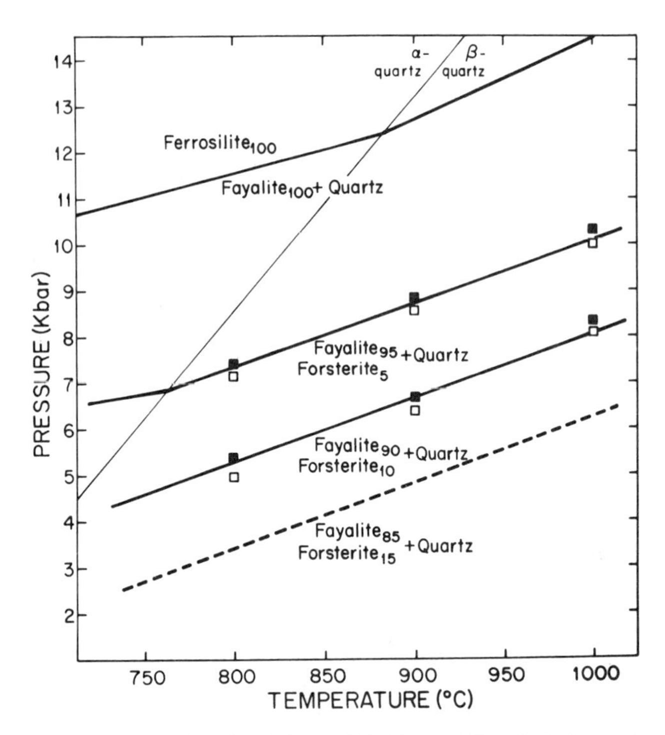

Fig. 5. *P–T* projection showing the effects of 5, 10 and 15 mol. % forsterite solution on stability of fayalitic olivine + quartz . □ indicates that olivine + quartz + orthopyroxene reacted completely to olivine + quartz. ■ indicates that olivine + quartz reacted to olivine + quartz + orthopyroxene. The dashed curve is calculated. α-β quartz from Cohen and Klement (1967).

first investigated by Green and Hibberson (1970) in a reconnaissance fashion, but neither the dP/dT slope or the absolute pressure of the equilibrum relation were known prior to our studies. The equilibrium relations forming the basis of the barometers (Figs. 6 and 7) are metastable and involve "two garnet" assemblages. This has been done because the individual activities of grossular and almandine can be calculated more easily than can the activity of Gr_1Alm_2. This greatly facilitates pressure calculations. Also shown in Figs. 6 and 7 are curves for constant $\log_{10}K$ for equilibria (13) and (12), respectively. It should be emphasized that in the following expressions for the equilibrium constant, K, for (13) and (12):

$$K_{(13)} = \frac{(a^3_{\mathrm{An}})(a^3_{\mathrm{Fa}})}{(a_{\mathrm{Gr}})(a^2_{\mathrm{Alm}})}, \qquad K_{(12)} = \frac{(a^3_{\mathrm{An}})(a^3_{\mathrm{Fs}})}{(a_{\mathrm{Gr}})(a^2_{\mathrm{Alm}})(a^3_{\mathrm{Qtz}})}$$

the activities of a_i of component i (An = anorthite, Fa = fayalite, Gr = grossular, Alm = almandine, Fs = ferrosilite, Qtz = quartz) are unity along the locus of the equilibrium curve for the pure phases. The $\log_{10}K$ curves have been calculated using available thermal expansion and compressibility data (Skinner, 1966; Smyth, 1975; Sueno *et al.*, 1976; Birch, 1966) and assuming an ideal volume of mixing for Ca–Fe garnets.

The accuracy of these barometers depends on accuracy in the calculation of the pressure–temperature loci of the metastable "two-garnet" equilibria, (12) and (13). The quality of these calculations depend on accuracy with which solution properties of Fe–Ca garnets are known. The solution properties of Fe–Ca garnets have been investigated experimentally (Cressey *et al.,* 1978; O'Neill and Wood, 1979) and deduced empirically by Ganguly and Kennedy (1974) and Ghent (1976). The experimental results of Cressey *et al.* indicate that Fe–Ca garnet mixing is substantially nonideal with activities being temperature dependent and a complex function of composition. The empirical studies suggest that the mixing is not appreciably nonideal. After a review of the available empirical and experimental data, Perkins (1979) obtained a temperature-dependent expression for W_{CaFe} ($W_{CaFe} = 1050 - 1.25T^{\circ}C$ cal/gm-atom) that for the metamorphic temperatures of most granulites gives $W_{CaFe} = 0$–300 cal/gm-atom, values in concert with other empirical studies and in agreement with the data of O'Neill and Wood (1979). Therefore, we have adopted the data of Perkins in our calculations. Use of an ideal solution model for Fe–Ca garnets results in a pressure-temperature loci of the "two-garnet" equilibria that are 400–500 bars above those shown in Figs. 6 and 7.

In pelitic rocks, assemblages containing ilmenite + Al_2SiO_5 + quartz and/or garnet + rutile are common at medium to high grades of metamorphism. This assemblage (hereafter referred to as GRAIL, for Garnet, Rutile, Al_2SiO_5, ILmenite, with excess quartz implied) is related by equilibrium (15), which forms the basis of a useful barometer (Fig. 2). The experiments using sillimanite have been carried out almost entirely in the stability field of kyanite. The location of the stable kyanite-bearing reaction can be accurately calculated and is shown in Fig. 8. The intersection of the breakdown of staurolite + quartz (Richardson, 1968; Rao and Johannes, 1979) with the GRAIL equilibrium generates additional reactions. These are only qualitatively located (Fig. 8) because of uncertanity in the entropy of staurolite (Anovitz and Essene, 1982). Such equilibria define the lower temperature stability of pyralspite garnet–rutile assemblages in the presence of H_2O vapor. Assemblages of staurolite–ilmenite–quartz–garnet–rutile and staurolite–rutile–quartz–ilmenite–Al_2SiO_5 are reported in a number of field areas (e.g., Ghent, 1975; Pigage, 1976; Fletcher and Greenwood, 1978; Holdaway, 1978; Tracy, 1978) and may themselves be useful barometers and/or thermometers. However, since the vapor-present equilibria are affected by the activity of H_2O, this variable must also be known to infer pressure and/or temperature data. (The quartz absent reaction staurolite + rutile + ilmenite = kyanite + ilmenite + V has been omitted from the diagram.)

Using the available volume thermal expansion and compressibility data (Lindsley, 1965; Robie *et al.,* 1966; Skinner, 1966; Birch, 1966) and the equilibrium curve in Fig. 8, a pressure–temperature–$\log_{10}K$[3] diagram for GRAIL equilibria has been calculated (Fig. 9). The shallow dP/dT slopes of the curves for constant $\log_{10}K$ enhance the utility of these equilibria for barometry. The GRAIL assemblage is multivariant as the result of solid–solution in garnet and, to a much

[3] K, the equilibrium constant is calculated $K = (a_{Il}^3)(a_{Al2SiO5})(a_{Qz}^2)/(a_{Alm})(a_{Ru}^3)$, where a_i is the activity of component i (Il = ilmenite, Qz = quartz, Alm = almandine, Ru = rutile). Unit activity refers to the pure phase along the univariant equilibrium curve.

Table 1. Pressures inferred from orthopyroxene + olivine + quartz geobarometry.

Locality	Reference	Sample No.	Composition	T(°C)	Pressure (kbar)[a]
Adirondacks					
Porter Mountain	Jaffe *et al*. (1978)	Po-17	$Fs_{91}En_5Rh_2Wo_2$	750	≥ 7.8
Saranac Lake	Bohlen *et al*. (1980)	SL-26	$Fs_{90}En_6Rh_2Wo_2$	760	≥ 7.8
Blue Mountain Lake	Bohlen *et al*. (1980)	IL-8	$Fs_{89}En_8Rh_2Wo_1$	740	≥ 7.5
Benson Mines	Bohlen (unpublished data)	78-Ben-3	$Fs_{89}En_9Rh_1Wo_1$	700	≥ 7.0
Au Sable Forks	Bohlen and Essene (1978)	76-AS-1	$Fa_{95}Fo_3Te_2$	700	≤ 7.5
Wanakena	Bohlen and Essene (1978)	NF-4	$Fa_{96}Fo_2Te_2$	700	≤ 8.0
Greenland					
Location unknown	Ramberg and DeVore (1951)	XYZ	$Fs_{84}En_{14}Rh_1Wo_2$	(750)[b]	≥ 5.0
Quvssagssat	Frisch and Bridgwater (1976)	49,081	$Fs_{77}En_{15}Rh_6Wo_2$ $Fa_{91}Fo_4Te_5$	800	4.5
China					
Hobie Province	Zhang *et al*. (pers. comm.)	$79\text{-}L_{3-4}$	$Fs_{86}En_{12}Rh_0Wo_2$	750	6.0
Je-ho-shen	Kuno (1954)	23	$Fs_{86}En_{12}Rh_0Wo_2$	(750)	≥ 6.0
Labrador					
Labrador Trough	Klein (1978)	6-MR	$Fs_{94}En_5Rh_1Wo_1$	700	≥ 8.8

Nain	Berg (1977)	2-1716	$Fs_{72}En_{20}Rh_5Wo_3$ $Fa_{88}Fo_8Te_5$	840	3.4
Nain		74-18	$Fs_{73}En_{23}Rh_1Wo_3$ $Fa_{88}Fo_{11}Te_1$	840	3.2
Nain		74-15x	$Fs_{77}En_{21}Rh_0Wo_2$ $Fa_{92}Fo_8Te_1$	750	3.2
Nain		2-1052	$Fs_{78}En_{18}Rh_2Wo_2$ $Fa_{93}Fo_6Te_2$	715	3.3
Nigeria					
Bauchi	Oyawoye and Makanjuola (1972)	BS-43	$Fs_{83}En_{11}Rh_2Wo_4$ $Fa_{91}Fo_4Te_3$	900	6.0
Norway					
Lofoten	Ormaasen (1977)	406	$Fs_{92}En_2Rh_5Wo_2$	800	9.5
Lofoten–Vesterålen	Griffin and Heier (1969)	107	$Fs_{90}En_8Rh_0Wo_2$	(800)	$\geq$7.8
Lofoten–Vesterålen	Griffin *et al.* (1974)	720	$Fs_{79}En_{14}Rh_5Wo_2$ $Fa_{92}Fo_4Te_4$	900	6.7
Lofoten–Vesterålen	Krogh (1977)		$Fs_{75}En_3Rh_{22}Wo_0$	800	8.0

[a]Pressure estimates are based on pyroxene and olivine (+ quartz) compositions listed in this table with adjustments for minor additional components.

[b]Temperatures in parentheses are estimated.

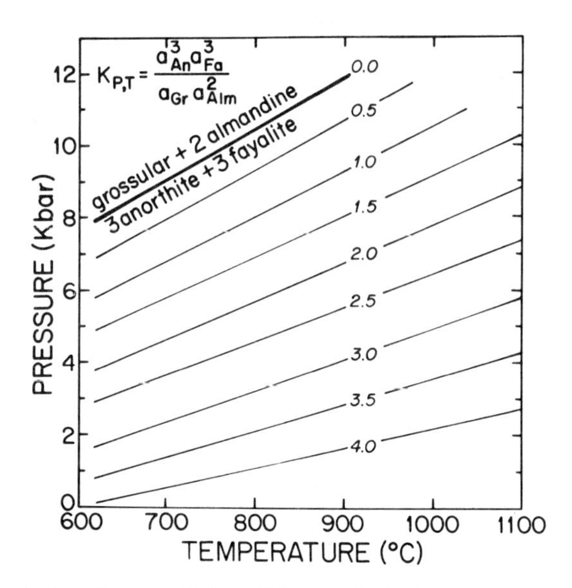

Fig. 6. Geobarometer based on coexisting olivine + plagioclase + garnet. $\mathrm{Log}_{10}K$ curves are shown for equilibrium constant as in upper left.

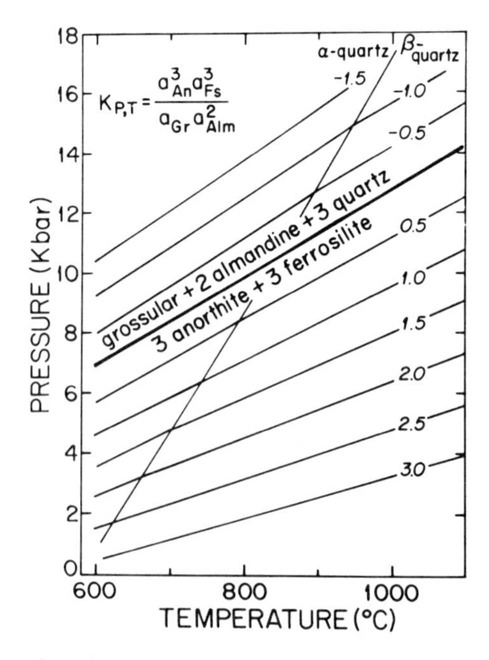

Fig. 7. Geobarometer based on coexisting orthopyroxene + plagioclase + garnet + quartz. $\mathrm{Log}_{10}K$ curves are shown for equilibrium constant as in upper left.

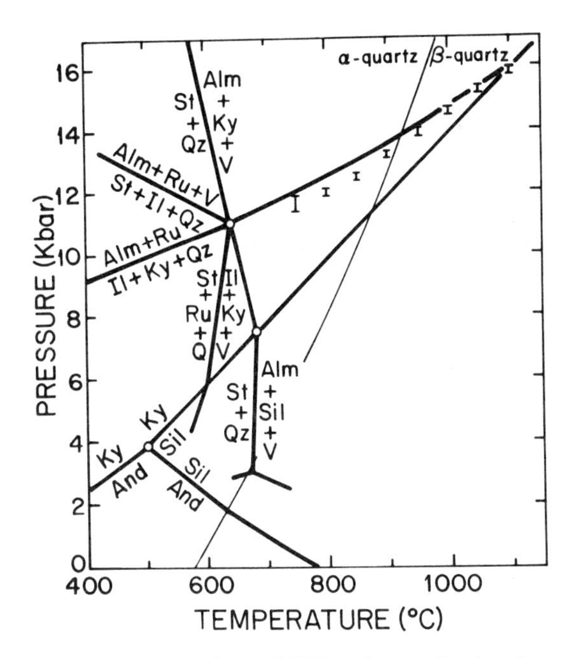

Fig. 8. P–T projection showing relative stabilities of staurolite-bearing assemblages and the GRAIL assemblage. Staurolite equilibria are from Richardson (1968), Rao and Johannes (1979). Al_2SIO_5 phase relations from Holdaway (1971). α-β quartz from Cohen and Klement (1967).

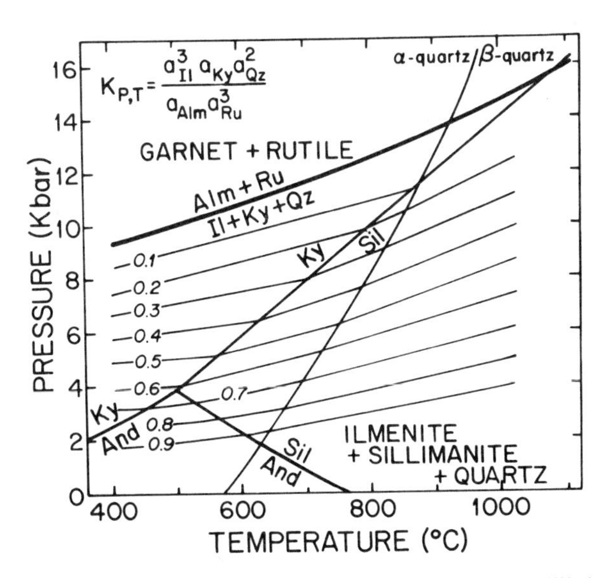

Fig. 9. The GRAIL geobarometer. $Log_{10}K$ curves are shown for equilibrium constant as in upper left. Other curves as in Fig. 8.

lesser extent, in ilmenite. Thus, to a first approximation, pressures inferred from Fig. 9 are a function of garnet composition. As a result the GRAIL geobarometer should be accurate and precise since pressures can be determined from an analysis of garnet for its major component, almandine, and errors from long extrapolations from the end member equilibrium boundary or long compositional extrapolations are minimized.

Figure 9 can be used to determine metamorphic pressures given (1) the equilibrium coexistence of the GRAIL assemblage, (2) the composition of the phases, (3) appropriate solution models for the activity of end member components in impure minerals, (4) a rough estimate of equilibration temperature. Conversely, application of the GRAIL barometer in terranes for which metamorphic pressures are well determined by other barometers will allow the activities of almandine to be deduced empirically. This, in turn, will allow critical evaluation and/or empirical adjustment of available garnet solution models.

Geobarometry

The solution properties of garnet are potentially the greatest obstacle to accurate geobarometry. In most pelitic rocks of high metamorphic grade, the garnets are essentially almandine–pyrope solutions with spessartine, grossular, and andradite components, together comprising less than 5–15 mol. %. In metaluminous rocks garnet compositions are more variable but generally are well described by the four component system of Fe–Mg–Ca–Mn garnet. Empirical determinations of W^{gt}_{FeMg} for garnet yield 2580, 2979 $\pm$ 185, 2580 $\pm$ 140 and 1509 $\pm$ 1392 cal/gm-atom (Saxena, 1968a; Ganguly and Kennedy, 1974; Oka and Matsumoto, 1974; Dahl, 1980, respectively). Experimentally determined values are substantially lower. Kawasaki and Matsui (1977) determined a value of 2120 $\pm$ 350 cal/gm-atom at high pressure and temperature. O'Neill and Wood (1979) inferred that Fe–Mg garnet mixing is significantly more ideal at 1000°C than it is in olivine. Heat of solution measurements on olivines by Wood and Kleppa (1981) indicate that W_{FeMg} for olivines is in the range of 1200 cal/gm-atom for Fe-rich compositions. This implies that W_{FeMg} for garnets must be in the range of 500–1000 cal/gm-atom. If one accepts Sack's (1980) value of W_{FeMg} = 800 cal/gm-atom for olivines, then W_{FeMg} for garnets of 0–300 cal/gm-atom are required by the experimental data. After a review of the available data, Perkins (1979) proposed a T-dependent expression for W^{gt}_{FeMg} (3480–1.2T°C) that yields W^{gt}_{FeMg} of 2500 $\pm$ 100 cal/gm-atom for granulite temperatures. However, in light of the recent experimental data, this expression has been rejected in favor of an ideal mixing model (W_{FeMg} = 0) by Newton and Perkins (1982), although they noted that such a choice is at the lower limit of permissible values. An additional empirical test of W^{gt}_{FeMg} can be made by comparing GRAIL barometry with estimates in terranes for which pressures have been reliably established. Choice of W^{gt}_{FeMg} below 1000–1500 cal/gm-atom yields sillimanite pressures for kyanite-bearing rocks and in a few cases, andalusite pressures for some sillimanite-bearing rocks. In terranes where pressures are well known, such as the Adirondacks (see Bohlen *et al.*, 1983b, for a review), empirical evaluations of W^{gt}_{FeMg} give values of 2200–2500 cal/gm-atom. Therefore, we have adopted the Perkins model for almandine-pyrope since it is consistent with the bulk of empirical data. It should be noted, however, that our conclusion regarding W^{gt}_{FeMg} is based primarily on Fe-rich compositions, whereas the data of O'Neill and Wood (1979) were determined for Mg-rich garnets. The apparent discrepancy between the two sets of Fe–Mg mixing properties could be an indication of marked asymmetry of these parameters in Fe–Mg garnets. Spessartine is a significant component of some garnets in peraluminous and metaluminous rocks, but there are few data on the mixing of Fe–Mn and Mg–Mn garnets. For those garnets with significant Mn we have assumed that Fe–Mn garnets mix ideally (Ganguly and Kennedy, 1974) and that interactions of Ca and Mg with Mn are the same as those for Fe. In consideration of the above and the evaluation of Ca–Fe garnet mixing discussed earlier, we have adopted the Perkins model for garnet solutions. In the pressure determinations that follow, we have calculated the activities of almandine and grossular components in garnet as $(\gamma X_{Alm})^3$ and $(\gamma X_{Gr})^3$, where:

$$\ln \gamma^{gt}_{Gr} = W_{CaFe} X^2_{Fe} + W_{CaMg} X^2_{Mg} + (W_{CaFe} - W_{FeMg} + W_{CaMg}) X_{Mg} X_{Fe},$$
$$\ln \gamma^{gt}_{Alm} = W_{CaFe} X^2_{Ca} + W_{FeMg} X^2_{Mg} + (W_{CaFe} - W_{CaMg} + W_{FeMg}) X_{Ca} X_{Mg},$$

and

$$W_{CaFe} = 1050 - 1.2T^{\circ}C \text{ cal/gm-atom},$$
$$W_{CaMg} = 4180 - 1.2T^{\circ}C \text{ cal/gm-atom},^{4}$$
$$W_{MgFe} = 3480 - 1.2T^{\circ}C \text{ cal/gm-atom}.$$

In addition to the activity model for garnet, pressures given in Tables 2 and 3 have been calculated using the activity models of Engi (1980), Orville (1972) and Wood and Banno (1973) for olivines, plagioclase, and orthopyroxenes, respectively. Calculated pressures are somewhat dependent on choice of activity models. Use of Saxena's (1973) model for orthopyroxene reduces the calculated pressures by roughly 0.3–0.5 kbar. Use of the garnet solution model preferred by Newton and Perkins (1982) results in calculated pressures only slightly lower than those given in Table 3. This is because the choice of ideal mixing of Ca–Fe garnets results in a higher calculated pressure locus for the "two-garnet" equilibrium in Fig. 7. The higher pressure ($\sim$ 0.4 kbar) largely compensates for the larger $\log_{10}K$ that results from assumption of ideal mixing for both Ca–Fe and Mg–Fe garnet components.

For GRAIL assemblages the compositions of quartz, Al_2SiO_5, and rutile differ little from those of end member compositions. Rutile containing Nb and Fe has

[4]Newton and Haselton (1981) have derived a different function for W_{CaMg} than that determined by Perkins ($W_{CaMg} = 3300 - 1.5T(K)$) which yields slightly lower values for W_{CaMg}. However, since the majority of garnets have relatively low amounts of Ca and/or Mg components, such differences do not affect the calculated pressures significantly.

Table 2. Pressures inferred from olivine + plagioclase + garnet geobarometry.

Locality	Reference	Sample No.	Tempera- ture (°C)	Pressure (kbar)	Average (kbar)	Other barometers (kbar)
Adirondacks,	Johnson and	187-1, LL-1	700 760	7.8 8.0		
New York	Essene				7.5	7.6[b,c]
	(1982)	187-4, 233-1	700 700	7.5 6.5[a]		
Nain Complex,	Berg (1977)		825	2.5	—	3.2[b]
Labrador						
South Harris,	Wood (1975)	16, 17	830, 830	8.8, 8.7	8.6	8.2[d]
Scotland		20	830	8.2		

[a]This pressure is for a sample collected in the extreme SE Adirondacks and may be lower grade than the other assemblages.

[b]Orthopyroxene–olivine–quartz (Bohlen and Boettcher, 1982).

[c]Akermanite breakdown (Valley and Essene, 1980).

[d]Olivine–plagioclase–garnet based on CMAS (Johnson and Essene, 1982).

been reported, but application of an ideal solution model should not lead to significant errors in calculated pressures. Ilmenite may contain appreciable proportions of hematite and pyrophanite. Ilmenite–hematite solutions are probably substantially nonideal at temperatures of medium metamorphic grades becoming nearly ideal under granulite facies conditions (Anderson and Lindsley, 1981). Hematite-rich ilmenites may not be compatible with almandine-rich garnets, the latter being unstable with respect to oxide–aluminum silicate assemblages at the relatively high f_{O_2} required by hemo–ilmenite. Fe^{3+} enriched ilmenite coexisting with almandine-rich garnet probably results from retrograde alteration. In the cases examined (Table 4), ilmenite in the GRAIL assemblage contains less than 15 mol % hematite and pyrophanite combined and an ideal solution model has been adopted for these and other less significant components.

Applications of the garnet barometers, Figs. 6, 7, and 9, to natural assemblages are summarized in Tables 2, 3, and 4, respectively. Unfortunately, a number of terranes had to be omitted from our analysis because only "representative" analyses of minerals were published and not compositions of coexisting phases from equilibrium assemblages. For GRAIL assemblages commonly no distinction is made between minor amounts of oxides, sulfides, etc., in the published modes. The utility of published modal data and mineral analyses would be enhanced greatly if *all* phases in a rock were clearly reported along with their textural relationships and mineral analyses were from equilibrium assemblages. Equilibrium coexistence of the GRAIL assemblage is sometimes difficult to demonstrate. In many rocks one or more of the phases, usually rutile, are present at levels of less than one modal percent. Also ilmenite may have oxidized during retrogression and/or weathering, resulting in the formation of rutile or anatase. Rutile has also been noted as a product of retrogression of titaniferous biotite. However, such retro-

Table 3. Pressures inferred from orthopyroxene + plagioclase + garnet + quartz geobarometry.

Locality	Reference	Sample No.[d]	Temperature (°C)	Pressure (kbar)	Av. (kbar)	Other barometers (kbar)
Nain Complex, Labrador	Berg (1977)	KI-3909, 2893C, 2893R	800, 750, 750	3.3, 2.9, 2.8	3.1	3.2[a]
Adirondack Highlands, New York	Bohlen (1979)	BM-2, BM-13, ET-1, ET-15,	750, 750, 700, 700	7.5, 7.4, 7.6, 7.5	7.6	7.6[a,b]
	Bohlen and Essene	ET-24, IN-2, IN-11, LL-6,	700, 720, 720, 750	7.1, 7.4, 8.2, 7.8		
	(1979, 1980) and	MM-4, MM-18, PH-4, SL-5,	750, 750, 700, 750	7.1, 7.8, 7.4, 7.5		
	unpublished data	SL-26, SR-29, SR-31, W-9	750, 750, 750, 700	8.1, 7.6, 7.8, 8.2		
Adirondack Lowlands, New York	Stoddard (1976)	74-C-248A, -252, -253E	730, 730, 730	6.6, 7.6, 6.8	7.0	5.9[c]
Furua Complex, Tanzania	Coolen (1980)	MF-283.2, -268.1, ZC-8	800, 800, 800,	10.2, 10.1, 10.0,	10.2	
		MF-266.3, -276, DMa-40	800, 800, 800,	10.4, 10.1, 9.6,		
		C-352.2, -247.1, -311.1	800, 800, 800, 800	11.4, 9.8, 10.2,		
		-311.4		10.6		
Ruby Range, Montana	Dahl (1980)	RMK27-3, RMK-51C, RMS-22C	750, 750, 750	7.5, 7.7, 7.4	7.5	8.2[c]
Molodezhnaya Station, E. Antarctica	Grew (1981)	341B, 342, 342A	700, 700, 700	5.7, 5.7, 5.5	5.6	5.5[c]
Inarijärvi, Finland	Hörmann *et al.* (1980)	47-III	750	7.6		
Nilgiri Hills, S. India	Janardhan *et al.* (1982)	5-2, 11-1C, 11-3, GN-4A	820, 760, 850, 880	8.8, 7.7, 9.0, 9.2	8.7	
Labwor Hills, Uganda	Nixon *et al.* (1973)	AR-51	950	6.5		
Doubtful Sound, New Zealand	Oliver (1977)	DS-36468, DS-36461	750, 750	9.6, 11.2	10.4	
Otter Lake Area, Quebec	Perkins *et al.* (1982)	RK-1, RK-2, DD-17, DD-11	700, 700, 700, 700	8.2, 8.1, 7.8, 7.0	8.0	
		A-12, DL-2, Q-11	700, 700, 700	8.1, 8.4, 8.7		
Madras, S. India	Weaver *et al.* (1978)	MP-44, MP-72	800, 800	7.0, 7.0	7.0	
Buksefjorden Area, S. W. Greenland	Wells (1979)	174018, 174031, 174087	800, 800, 800,	8.0, 8.7, 8.3,	8.0	
		174090, 174102C	800, 800	7.8, 7.4		

[a] Orthopyroxene–olivine–quartz (Bohlen and Boettcher, 1982).

[b] Akermanite breakdown (Valley and Essene, 1980).

[c] Plagioclase–garnet–Al_2SiO_5–quartz (Newton and Haselton, 1981).

[d] C and R after sample number refer to pressure calculated from garnet core or rim composition. Sample No., temperature and pressure are listed in corresponding column and row positions.

Table 4. Pressures inferred from the GRAIL barometer.

Locality	Reference	Sample No.[a]	Observed Al–Silicate	Temperature (°C)	Pressure (kbar)	Other barometers (kbar)
Lake Bonaparte, Benson Mines, Adirondacks	Bohlen, unpublished	78-LB-20, 78-Ben-1	Sill	730, 730	7.5, 7.8	7.5[b]
Western Maine	Evans and Guidotti (1966)	9, 24	Sill	600, 600	6.7, 7.5	
Quesnel Lake, British Columbia	Fletcher and Greenwood (1979)	6, 9, 13	Ky, Ky-Sill, Ky-Sill	680, 680, 680	8.4, 8.4, 7.7	7.6[e]
Esplanade Range, British Columbia	Ghent (1975)	CV-150	Ky	540	7.7	7.9[c]
South Island, New Zealand	Hattori (1967)	11468	Ky-Sill	620	6.3	6.5[e]
Northen Idaho Batholith	Hietanen (1969)	2127	Ky	650	8.5	
Funeral Mts., Death Valley, California	Labotka (1980)	FML-110	Ky-Sill	660	7.3	7.2[e]
Kilbourne Hole, New Mexico	Padovani and Carter (1977)	4, 18, 51, 54, 57, 62, 68	Sill	800, 800, 800, 800, 800, 800, 800	7.2, 7.3, 8.0, 7.3 8.1, 7.8, 7.9	
Yale, British Columbia	Pigage (1976)	3, 4, 5	Ky, Ky, Sill	600, 600, 600	7.8, 6.9, 8.0	7.3[e]
Ivrea Zone, N. Italy	Schmidt and Wood (1976)	SD-121, SD-139, SD-430E	Sill	710, 710, 710	7.2, 7.0, 7.4	5.8[c]
Colton, New York	Stoddard (1976)	C-74A, C-79B1, C-78B2 C-74B, C-79A, C-74 C-629, C-532, C-529	Sill	730, 730, 730 730, 730, 730 730, 730, 730	7.5, 7.5, 7.6 7.3, 7.4, 7.1 7.2, 7.1, 7.0	5.8[c], 7.0[d]
Central Massachusetts	Tracy (1978)	871, L11y, C26A, M34, 595C	Sill	675, 675, 675 675, 675	7.0, 6.3, 7.2 6.5, 5.3	6.8[c]

[a]Sample No., Al–silicate, temperature and pressure are listed in corresponding column and row positions.

[b]Orthopyroxene–olivine–quartz (Bohlen and Boettcher, 1982).

[c]Plagioclase–garnet–Al_2SiO_5 –quartz (Newton and Haselton, 1981).

[d]Orthopyroxene–plagioclase–garnet-quartz, see Table 3.

[e]Kyanite–sillimanite at known temperature (Richardson *et al.*, 1968).

grade development of TiO_2 is usually recognizable by textural examination. Additionally garnets in medium grade and some upper amphibolite grade rocks exhibit growth zoning whereas those in high-grade paragneisses are usually broadly homogeneous aside from retrograde rim compositions. Therefore, accurate geobarometry requires careful assessment of the garnet compositions in equilibrium with other phases. Zoned garnets that have remained saturated with ilmenite–Al_2SiO_5–quartz–rutile might conceivably yield information on prograde, peak, and retrograde metamorphic pressures (see below). Garnet zonation, the bane of many petrologists, may ultimately be a boon to those attempting to determine pressure–temperature paths of metamorphism.

Inspection of Tables 2, 3, and 4 indicates that these barometers yield accurate pressures that agree well with a variety of other well-calibrated barometers. In terranes where a large number of phase assemblages are available, precision and accuracy appear to be excellent. For example, in the Adirondack Highlands, phase compositions span a broad range (garnet: $Alm_{52-73}Py_{2-30}Gr_{16-25}$; plagioclase: An_{72-17}; orthopyroxene Fs_{42-90}), but pressures range only from 7.1–8.2 kbar (Table 5). Some of the differences may result from real variations in pressure within the terrane. Pressures calculated from Fig. 7 are generally lower than those calculated from the equivalent barometer in the CMAS system (Newton and Perkins, 1982) by as much as 2 kbar although more typical values range between 0.5–1.0 kbar (Table 5). We suggest that pressures calculated from our barometer (Fig. 7) should be more accurate especially for iron-rich compositions. Our direct experimental calibration involves substantially less uncertainty than calculation of the relevant equilibria from thermochemical data. However, as knowledge of Ca–Fe–Mg solution properties in garnets and pyroxenes improve, both experimental and thermodynamically based barometers will yield better, and, one hopes, convergent pressure data.

The pressures calculated for upper amphibolite and granulite grade metamorphics using the experimentally calibrated barometers discussed here (Tables 1–4) are generally lower than previous estimates. Nearly all granulite terranes examined equilibrated at pressures between 6 and 10 kbar, but there is a marked clustering of pressures at 7.5 ± 1 kbar. The exposed granulite belts may not be representative of the lower most continental crust, particularly in orogenic regions where possible modern analogs, such as the Himalayas or Peruvian Andes, suggest crustal thicknesses of over 60 km. Nearly all granulite terranes exhibit a supracrustal component implying operation of nappe/thrust style tectonics and perhaps, additionally, magmatic thickening (England and Richardson, 1977; Wells, 1980). Such thickened crust was widespread by late Archean time since many exposed, old granulite terranes exhibiting pressures of 7–10 kbar are underlain by 30–50 km of (apparently) continental crust. In these cases, latter magmatic thickening can be ruled out, since the thermal effects of such underplating would produce later metamorphisms. Alternatively, passive tectonic thickening such as that proposed by O'Hara (1977) seems unlikely. The consistent pressures of 7.5 ± 1 kbar of granulites of widely varying ages suggests that some recurrent tectonic process might have been operative since the early Proterozoic. Newton and Perkins (1982) first noted the clustering of inferred peak metamorphic pres-

Table 5. Comparison of metamorphic pressures.

Locality	Sample No.	Newton and Perkins 1982 (kbar)	This study[a] (kbar)
Nain Complex	KI3909	3.9	3.3
	2893C	3.6	2.9
	2893R	1.1	2.8
Adirondack Highlands	ET-1	9.8	8.1
	ET-24	9.3	7.6
	IN-11	8.5	8.2
	SL-5	9.9	7.8
Adirondack Lowlands	74-C-248A	6.6	6.6
	-252	5.6	7.6
	-253E	6.3	6.8
Furua Complex	MF-283.2	10.2	10.2
	-268.1	10.5	10.1
	DMa-40	10.2	9.6
	C-247.1	11.0	10.2
	-311.1	10.6	10.2
	-352.2	12.7	11.4
Ruby Range	RMK-27-3	8.0	7.5
Invaijärvi	47-III	7.3	7.4
Nilgiri Hills	5-2	9.5	8.8
	11-1	7.4	7.7
	11-3E	8.3	9.0
	GN-4A	9.5	9.2
Labwor Hills	AR-51	8.1	6.5
Doubtful Sound	DS-36461	12.4	11.2
Otter Lake	RK-2	5.2	7.9
	DD-17	6.9	7.6
	DD-11	7.4	6.8
	A-12	6.6	7.9
	DL-2	8.1	8.2
Madras	MP-44	7.1	7.0
	MP-72	7.7	7.0
Buksefjorden	174087	7.6	8.3
	174102	7.5	7.4

[a]Pressures listed here have been adjusted for the temperatures used by Newton and Perkins and will not necessarily agree with those in Table 3.

sures and proposed that continental-scale overthrusting such as that in progress under the Tibetan Plateau was the principal tectonic process responsible for the formation of granulite terranes.

It is important to note that inferred pressures for upper amphibolite grade and transitional amphibolite–granulites (Esplanade Range, New Zealand, Funeral Mountains, Yale-B.C., central Massachusetts, Adirondack Lowlands) are not much different from pressures inferred in granulite and garnet–granulite terranes.

This suggests that formation of granulite terranes is dependent on an increase in temperature with little or only modest increases in pressure and implies addition of heat, perhaps by intrusion of magmas, at the base of, or into, the terrane. A thermal pulse such as that arising from addition of magma into the base of the crust is also implied from the retrograde pressure–temperature paths inferred using the GRAIL and ferrosilite–anorthite–garnet–quartz barometers. As noted above, at high grades of metamorphism, garnet compositions are broadly homogeneous with a thin compositionally zoned rim, generally believed to result from retrograde adjustments of mineral compositions after the peak of metamorphism. If it is assumed that garnet rim compositions were in equilibrium with the other phases in the rock during the initial stages of retrogression, the retrograde pressure path can be evaluated. Table 6 shows the $\log_{10}K$ values calculated from GRAIL and orthopyroxene–plagioclase–garnet–quartz assemblages in terranes for which garnet core and rim compositions are known. In every case the calcu-

Table 6. Pressures inferred from garnet core and rim compositions.

Locality	Barometer	Sample No.	Temperature (°C)	Log K	Pressure (kbar)
Adirondacks	GRAIL	C-74Ac[a]	730	0.37	7.5
		C-74Ar[a]	730	0.28	8.4
		C-79B1c	730	0.37	7.5
		C-79B1r	730	0.30	8.2
		C-529c	730	0.41	7.0
		C-529r	730	0.37	7.5
Adirondacks	FAGS[b]	74-C-248c	730	0.83	6.6
		C-248r	730	0.71	7.0
Western Maine	GRAIL	9c	600	0.37	6.7
		9r	600	0.24	8.1
		24c	600	0.30	7.5
		24r	600	0.29	7.6
Central Massachusetts	GRAIL	871c	675	0.38	7.0
		871r	675	0.36	7.2
		L11yc	675	0.46	6.3
		L11yr	675	0.46	6.3
Furua Complex	FAGS	DMa-40c	800	0.21	9.6
		DMa-40r	800	−0.13	10.2
		C-247.1c	800	0.18	9.8
		C-247.1r	800	−0.45	11.1
		C-311.1c	800	−0.13	10.2
		C-311.1r	800	−0.59	11.6
Buksefjorden	FAGS	174102c	800	0.79	7.4
		174102r	800	0.58	8.2

[a] c and r refer to core and rim garnet compositions, respectively.

[b] Ferrosilite–anorthite–garnet–silica (quartz).

lated $\log_{10}K$ is smaller for the rim than for the broadly homogeneous internal portions of the garnets. Using Figs. 7 and 9, this requires either an increase in pressure in the latest moments of metamorphism or cooling of the terrane isobarically or, at least, with a decrease in pressure at a rate less than the dP/dT slope of the curves of constant $\log_{10}K$. In the GRAIL barometer, the curves of constant $\log_{10}K$ are not appreciably pressure dependent and, therefore, the garnet rim compositions in such terranes as western Maine and central Massachusetts imply appreciable cooling before pressure decreased significantly. There are a number of terranes that appear to have initially cooled isobarically or nearly so, implying that such a retrograde path is common to high-grade metamorphic terranes. The inferred retrograde pressure–temperature paths are consistent with diminishing heat supply during the waning stages of metamorphism. This further suggests, following the thermal–tectonic modeling of England and Richardson (1977) and Wells (1980) that addition of magma into the crust may be an essential component of the heat budget of most high-grade terranes. Variable amounts of magmatic heating may explain the large differences in average geothermal gradients (20–50°C/km) inferred for granulites.

Acknowledgments

This research was supported by National Science Foundation Grants EAR82-06268 to SRB and NSF EAR78-16413 to ALB, and Australian Research Grants to VJW. We thank Lois Koh for drafting the line drawings and Shirley King for typing the manuscript. We also thank S. K. Saxena and A. B. Thompson for their reviews of the manuscript.

References

Akimoto, S., Matsui, Y., and Syono, Y. (1976) High pressure crystal chemistry of orthosilicates and the formation of the mantle transition zone, in *The Physics and Chemistry of Minerals and Rocks,* edited by R. J. G. Strens, pp. 327–363, Wiley, New York.

Anderson, D. J., and Lindsley, D. H. (1981) A valid Margules formulation for an asymmetric ternary solution: revision of the olivine–ilmenite thermometer, with applications, *Geochim. Cosmochim. Acta* **45**, 847–853.

Anovitz, L. M., and Essene, E. J. (1982) Phase equilibria in the system Fe–Al–Si–O–H (abstr.), *Geolog. Soc. Amer. Abstr. Progr.* **14**, 434.

Berg, J. H. (1977) Regional geobarometry in the contact aureoles of the anorthositic Nain Complex, Labrador, *J. Petrol.* **18**, 399–430.

Birch, F. (1966) Compressibility: elastic constants, in *Handbook of Physical Constants,* edited by S. P. Clark, Jr., Geological Society of America Memoirs, Boulder, Colorado, **97**, 97–174.

Boettcher, A. L., Windom, K. E., Bohlen, S. R., and Luth, R. W. (1981) A low-friction, low- to high-temperature furnace-sample assembly for piston-cylinder, high-pressure apparatus, *Rev. Scient. Instr.* **52**, 1903–1904.

Bohlen, S. (1979) The pressure, temperature and fluid composition of Adirondack metamorphism as determined in orthogneisses, Adirondack Mountains, New York, Unpublished Ph.D. Thesis, The University of Michigan.

Bohlen, S. R., and Boettcher, A. L. (1982) Experimental investigations and geological applications of orthopyroxene geobarometry, *Amer. Minerol.* **66**, 951–964.

Bohlen, S. R., and Essene, E. J. (1978) The significance of metamorphic fluorite in the Adirondacks, *Geochim. Cosmochim.* Acta **42**, 1669–1678.

Bohlen, S. R., Essene, E. J., Boettcher, A. L. (1980a) The investigation and application of olivine-quartz-orthopyroxene barometry, *Earth Planet. Sci. Lett.* **47**, 1–10.

Bohlen, S. R., Wall, V. J., and Boettcher, A. L. (1983a) Experimental investigation and application of garnet granulite equilibria. *Contr. Mineral. Petrol.,* in press.

Bohlen, S. R., Wall, V. J., and Boettcher, A. L. (1983b) Experimental investigations and geological applications of equilibria in the system FeO-TiO_2-Al_2O_3-SiO_2-H_2O, *Amer. Mineral.,* in press.

Bohlen, S. R., Boettcher, A. L., Dollase, W. A., and Essene, E. J. (1980b) The effect of manganese on olivine-quartz-orthopyroxene stability, *Earth Planet. Sci. Lett.* **47**, 11–20.

Bostwick, T. R. (1976) The effect of Mn on the stability and phase relations of iron-rich pyroxenes, M.S. Thesis, State University of New York, Stony Brook.

Buddington, A. F. (1965) The origin of three garnet isograds in Adirondack gneisses, *Mineral. Mag.* **34**, 71–81.

Buddington, A. F. (1966) The occurrence of garnet in the granulite facies terrane of the Adirondack Highlands: a discussion, *J. Petrol.* **7**, 331–335.

Carmichael, D. M. (1978) Metamorphic bathozones and bathograds: A measure of post-metamorphic uplift and erosion on a regional scale, *Amer. J. Sci.* **278**, 769–797.

Clark, S. P. (1959) Effect of pressure on the melting points of eight alkali halides, *J. Chem. Phys.* **31**, 1526–1531.

Cohen, L. H., and Klement, W., Jr. (1967) High-low quartz inversion: determination to 35 kilobars, *J. Geophys. Res.* **72**, 4245–4251.

Coolen, J. J. M. M. M. (1980) Chemical petrology of the Furua granulite complex, southern Tanzania, GUA (Amsterdam) Paper 13, 1–258.

Cressey, G., Schmid, R., and Wood, B. J. (1978) Thermodynamic properties of almandine-grossular garnet solid solutions, *Contrib. Mineral. Petrol.* **67**, 397–404.

Currie, K. L. (1971) The reaction 3 cordierite = 2 garnet + 4 sillimanite + 5 quartz as a geologic thermometer in the Opinicon Lake Region, Ontario, *Contrib. Mineral. Petrol.* **33**, 215–226.

Dahl, P. S. (1980) The thermal-compositional dependence of Fe^{2+}-Mg^{2+} distributions between coexisting garnet and pyroxene: Applications to geothermometry, *Amer. Mineral.* **65**, 852–866.

DeWaard, D. (1965) The occurrence of garnet in the granulite facies terrane of the Adirondack Highlands, *J. Petrol.* **6**, 154–191.

DeWaard, D. (1967) The occurrence of garnet in the granulite facies terrane of the Adirondack Highlands and elsewhere, an amplification and reply, *J. Petrol.* **8**, 210–232.

Engi, M. (1980) Theoretical analysis of exchange equilibria between a solid solution and an electrolyte fluid undergoing partial association (abstr.), *Geol. Soc. Amer. Abstr. Progr.* **12**, 422.

England, P. C., and Richardson, S. W. (1977) The influence of erosion upon the mineral facies of rocks from different metamorphic environments, *Quart. J. Geol. Soc. London* **134**, 201–213.

Essene, E. J. (1982) Geologic thermometry and barometry, in *Reviews of Mineralogy, Vol. 10,* edited by J. M. Ferry, 153–206. Mineralogical Soc. of America, Washington, D.C.

Evans, B. W., and Guidotti, C. V. (1966) Thc sillimanitc-potash fcldspar isograd in wcstern Maine, U.S.A., *Contrib. Mineral. Petrol.* **12**, 25–62.

Fletcher, C. J. W., and Greenwood, H. J. (1978) Metamorphism and structure of Penfold Creek area, near Quesnel Lake, British Columbia, *J. Petrol.* **20**, 743–794.

Frisch, T., and Bridgwater, D. (1976) Iron- and manganese-rich minor intrusions emplaced under late orogenic conditions in the Proterozoic of south Greenland, *Contrib. Mineral. Petrol.* **57**, 25–48.

Ganguly, J., and Kennedy, G. C. (1974) The energetics of natural garnet solid solution. I. Mixing of the aluminosilicate end members, *Contrib. Mineral. Petrol.* **48**, 137–148.

Ghent, E. (1975) Temperature, pressure, and mixed-volatile equilibria attending metamorphism of staurolite-kyanite-bearing assemblages, Esplanade Range, British Columbia, *Geol. Soc. Amer. Bull.* **86**, 1654–1660.

Ghent, E. A. (1976) Plagioclase-garnet-Al_2SiO_5-quartz: A potential geothermometer-geobarometer, *Amer. Mineral.* **61**, 710–714.

Ghent, E. A., Robbins, D. B., and Stout, M. Z. (1979) Geothermometry, geobarometry and fluid compositions of metamorphosed calcsilicates and pelites, Mica Creek, British Columbia, *Amer. Mineral.* **64**, 874–885.

Green, D. H., and Ringwood, A. E. (1967) An experimental investigation of the gabbro to eclogite transformation and its petrological applications, *Geochim. Cosmochim. Acta* **31**, 767–833.

Green, D. H., and Hibberson, W. (1970) The instability of plagioclase in peridotite at high pressure, Lithos **3**, 209–221.

Green, T. H. (1970) High-pressure experimental studies on the mineralogical constitution of the lower crust, *Phys. Earth Planet. Interiors* **3**, 441–450.

Grew, E. S. (1981) Granulite-facies metamorphism at Molodezhnaya Station, East Antarctica, *J. Petrol.* **22**, 297–336.

Griffin, W. L., and Heier, K. S. (1969) Parageneses of garnet in granulite-facies rocks, Lofoten-Vesterålen, Norway, *Contrib. Mineral. Petrol.* **23**,89–116.

Griffin, W. L., Heier, K. S., Taylor, P. N., and Weigand, P. W. (1974) General geology, age and chemistry of the Raftsund mangerite intrusion, Lofoten-Vesterålen, *Norges Geolog. Undersøkelse* **312**, 1–30.

Harley, S. L., and Green, D. H. (1983) Garnet-orthopyroxene barometry for granulites and peridotites, *Nature,* **300**, 697–701.

Hattori, H. (1967) Occurrence of sillimanite-garnet-biotite gneisses and their significance in metamorphic zoning in South Island, New Zealand, *New Zealand J. Geol. Geophys.* **10**, 269–299.

Hays, J. F., Bell, P. M. (1973) Albite-jadeite-quartz equilibrium: A hydrostatic determination. *Trans. Amer. Geophys. Union* **54**, 482.

Hensen, B. J. and Green, D. H. (1971) Experimental study of the stability of cordierite and garnet in pelitic compositions at high pressures and temperatures. I. Compositions with excess aluminosilicate, *Contrib. Mineral. Petrol.* **33**, 309–330.

Hensen, B. J., and Green, D. H. (1972) Experimental study of the stability of cordierite and garnet in pelitic compositions at high pressures and temperatures. II. Compositions without excess alumino-silicate, *Contrib. Mineral. Petrol.* **35**, 331–354.

Hietanen, A. (1969) Distribution of Fe and Mg between garnet, staurolite and biotite in aluminum-rich schist in various metamorphic zones north of the Idaho Batholith, *Amer. J. Sci.* **267**, 422–456.

Holdaway, M. J. (1971) Stability of andalusite and the aluminum silicate phase diagram, *Amer. J. Sci.* **271**, 97–131.

Holdaway, M. J. (1978) Significance of chloritoid-bearing and staurolite-bearing rocks in the Picuris Ringe, New Mexico, *Geol. Soc. Amer. Bull.* **89**, 1404–1414.

Holdaway, M. J., and Lee, S. M. (1977) Mg-Fe corderite stability in high grade pelitic rocks based on experimental, theoretical and natural observations, *Contrib. Mineral. Petrol.* **63**, 175–198.

Hörmann, P. K., Raith, M. Raase, P., Akermand, D., Seifert, F. (1980) The granulite complex of Finnish Lappland: Petrology and metamorphic conditions in the Iralojoki-Inarijärvi area, *Geol. Survey Finland Bull.* **308**, 1–95.

Ito, K., and Kennedy, G. C. (1971) An experimental study of the basalt-garnet-granulite-eclogite transition, in *The Structure and Physical Properties of the Earth's Crust*, edited by J. C. Heacock, **14**, 303–314. American Geophysical Union Monograph, Washington, D.C.

Jaffe, H. W., Robinson, P., and Tracy, R. J. (1978) Orthoferrosilite and other iron-rich pyroxenes in microperthite gneiss of the Mount Marcy Area Adirondack Mountains, *Amer. Mineral.* **63**, 1116–1136.

Janardhan, A. S., Newton, R. C., Hansen, E. C. (1982) The transition from amphibolite facies gneiss to charnockite in southern Karnataka and northern Tamil Nadu, India, *Contrib. Mineral. Petrol.* **79**, 130–149.

Johannes, W. (1978) Pressure comparing experiments with NaCl, AgCl, talc, pyrophyllite assemblies in a piston-cylinder apparatus, *Neues, Jahr. Mineral. Monat.* **2**, 84–92.

Johnson, C. A., Essene, E. J. (1982) The formation of garnet in olivine-bearing metagabbros from the Adirondacks, *Contrib. Mineral. Petrol.* **81**, 240–251.

Kawasaki, T., and Matsui, Y. (1977) Partitioning of Fe^{2+} and Mg^{2+} between olivine and garnet, *Earth Planet. Sci. Lett.* **37**, 159–166.

Klein, C. (1978) Regional metamorphism of Proterozoic iron formation, Labrador Trough, Canada, *Amer. Mineral.* **63**, 898–912.

Kretz, R. (1964) Analysis of equilibrium in garnet-biotite-sillimanite gneisses from Quebec, *J. Petrol.* **5**, 1–20.

Krogh, E. J. (1977) Origin and metamorphism of iron formations and associated rocks, Lofoten-Vesterålen, N. Norway. I. The Vestpolltind Fe-Mn deposit, *Lithos* **10**, 243–255.

Kuno, H. (1954) Study of orthopyroxenes from volcanic rocks, *Amer. Mineral.* **39**, 30–46.

Labotka, T. C. (1980) Petrology of a medium-pressure regional metamorphic terrane, Funeral Mountains, California, *Amer. Mineral.* **65**, 670–689.

Lindsley, D. H. (1965) Iron-titanium oxides, *Carnegie Institution of Washington Year Book,* **64**, 144–148.

Lonker, S. W. (1981) The *P-T-X* relations of the cordierite-garnet-sillimanite-quartz equilibrium, *Amer. J. Sci.* **281**, 1056–1090.

Martignole, J., and Sisi, J. C. (1981) Cordierite-garnet-H_2O equilibrium: A geological thermometer, barometer and water fugacity indicator, *Contrib. Mineral. Petrol.* **77**, 38–46.

Newton, R. C., and Haselton, H. T. (1981) Thermodynamics of the garnet-plagioclase-Al_2SiO_5-quartz geobarometer, in *Thermodynamics of Minerals and Melts*, edited by R. C. Newton, A. Navrotsky, and B. J. Wood, pp. 129–145, Springer-Verlag, New York.

Newton, R. C., and Perkins, D., III (1982) Thermodynamic calibration of geobarometers for charnockites and basic granulites based on the assemblages garnet-plagioclase-orthopyroxene (clinopyroxene)-quartz with applications to high grade metamorphism, *Amer. Mineral.* **67**, 203–222.

Newton, R. C., and Wood, B. J. (1979) Thermodynamics of water in cordierite and some petrologic consequences of cordierite as a hydrous phase, *Contrib. Mineral. Petrol.* **68**, 391–405.

Nixon, P. H., Reedman, A. J., and Burns, L. K. (1973) Sappirine-bearing granulites from Labwor, Uganda, *Mineral. Mag.* **39**, 420–428.

O'Hara, M. J. (1975) Great thickness and high geothermal gradient of archaen crust: the Lewisian of Scotland (abstr.), *International Conference on Geothermometry and Geobarometry,* Pennsylvania State University, University Park.

O'Hara, M. J. (1977) Thermal history of excavation of Archaen gneisses from the base of the continental crust. *Geol. Soc. London J.* **134**, 185–200.

Oka, Y., and Matsumoto, T. (1974) Study on the compositional dependence of the apparent partition coefficient of iron and magnesium between coexisting garnet and clinopyroxene solid solutions, *Contrib. Mineral. Petrol.* **48**, 115–121.

Oliver, G. J. H. (1977) Feldspathic hornblende and garnet granulites and associated anorthosite pegmatites from Doubtful Sound, Fiordland, New Zealand, *Contrib. Mineral. Petrol.* **65**, 111–121.

O'Neill, H. St. C., and Wood, B. J. (1979) An experimental study of Fe-Mg partitioning between garnet and olivine and its calibration as a geothermometer, *Contrib. Mineral. Petrol.* **70**, 59–70.

Ormaasen, D. E. (1977) Petrology of the Hopen mangerite-charnockite intrusion, Lofoten, north Norway, *Lithos,* **10**, 291–310.

Orville, P. M. (1972) Plagioclase cation exchange equilibria with aqueous chloride solution: Results at 700 °C and 2000 bars in the presence of quartz, *Amer. J. Sci.* **72**, 234–272.

Oyawoye, M. O., and Makanjuola, A. A. (1972) Bauchite: a fayalite-bearing quartz monzanite, *International Geological Congress 24th Session,* Section 2, 251–266.

Padovani, E. R., and Carter, J. J. (1977) Aspects of the deep crustal evolution beneath south central New Mexico, *Amer. Geophysical Union Monograph,* **20**, 19–55.

Perkins, D., III. (1979) Application of new thermodynamic data to mineral equilibria, Unpublished Ph.D. thesis, The University of Michigan.

Perkins, D., III, Essene, E. J., Marcotty, L. A. (1982) Thermometry and barometry of some amphibolite-granulite facies rocks from the Otter Lake area, southern Quebec, *Canad. J. Earth Sci.* **19**, 1759–1774.

Perchuk, L. L. Podlesskii, K. K., and Aranovich, L. Ya. (1981) Calculation of thermodynamic properties of end-member minerals from natural parageneses, in *Thermodynamics of Minerals and Melts,* edited by R. C. Newton, A. Navrotsky, and B. J. Wood, pp. 111–129, Springer-Verlag, New York.

Phillips, G. N., and Wall, V. J. (1981) Evaluation of prograde regional metamorphic conditions: Their implications for the heat source and water activity during metamorphism in the Willyama Complex, Broken Hill, Australia, *Bull. Mineral.* **104**, 801–810.

Pigage, L. C. (1976) Metamorphism of the Settler schist, southwest of Yale, British Columbia, *Canad. J. Earth Sci.* **13**, 405–421.

Podpora, C., Lindsley, D. H. (1979) Fe-rich pigeonites: minimum temperatures of stability in the Ca-Mg-Fe quadrilateral, *EOS Trans. Amer. Geophys. Union* **60**, 420–421.

Ramberg, H., and DeVore, G. (1951) The distribution of Fe^{++} and Mg^{++} in coexisting olivines and pyroxenes, *J. Geol.* **59**, 193–210.

Rao, B. B., and Johannes, W. (1979) Further data on the stability of staurolite + quartz and related assemblages, *Neues Jahr. Mineral. Monat.* **10**, 437–447.

Reinhardt, E. W. (1968) Phase relations in cordierite-bearing gneisses from the Gananoque area, Ontario, *Canad. J. Earth Sci.* **5**, 455–482.

Richardson, S. W. (1968) Staurolite stability in part of the system Fe-Al-Si-O-H, *J. Petrol.* **9**, 467–488.

Richardson, S. W., Bell, P. M., and Gilbert, M. C. (1968) Kyanite-sillimanite equilibrium between 700 and 1500 °C, *Amer. J. Sci.* **266**, 513–541.

Ringwood, A. E., Green, D. H. (1966) An experimental investigation of the gabbro-eclogite transformation and some geophysical implications, *Tectonophysics* **3**, 383–427.

Robie, R. A., Bethke, P. M. Toulmin, M. S., and Edwards, J. L. (1966) X-ray crystallographic data, densities and molar volumes of minerals, in *Handbook of Physical Constants,* edited by S. P. Clark, Jr., vol. 97, pp, 27–74. Geological Society of America Memoirs, Boulder, Colorado.

Sack, R. O. (1980) Some constraints on the thermodynamic mixing properties of Fe-Mg orthopyroxenes and olivines, *Contrib. Mineral. Petrol.* **71**, 237–246.

Saxena, S. K. (1968a) Distribution of iron and magnesium between coexisting garnet and clinopyroxene in rocks of varying metamorphic grade, *Amer. Mineral.* **53**, 2018–2024.

Saxena, S. K. (1968b) Chemical study of phase equilibria in charnockites, Varberg, Sweden, *Amer. Mineral.* **53**, 1674–1695.

Saxena, S. K. (1973) *Thermodynamics of Rock-Forming Crystalline Solutions,* Springer-Verlag, New York.

Saxena, S. K. (1977) The charnockite geotherm, *Science* **198**, 614–617.

Schmid, R., and Wood, B. J. (1976) Phase relationships in granulitic metapelites from the Ivrea-Verbano zone (Northern Italy), *Contrib. Minteral. Petrol.* **54**, 255–279.

Skinner, B. J. (1966) Thermal expansion, in *Handbook of Physical Constants,* edited by S. P. Clark, Jr., vol. 97, pp. 75–96. Geological Society of American Memoirs, Boulder, Colorado.

Smyth, J. R. (1975) High-temperature crystal chemistry of fayalite, *Amer. Mineral.* **60**, 1092–1097.

Stoddard, E. F. (1976) Granulite facies metamorphism in the Colton-Rainbow Falls area, northwest Adirondacks, New York, Ph.D. Thesis, University of California, Los Angeles.

Sueno, S., Cameron, M., and Prewitt, C. T. (1976) Orthoferrosilite: high temperature crystal chemistry, *Amer. Mineral.* **61**, 39–53.

Thompson, A. B. (1976) Mineral reactions in pelitic rocks, II. Calculation of some P-T-X (Fe-Mg) phase relations, *Amer. J. Sci.* **276**, 425–454.

Tracy, R. J. (1978) High grade metamorphic reactions and partial melting in pelitic schist, west-central Massachusetts, *Amer. J. Sci.* **278**, 150–178.

Tracy, R. J., Robinson, P., and Thompson, A. B. (1976) Garnet composition and zoning in the determination of temperature and pressure of metamorphism, central Massachusetts, *Amer. Mineral.* **61**, 762–775.

Turnock, A. C., Lindsley, D. H., and Grover, J. E. (1973) Synthesis and cell parameters of Ca-Mg-Fe pyroxenes, *Amer. Mineral.* **58**, 50–59.

Valley, J. W. and Essene, E. J. (1980) Akermanite in the cascade slide xenolith and its significance for metamorphism in the Adirondacks, *Contrib. Mineral Petrol.* **74**, 143–152.

Weaver, B. L., Tarney, J., Windley, B. F., Sugavanam, E. B., Venkata, Rao V. (1978) Madras granulites: geochemistry and P-T conditions of crystallization, in *Archaean Geochemistry,* edited by B. F. Windley and S. M. Naqvi, pp. 177–204, Elsevier, Amsterdam.

Weisbrod, A. (1973) Refinements of the equilibrium conditions of the reaction Fe cordierite = almandine + quartz + sillimanite + H_2O. *Carnegie Institution Washington Year Book* **72**, 515–522.

Wells, P. R. A. (1979) Chemical and thermal evolution of Archaean sialic crust, southern West Greenland, *J. Petrol.* **20**, 187–226.

Wells, P. R. A. (1980) Thermal models for the magmatic accretion and subsequent metamorphism of continental crust, *Earth Planet. Sci. Lett.* **46**, 253–265.

Wood, B. J. (1975) The influence of pressure temperature and bulk composition on the appearance of garnet in orthogneisses—an example from South Harris, Scotland, *Earth Planet. Sci. Lett.* **26**, 299–311.

Wood, B. J., and Banno, S. (1973) Garnet-orthopyroxene and orthopyroxene-clinopyroxene relationships in simple and complex systems, *Contrib. Mineral. Petrol.* **42**, 109–124.

Wood, B. J., and Kleppa, O. J. (1981) Thermochemistry of forsterite-fayalite olivine solutions, *Geocheim. Cosmochim. Acta* **45**, 529–534.

Chapter 6
The Cordierite–Garnet–Sillimanite–Quartz Equilibrium: Experiments and Applications

L. Ya. Aranovich and K. K. Podlesskii

Symbols

Thermodynamic Parameters

T: temperature, °K
t: temperature, °C
P: pressure, bar
ΔG(i): Gibbs free energy change of ith reaction at temperature T and 1 bar, cal·mol^{-1}
ΔS(i): entropy change of ith reaction at temperature T and 1 bar, cal·k^{-1} mol^{-1}
ΔV(i): volume change of solids in ith reaction, cal·bar^{-1} · mol^{-1}
X_{Mg}: mole fraction of Mg in mineral; $X_{Mg} = (Mg)/(Mg + Fe)$
N_{Mg}: $X_{Mg} \cdot 100$

Minerals

Alm: almandine
And: andalusite
Cor_{Fe}: Fe–cordierite
Cor_{Mg}: Mg–cordierite
Cor: cordierite solid solution of Fe and Mg end members
En: enstatite
Gr: garnet solid solution of pyrope and almandine
Gros: grossularite
Her: hercynite
Ky: kyanite
Pl: plagioclase
Pyr: pyrope
Spes: spessartine
Qz: quartz
Saph: sapphirine
Sil: sillimanite
Ta: talc

Introduction

The paragenesis of cordierite–garnet–sillimanite–quartz is widespread in regionally metamorphosed high-grade pelitic rocks and in hornfelses of contact aureoles. Since Korzhinskii (1936) considered it as a potential indicator of metamorphic conditions, this assemblage has been of great interest to petrologists. Qualitative thermodynamic analysis by Marakushev (1965) showed that Cor + Gr + Sil + Qz paragenesis could be treated as divariant assemblage in the FeO–MgO–Al_2O_3–SiO_2 system, the Fe–Mg minerals compositions having been defined by the P–T conditions. This approach allowed P–T estimates to be obtained from the intersection of isopleths of Mg/(Mg + Fe) of coexisting cordierite and garnet on a P–T diagram, but unfortunately the experimental and theoretical studies provided conflicting calibrations of the diagram. In his review of the literature on the subject, Lonker (1981) has listed at least eight studies which disagree with each other concerning the effects of pressure, temperature, and water fugacity on the Cor–Gr–Sil–Qz equilibrium.

P–T–X_{Mg} relations of the Cor–Gr–Sil–Qz equilibrium can be described by two reactions:

$$\underset{Py}{\tfrac{1}{3}Mg_3Al_2Si_3O_{12}} + \underset{Sil}{\tfrac{2}{3}Al_2SiO_5} + \underset{Qz}{\tfrac{4}{5}SiO_2} = \underset{Cor_{Mg}}{\tfrac{1}{2}Mg_2Al_4Si_5O_{18}}, \tag{1}$$

$$\underset{Alm}{\tfrac{1}{3}Fe_3Al_2Si_3O_{12}} + \underset{Sil}{\tfrac{2}{3}Al_2SiO_5} + \underset{Sil}{\tfrac{4}{5}SiO_2} = \underset{Cor_{Fe}}{\tfrac{1}{2}Fe_2Al_4Si_5O_{18}}. \tag{2}$$

One of these reactions can be substituted by the exchange reaction:

$$\underset{Py}{\tfrac{1}{3}Mg_3Al_2Si_3O_{12}} + \underset{Cor_{Fe}}{\tfrac{1}{2}Fe_2Al_4Si_5O_{18}} = \underset{Alm}{\tfrac{1}{3}Fe_3Al_2Si_3O_{12}} + \underset{Cor_{Mg}}{\tfrac{1}{2}Mg_2Al_4Si_5O_{18}}. \tag{3}$$

Principal disagreements have arisen concerning the dP/dT slopes of reactions (1) and (2) on the P–T diagram, not to mention the positions of the reactions. While calculations based on the calorimetric data (Hutcheon *et al.*, 1947; Martignole and Sisi, 1981) and the data from natural parageneses (Perchuk, 1977; Perchuk *et al.*, 1981) gave convincing evidence of the positive slopes, the experiments could not provide an unambiguous solution. All experimental determinations of reaction (2) in the Mg-free system indicated a negative slope (Richardson, 1968; Weisbrod, 1973a,b,c; Holdaway and Lee, 1977). In the investigation of the stability limit of Mg–cordierite (Newton, 1972) a positive slope for the metastable reaction (1) was assumed and the experiments by Currie (1971, 1974) supported such slopes [negative for reaction (2) and positive for reaction (1)]. The positions of reactions (1) and (2) cannot be determined from the results of Hensen and Green (1971, 1973) because of large compositional uncertainties.

The influence of water content in cordierite on the activities of Cor_{Mg} and Cor_{Fe} has also become the subject of controversy. Consideration of the effect of water on the cordierite-bearing equilibria has led Newton and Wood (1979) to conclude that the distribution of Fe and Mg between cordierite and garnet is nearly independent of temperature. This conclusion is contrary not only to the other studies concerning the effect of water (Kurepin, 1979; Lonker, 1981; Martignole and Sisi, 1981), but also to the evidence from natural occurrences.

It should be noted that direct experimental determinations of the equilibrium relations of cordierite and garnet in the divariant field are insufficient and that only one point (1000 °C, 9 kbar) has been obtained from reversal experiments (Hensen, 1977). The most recent experimental results on the Cor–Gr–Sil–Qz equilibrium were obtained more than 7 years ago, and since that time conflicting schemes continue to appear (Lonker, 1981; Martignole and Sisi, 1981). It is clear, therefore, that without additional experiments the problem cannot be solved.

The purpose of this chapter is to present new experimental results on the P–T–X_{Mg} relations of the Cor–Gr–Sil–Qz assemblage in the presence of pure water or a water–carbon dioxide mixture and to propose an appropriate thermodynamic description that allows an application of the data on the iron–magnesium ratio of the garnet and cordierite to the geothermometry and barometry of metapelites.

Experimental Methods and Results

We chose to investigate experimentally the Cor–Gr–Sil–Qz equilibrium at 700 and 750 °C in the pressure range of 4–8 kbar. The objective was to determine, as accurately as possible, the equilibrium compositions of coexisting cordierite and garnet. To obtain information on the influence of the fluid-phase composition, the experiments were conducted in pure H_2O and in an H_2O–CO_2 mixture (H_2O:CO_2 = 1:1).

Starting Materials

Except for a few runs in which synthetic garnets and cordierites were used, the experimental charges were prepared from natural minerals. All minerals were analyzed by electron microprobe to ensure distinction between the starting compositions and those of the run products. The details of the analytical procedure and the analyses have been given by Aranovich and Podlesskii (1982a). The natural garnets corresponded to solid solutions of pyrope and almandine end members with minor impurities of the other components (less than 6 mol% Gros and less than 3 mol% Spes) and the cordierites corresponded to the Cor_{Fe}–Cor_{Mg} series. The synthetic minerals used, as well as the sillimanite and quartz, contained no impurities.

The garnets, both natural and synthetic, were found inhomogeneous with regard to their Mg/(Mg + Fe) ratio, although the average microprobe analyses corresponded to the bulk chemical determinations. The inhomogeneity was undetectable by X-ray powder diffraction. It could be seen from Fig. 1 that a sample of a garnet of considerable inhomogeneity ($18 \leq N_{Mg} \leq 37$) yielded quite sharp reflection, comparable to that of pure silicon used as an internal standard.

The starting minerals were mixed to produce the charges of cordierite stoichi-

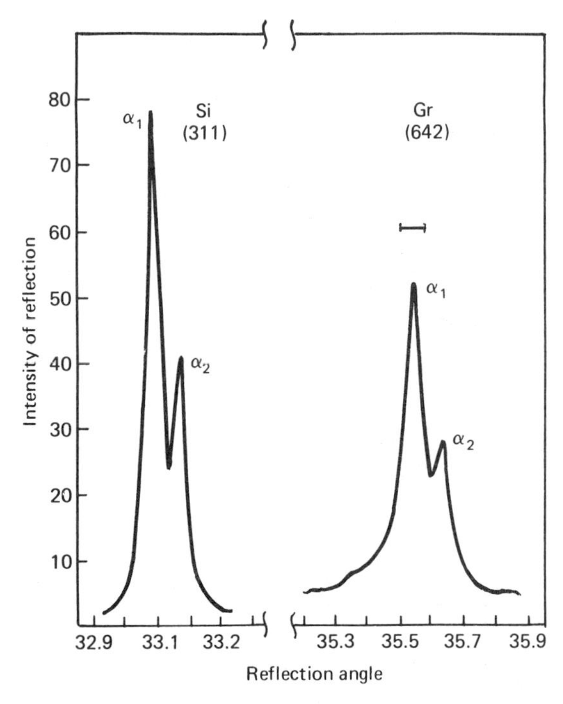

Fig. 1. Reflection (642) of natural garnet ($18 \leq N_{Mg} \leq 37$) and reflection (311) of Si (CoK_{α}: radiation). Bars show a range of reflection angle values that should correspond to a range of the garnet compositions.

ometry of various N_{Mg}. Small amounts of quartz over the stoichiometry were used to saturate the fluid phase with SiO_2.

Apparatus, Capsules, and Run Procedures

All experiments were carried out in Tuttle-type high-pressure vessels (Ivanov *et al.*, 1977). The temperature measurements are accurate to within ±5°C, and the accuracy of the given pressures is to ±50 bars. To maintain oxygen fugacity corresponding to the value defined by the QFM buffer, a double capsule was used. Because gold had been reported sufficiently permeable to hydrogen at the conditions in our experiments (Popp and Frantz, 1978), we used only Au capsules to avoid Fe losses from the charges. For every run the charge of minerals (20 ± 0.5 mg) ground in acetone was sealed in a 3 × 0.1 × 25-mm Au capsule together with 10 ± 0.5 mg H_2O or oxalic acid. The inner capsule was loaded into a 5.4 × 0.2 × 50-mm Au outer capsule containing QFM buffer and sufficient H_2O. The outer capsule was sealed and placed into the high-pressure vessel together with an Ni open capsule containing ∽500 mg QFM buffer. Mechanical shaking of the capsules during the runs was used to promote reaction rates in several experiments at 700°C and 4 kbar. Duration of the runs was no less than 120 hours. Some run products were reused in the experiments at the same P and T after they had been analyzed by X-ray powder diffraction and electron microprobe and ground in acetone. The details of the experimental conditions are given by Aranovich and Podlesskii (1982a).

Run Products

All run products were analyzed optically and by X-ray diffraction and electron microprobe. Microprobe analyses were obtained to reveal the composition changes of the cordierites and garnets, while the other methods allowed to check the phase composition of the charges. The results of the investigation of the products are presented in Fig. 2. Because we tried to use the charges of bulk compositions corresponding to a divariant field in which all the four phases are stable, disappearance of phases was observed only twice. Grinding and reusing the run products facilitated the disappearance of cordierite at 700°C, 6 kbar, and Cor_{20} bulk composition and of garnet and sillimanite at 700°C, 5 kbar, and Cor_{60} bulk composition.

Although the runs were of long duration (120–672 hours), considerable inhomogeneity and zonation of cordierite and garnet was encountered in all run products, despite grinding and rerunning the charges in specific cases. Grains of the starting composition were always found by microprobe, and in many cases we did not manage to discriminate between starting minerals and run products. The starting minerals typically occurred as zoned grains and, as a rule, the rim compositions represented a closer approach to equilibrium than the core compositions. Unzoned small grains (less than 10 μk) with compositions corresponding to those of the rims always occurred along with larger-zoned grains. Minerals whose starting compositions were near to the equilibrium value moved closer to complete equilibration. The higher temperatures and pressures favored more rapid changes in the mineral compositions.

Experimental Results[1]

The experiments gave no evidence for the influence of the fluid composition on the Cor–Gr–Sil–Qz equilibrium. The results for aqueous and H_2O–CO_2 fluids practically coincide with regard to N_{Mg} values of coexisting cordierite and garnet.

The equilibrium N_{Mg} values are given in Table 1 and Fig. 2. These estimates are based on the compositional reversals and the phase relations at given bulk compositions. In the cases in which compositional reversals were not achieved, product N_{Mg} values corresponding to the maximum changes from starting compositions were accepted as the equilibrium values. For example, in experiments at 700°C and 4 kbar the garnet equilibrium composition was determined from compositional reversal ($N_{Mg} = 10 \pm 1$). To estimate the cordierite equilibrium composition, we considered the stability of all four phases at Cor_{40} bulk composition,

[1]The table is available upon request from the authors.

Table 1. Experimentally determined equilibrium cordierite and garnet compositions and related values.

t (°C)	P (kbar)	N_{Mg}^{Gr}	N_{Mg}^{Cor}	K_{Mg}	ln K_{Mg}	K_D
700	4	10	44	4.4	1.4816	7.071
	5	15	55	3.667	1.29928	6.926
	6	24	69	2.875	1.05605	7.048
	8	51	89	1.745	0.55681	7.774
750	5	12	44	3.667	1.29928	5.762
	8	44	83	1.886	0.63465	6.214

which lies in the divariant field. Here N_{Mg}^{Cor} must be greater than 40. Compositional changes of magnesian cordierite indicate that the equilibrium $N_{Mg}^{Cor} \leq 44$. In Table 1 the accepted value is 44 ± 2.

The experimental evidence points to the fact that the equilibrium N_{Mg}^{Cor} and N_{Mg}^{Gr} tends to decrease as the temperature decreases. It is seen from Fig. 2 that the starting cordierite ($N_{Mg}^{Cor} = 86$) has become more magnesian ($N_{Mg}^{Cor} = 89$) at 700°C and 8 kbar, while at 750°C and 8 kbar the same cordierite has become more Fe-rich ($N_{Mg}^{Cor} = 83$). In experiments at 5 kbar and 700°C we observed the change of Cor_{59} to Cor_{68}, while at 5 kbar and 750°C Cor_{59} changed to Cor_{46}.

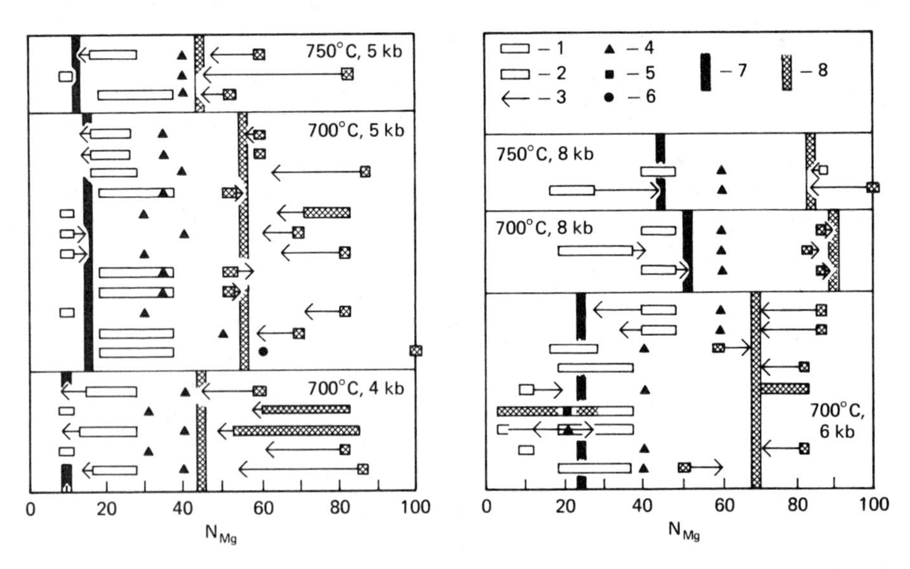

Fig. 2. Results of experiments on the Cor–Gr–Sil–Qz equilibrium at 700 and 750°C. (1) Range of the starting garnet compositions. (2) Range of the starting cordierite compositions. (3) Changes of the compositions outside the starting ranges. (4) Bulk charge composition at which all four phases remain stable in the run products. (5) Bulk composition at which cordierite disappeared. (6) Bulk composition at which garnet disappeared. (7) Garnet composition accepted as equilibrium one. (8) Cordierite composition accepted as equilibrium.

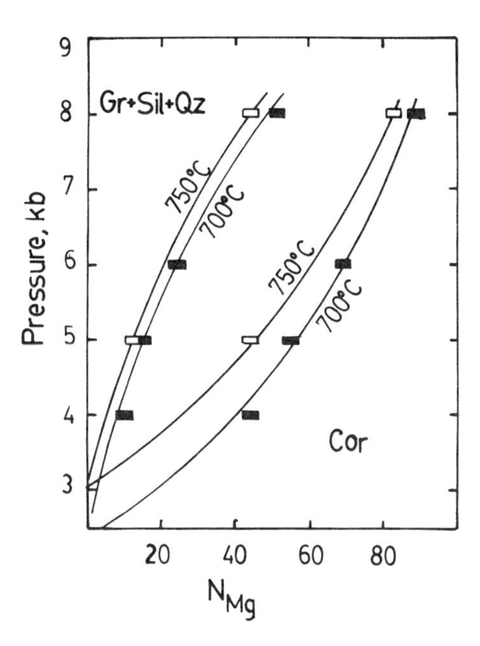

Fig. 3. P–N_{Mg} diagram of the Cor–Gr–Sil–Qz stability at 700 and 750°C. The four field boundaries are drawn according to Eqs. (5) and (16).

The equilibrium relations of coexisting cordierite and garnet are described by divariant P–N_{Mg} loops shown in Fig. 3. Figure 3 demonstrates that our data indicate positive slopes of isopleths of the coexisting minerals on a P–T diagram which oppose the determinations of the Fe end member reactions made by Richardson (1968), Weisbrod (1973a,b,c), and Holdaway and Lee (1977), and the calibrations made by Thompson (1976) and Hensen and Green (1973), but which are in accord with the predictions of Hutcheon *et al.* (1974), Perchuk *et al.* (1981), and Martignole and Sisi (1981).

Calculated Equilibria

The experimentally determined data of Table 1 indicate that K_D decreases with increasing temperature, the effect of pressure on Fe–Mg exchange between coexisting garnet and cordierite being very small. These relations can be described in terms of the equilibrium conditions of reaction (3):

$$RT \ln K_D + P\,\Delta V(3) + \Delta G_T(3) = 0. \tag{4}$$

Using $\Delta V(3) = -0.03535$ cal·bar^{-1}·mol.$^{-1}$ (Perchuk *et al.*, 1981), we obtained $\Delta G_T(3)$ from the data of Table 1. Within the accuracy of the experimental determinations, $\Delta G_T(3)$ values at 700 and 750°C are -3618 ± 275 cal·mol.$^{-1}$ and -3412 ± 293 cal·mol.$^{-1}$, respectively. These results agree very well with the values -3539 and -3390 cal·mol^{-1}, respectively, calculated by Lavrent'eva and Perchuk (1981) from recent data on reaction (3) over a temperature range of 600–1000°C at 6 kbar pressure in the presence of complex fluid. Both our data and the data of Lavrent'eva and Perchuk (1981) for a wide range of temperatures and compositions suggest that the fluid composition has no effect on Fe–Mg exchange between cordierite and garnet. These data also imply that the binary Pyr–Alm solid solution is ideal if the cordierite is also considered ideal (Engi, 1978). We used the data of Table 1 and the values calculated from the data of Lavrent'eva and Perchuk (1981) to fit the $\Delta G_T(3)$ temperature dependence by linear regression, and obtained the equation:

$$\Delta G_T(3) = -6134(\pm 282)^2 + 2.67(\mp 0.26)T \tag{5}$$

with the regression coefficient $r^2 = 0.983$. The constants in Eq. (5) are in good agreement with ΔH° and ΔS° thermodynamic properties of exchange reaction (3), calculated by Holdaway and Lee (1977) from the data on natural occurrences.

The data on ln $\overline{K}_{Mg}$ (Table 1) reveal a pronounced effect of fluid pressure on the Cor–Gr–Sil–Qz equilibrium. Based on this data we calculated the average values of $(\partial RT \ln \overline{K}_{Mg}/\partial_P)_T$ for 700 and 750°C and obtained 0.44 ± 0.06 and 0.45 ± 0.07 cal·bar^{-1}·mol^{-1}, respectively. If both Cor_{Fe}–Cor_{Mg} and Pyr–Alm are considered as ideal, with no influence of any other components, application of the thermodynamic equilibrium constraints for reaction (1) result in a $\Delta V(1)$ which is equal to these values. Comparison of the calculated values to a $\Delta V(1)$ value of 0.64164 cal·bar^{-1}, from Perchuk *et al.* (1981), shows a difference of $\backsim$0.2 cal·bar^{-1}. This difference is consistent with a ΔV_{H_2O} value calculated by Holdaway and Lee (1977) for the end member reaction Cor_{Fe} = Alm + Sil + Qz, and arises from the effect of the fluid content of cordierite on the stability of the mineral.

Appreciable stabilization of cordierite due to introduction of water molecules into the structural channels has long been recognized by many petrologists (Lepzin, 1969; Perchuk, 1969; Weisbrod, 1973b; Holdaway and Lee, 1977; Kurepin, 1979; Newton and Wood, 1979; Martignole and Sisi, 1981), who developed different models to describe this effect. In this study we used the model by Kurepin (1979) with regard to independence of water content in cordierite on its Mg/Fe ratio (Gunter, 1977). According to the model, cordierite is assumed to be the ideal solution of "dry" and "wet" end members, i.e., Cor_{Mg}, Cor_{Fe}, $Cor_{Mg} \cdot H_2O$, $Cor_{Fe} \cdot H_2O$, and an activity of Cor_{Mg} is defined as

$$a_{CorMg} = X_{Mg}^{Cor} \cdot n^{1/2}, \tag{6}$$

where n denotes "dry" cordierite molecule content, defined by a reaction

$$Cor + H_2O = Cor \cdot H_2O. \tag{7}$$

The equilibrium conditions for the end member reactions (1) and (7) will

[2]Here and further on, the values in parentheses denote estimated ranges of uncertainty for the 0.95 probability level.

define the value of the water content of cordierite coexisting with garnet. sillimanite and quartz:

$$RT\,[\ln \overline{K}_{Mg} + \tfrac{1}{2}\ln n] + \Delta G_T(1) + P\,\Delta V(1) = 0, \tag{8}$$
$$RT\,[\ln (1 - n) - \ln n - \ln f_{H_2O}] + \Delta G_T(7) = 0. \tag{9}$$

Following several authors (Holdaway and Lee, 1977; Kurepin, 1979; Newton and Wood, 1979), we accepted $\Delta V(7)$ to be negligible. When applied to isothermal conditions and pressure values P_1 and P_2, Eqs. (8) and (9) can be easily converted to

$$\ln (\overline{K}_1 - \overline{K}_2) + \Delta V(1)(P_1 - P_2)/RT = \tfrac{1}{2}\ln (n_1/n_2), \tag{10}$$
$$\ln [f_{1(H_2O)}/f_{2(H_2O)}] = \ln [(1 - n_1)n_2/n_1(1 - n_2)], \tag{11}$$

where subscripts 1 and 2 denote the values corresponding to P_1 and P_2. Based on these equations, $(\partial RT \ln \overline{K}_{Mg}/\partial P)_T$ values (see above), $\Delta V(1)$ from Perchuk *et al.* (1981), and data on f_{H_2O} from Burnham *et al.* (1969), the water content of cordierite under the experimental conditions has been estimated. As shown in Table 2, the estimates are in good agreement with direct experimental determinations

Table 2. Comparison of direct determinations of water solubility in cordierite with the n values derived from experiments on the Cor–Gr–Sil–Qz equilibrium.

		n						
t (°C)	P (kbar)	1	2	3	4	5	6	7
700	1	0.68			0.61			0.63
	2	0.54	0.50		0.51			0.52
	3	0.45			0.45			0.43
	4	0.37			0.38			0.36
	5	0.30						0.29
	6	0.24						0.24
	7	0.20		0.17	0.31			0.20
	8	0.16						0.17
	9	0.12		0.15	0.26			0.14
	10	0.10	0.09					0.12
750	1	0.71				0.66	0.59	0.67
	2	0.63					0.81–0.62	0.56
	4	0.45			0.42			0.39
	5	0.37						0.33
	8	0.21						0.19
	10	0.15		0.12	0.20			0.14

1: According to our data using Eqs. (10) and (11). 2: Derived from Schreyer and Yoder (1964). 3: Derived from Mirwald and Schreyer (1977). 4: Derived from Mirwald *et al.* (1979). 5: Derived from Gunter (1977). 6: Derived from Duncan and Greenwood (1977). 7: According to Eq. (12).

of the water content. To facilitate extrapolation of the data, all the determinations of the water content, known from literature (Schreyer and Yoder, 1964; Gunter, 1977; Duncan and Greenwood, 1977; Mirwald and Schreyer, 1977; Mirwald *et al.*, 1979), have been fitted by least squares to yield the equation:

$$\ln n = 1.006(\pm 0.804) - 1252(\mp 783)/T - 0.18205(\mp 0.03319)P/T. \quad (12)$$

Significant ranges of uncertainty for the parameters of Eq. (12) result from discrepances between different sets of the data due to quench problems, analytical errors, and pressure uncertainties (Lonker, 1981). However, the n values calculated using Eq. (12) agree closely with our estimates, which were obtained by simultaneous solution of Eqs. (10) and (11) (see Table 2).

It should be noted that the data points of Table 1, which were used to estimate n, had been obtained from the experiments both with pure water and the H_2O–CO_2 mixture. This may suggest that the amount of gas molecules entering the cordierite channels does not depend on the fluid composition, at least for fluids with $CO_2/H_2O \leq 1$, and is almost equal to the amount of H_2O under the same P–T conditions in a purely aqueous system. The determinations by Johannes and Schreyer (1981) provide support for this suggestion, though their data indicate that the content of gas molecules decreases with increases in the CO_2 content of the fluid. Despite the fact the that, in our opinion, Johannes and Schreyer have underdetermined the content of foreign molecules in cordierite in the presence of CO_2-rich fluids, their data are qualitatively consistent with the lower solubility of CO_2 molecules in the presence of pure carbonic acid, as determined by Aranovich *et al.* (1981). The latter data were also fitted by least squares to yield the equation for "dry" cordierite molecule content in the solid solution:

$$\ln n = 0.783(\pm 0.39) - 220(\mp 369)/T - 0.20242(\mp 0.02954)P/T. \quad (13)$$

It could be seen from Fig. 4 that a simple form of Eq. (13) is adequate to describe the dependence of the ln n on pressure and temperature, according to the data of Aranovich *et al.* (1981).

To derive the equilibrium P–T–X_{Mg} relations from the data of Table 1, estimation of $\Delta G_T(1)$ is necessary. This can be readily achieved by employing Eq. (8). The calculated $\Delta G_T(1)$ yields -4494 ± 351 and -4838 ± 402 cal·mol^{-1} for 700 and 750°C, respectively, and the reduction of these data leads to a linear equation of temperature dependence $\Delta G_T(1)$:

$$\Delta G_T(1) = 2387(\pm 2074) - 7.07(\mp 2.10)T, \quad (14)$$

with $r^2 = 0.98$.

There exists another way to estimate the $\Delta G_T(1)$ temperature dependence from the experimental data based on Eq. (8). Substitution of Eqs. (12) and (14) together with the $\Delta V(1)$ into Eq. (8) provides the equation:

$$RT \ln K_{Mg} + 1135(\pm 2857) - 6.07(\mp 2.90)T + 0.46077(\pm 0.03297)P = 0. \quad (15)$$

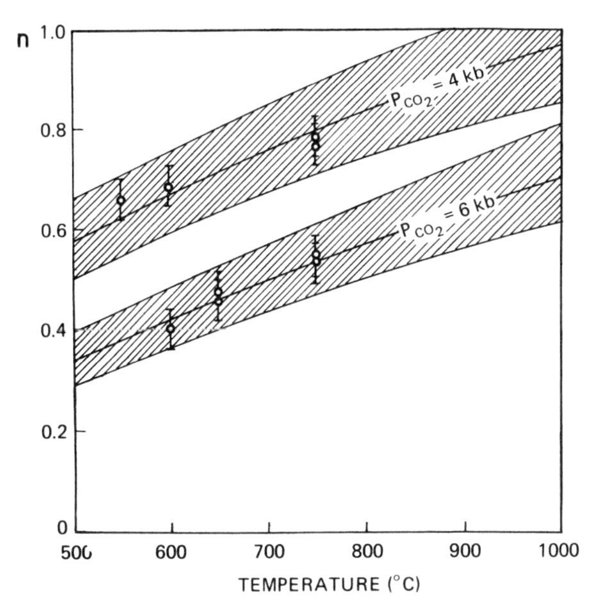

Fig. 4. Temperature dependence of "dry" cordierite content of the solid solution at $P = P_{CO_2}$. Dashed areas define uncertainty ranges corresponding to Eq. (13). Data from Aranovich *et al.* (1981).

But a similar equation can be derived directly from the data of Table 1 without estimating the $(dRT \ln \overline{K}_{Mg}/dP)$, n and $\Delta G_T(1)$ at 700 and 750°C. Regression by least squares yields:

$$RT \ln \overline{K}_{Mg} - 1201(\pm 1945) - 3.62(\mp 2.00)T + 0.45213(\pm 0.03086)P = 0. \quad (16)$$

By subtracting Eq. (12), $\Delta V(1)P$ and Eq. (8) from Eq. (16), we obtain

$$\Delta G_T(1) = 51(\pm 2728) - 4.62T(\mp 2.80). \quad (17)$$

Though coefficients of Eq. (14) and (17) fall within overlapping ranges of uncertainty, Eq. (17) is likely a better alternative for several reasons. First, it demonstrates an excellent agreement with $\Delta H_T(1)$ and $\Delta S_T(1)$ derived from thermochemical data. Calculations from the heats of solution at 970 K determined by Charlu *et al.* (1975) lead to $\Delta H_T(1) = -818 \pm 335$ cal, compared to our 51 ± 2728. Newton (personal communication) estimated $\Delta S_T(1) = 3.73$ cal/K, which was close to Martignole and Sisi's (1981) 4.02 ± 0.56 and our 4.62 ± 2.80. Second, since Eq. (16) is derived directly from the equilibrium compositions, it is probably devoid of some errors which may have been introduced into Eq. (15), and thus provides for a better approximation to the data of Table 1.

It should be noted that large ranges of uncertainty obtained for the coefficients of Eqs. (14)–(17) originated mainly from a narrow temperature range of our experiments and not from determination of the equilibrium compositions.

End Member Reactions

The assemblage Cor_{Mg}+Pyr + Al_2SiO_5 + Qz under "wet" ($P_{H_2O} = P$) and "dry" ($P_{fl} = 0$) conditions is metastable relative to other assemblages (Newton, 1972). Equation (14) allows the position of reaction (1) to be calculated for "dry" conditions. It is evident from Fig. 5 that the calculated position is topologically consistent with the stable breakdown reactions of anhydrous Cor_{Mg}. This position almost coincides with that calculated by Martignole and Sisi (1981) from thermochemical data.

To calculate the end member reaction (1) at hydrous (purely aqueous fluid present) conditions, Eq. (16) was employed. The calculations assume a position of the reaction curve shown in Fig. 6, where the stable cordierite breakdown reaction curves have also been plotted. It could be seen from the diagram that water pressure stabilizes cordierite by several kilobars compared to "dry" conditions, and that metastable reaction (1) occurs at higher pressures than the stable reactions to sillimanite, quartz and talc, or enstatite. Our data are in excellent agreement with the calculations by Martignole and Sisi (1981), who used the thermochemical data and a different approach to account for the cordierite hydration.

A combination of Eqs. (4) and (8) provides for calculation of the equilibrium conditions for reaction (2) at hydrous conditions and allows the estimation of its curve on the P–T plane. Using Eqs. (5), (17) and $V(2)$ we obtain:

$$RT \ln \overline{K}_{Fe} + 4933(\pm 2227) - 6.288(\mp 2.264)\,T + 0.487(\pm 0.031)P = 0. \quad (18)$$

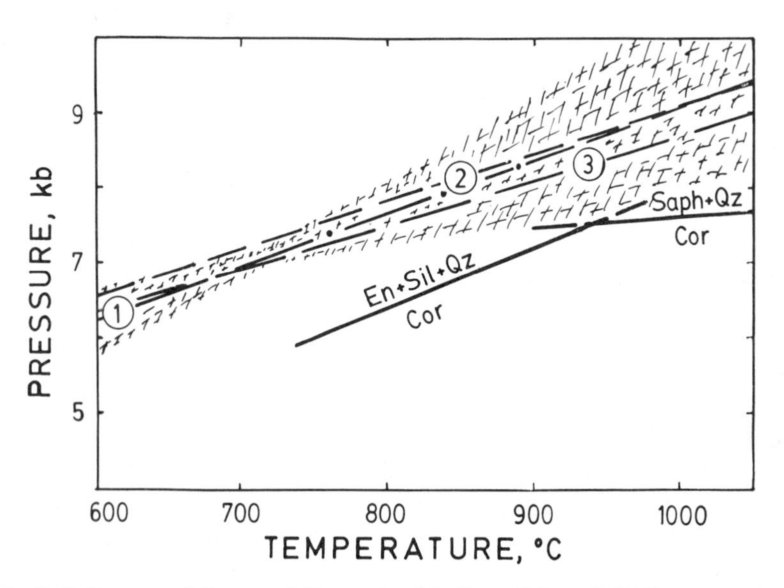

Fig. 5. P–T diagram of Cor_{Mg} stability under "dry" conditions. Solid lines according to Newton's data (personal communication). Patterned area denotes uncertainty ranges according to Eq. (16). Dashed lines refer to reaction (1). (1) Our calculation. (2) From Martignole and Sisi (1981).

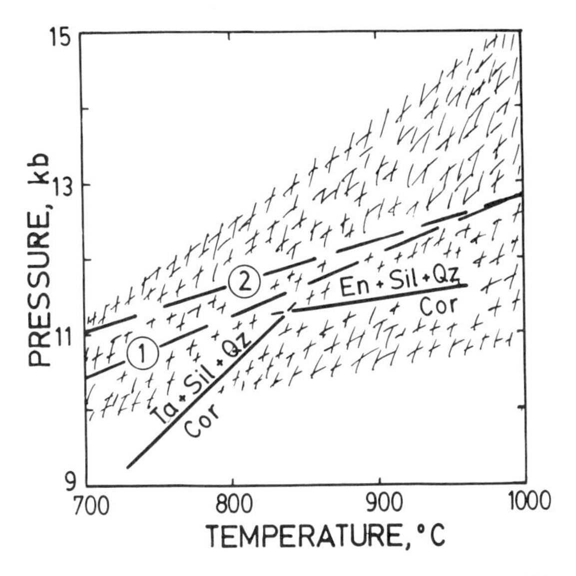

Fig. 6. *P–T* diagram of Cor_{Mg} stability under $P = P_{H_2O}$ conditions. Solid lines: data from Newton (1972) and Newton *et al.* (1974). Patterned area denotes uncertainty ranges according to Eq. (16). Dashed lines refer to reaction (1). (1) Our calculation. (2) According to Martignole and Sisi (1981).

The estimated position of the curve is shown in Fig. 7. It is evident that our results disagree, even qualitatively, with the positions obtained by Weisbrod (1973) and Holdaway and Lee (1977) from the experimental studies. As has been pointed out by Lonker (1981) and Martignole and Sisi (1981), a negative slope of the Fe end member reaction (2) curve on the *P–T* plane, deduced by Richardson (1968), Weisbrod (1973b), Holdaway and Lee (1977), is at variance with the calorimetric data, with the stable breakdown reactions of Cor_{Mg} bracketed experimentally, and with the data on dependence of K_D on temperature. Based on the evidence from natural occurrences, Korikovskii (1979) emphasizes that the negative slope is also inconsistent with the field studies. To explain these discrepancies Martignole and Sisi (1981) suggest that the cordierite starting materials in the available experiments on hydrous Cor_{Fe} stability "have not had enough time to reach equilibrium hydration, and hence, maximum pressure stability." Although this explanation may not be sufficient, there is little doubt that it is highly disputable that the experiments under discussion have reached equilibrium. In our opinion, the data of Richardson (1968), Weisbrod (1973b), and Holdaway and Lee (1977) are insufficient to support the negative slope. The plot of the experimental data by Richardson (1968) in Fig. 7 shows that there are ten runs which are consistent with our data and only one run (700°C, 3 kbar, SQ 46), in which a moderate increase of Cor was determined by X-ray diffraction, that disagrees. It should be noted that none of the experimenters who studied Fe end member reaction (2) managed to avoid tiny inclusions of hercynite in the synthetic minerals of starting charges and run products. It is likely that the inevitable hercynite inclusions have

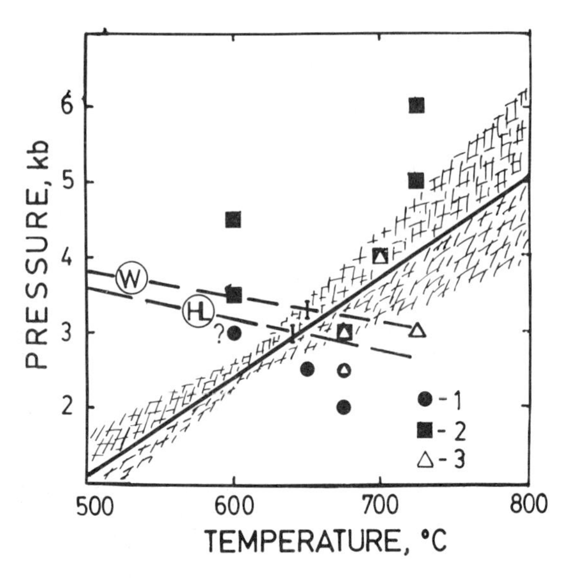

Fig. 7. P–T diagram of Cor_{Fe} stability under $P = P_{H_2O}$ conditions. Solid lines: our calculation. Dashed lines: (w) data from Weisbrod (1973b); (HL) data from Holdaway and Lee (1977). Data from Richardson (1968). (1) Cor growth (?—moderate). (2) Alm + Sil growth. (3) Her present.

distorted equilibrium relations of the minerals and have led to misinterpretation of the experimental results. Weisbrod (1973b) and Holdaway and Lee (1977) supported their investigations of the univariant end member reaction (2) curve by the data which were determined by an approach from divariant fields Fe–Mn and Fe–Mg. This method has been demonstrated to be the most reliable for bracketing univariant reactions (Schmid *et al.*, 1978), but as these points, 750°C, 3.3 ± 0. 1 kbar and 740°C, 3.0 ± 0.1 kbar, fall on our line in Fig. 7, they cannot be used to support the negative slope.

Fe–Mg Reactions

Simultaneous solution of Eqs. (4) and (8) for the mineral compositions provides the isothermal P–X_{Mg} sections for "dry" (ln n = 0) conditions (Fig. 8). Comparison of calculated P–N_{Mg} loops and experimentally determined points shows that the data of Hensen and Green (1971) and Hensen (1977) disagree with our calculations. The latter discrepancy may have arisen from erroneous estimation of the equilibrium mineral compositions by Hensen and Green (1971). Under close examination the average Mg/(Mg + Fe) ratio of the garnets, despite inhomogeneity of the grains, was found to be similar to that of the starting mixtures (for both C_{30} and C_{70} compositions). These relations resemble our data on the garnet synthesis at 1100°C and 30 kbar (Aranovich and Podlesskii, 1982a) and imply that the compositions accepted as being in equilibrium may have been produced during crystallization of the starting mixtures and not by equilibration.

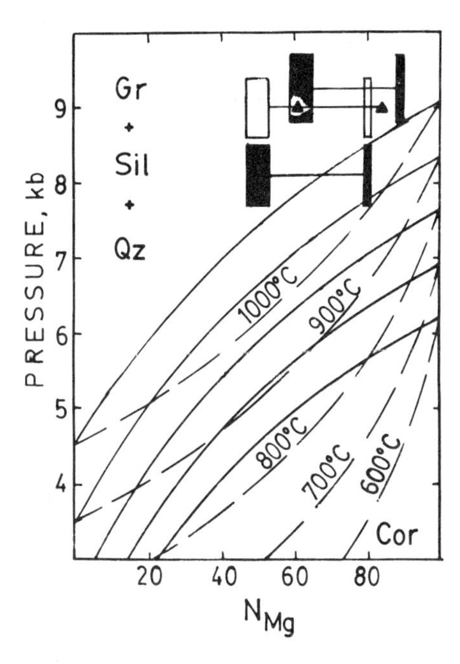

Fig. 8. P–N_{Mg} diagram of the Cor–Gr–Sil–Qz stability under "dry" conditions. Solid lines: Gr composition. Dashed lines: Cor composition. Triangles: data from Hensen (1977). Rectangles: data from Hensen and Green (1971) (dashed: 100 °C, open: 900 °C).

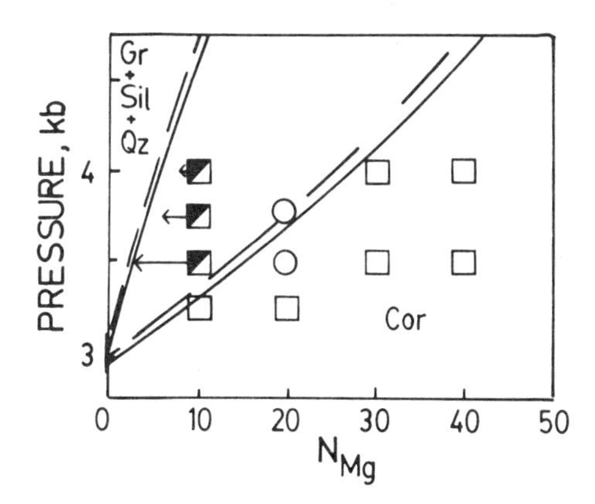

Fig. 9. P–N_{Mg} diagram of the Cor–Gr–Sil–Qz stability at 740 °C ($P = P_{H_2O}$). Solid line: our calculation. Dashed line and data from Holdaway and Lee (1977). Open boxes: Cor stability. Shaded boxes: Gr + Cor stability. Circles: uncertain mixtures. Arrows denote changes of the starting garnet compositions.

Although the reversal point (Hensen, 1977) has been obtained in experiments under low water activity conditions, addition of oxalic acid (5%) might have changed the compositional relations of the minerals according to Eq. (8), which is probably the reason why this point does not fall on the 1000° C loop in Fig. 8, while the $RT \ln K_D$ value agrees with Eq. (5). If the fluid content of cordierite from these experiments had been determined, we could know that for sure.

The results of calculation of the P–X_{Mg} sections for "wet" conditions ($\ln n$ defined by Eq. (12)) are incompatible with the experiments by Currie (1971) and agree with the 740° C isotherm derived by Holdaway and Lee (1977). Hensen's (1977) critical observation of the experiments by Currie showed that they had not been accurate enough to provide for correct thermodynamic data on the equilibrium in question, and because we agree with the criticism, we do not discuss them. The plot of the 740° C P–X_{Mg} section calculated with our data is shown in Fig. 9, in comparison with the isotherm derived by Holdaway and Lee (1977) from both experimental and natural data. It is evident that the calculated loop fit the limited experimental data of Holdaway and Lee (1977) in much the same manner as their own calculation.

Application to Geothermobarometry, and Discussion

The above calculations offer an instrument for measuring metamorphic temperatures and pressures, but to use it efficiently we need to know the "dry" cordierite molecule content of natural cordierites, and here some problems arise. (1) The data on the gas content of natural cordierites are difficult to obtain and, thus, are very limited. (2) The natural cordierite channels are filled not only with H_2O and CO_2, but also with other components of coexisting metamorphic fluid such as Ar, N_2, CO_2, CH_4, and C_nH_n (Beltrame *et al.*, 1976, Zimmermann, 1981), and the amount of these gases may range up to 30% of the total gas content. (3) The determined gas content may be related to the latter processes other than those of equilibration of the Cor–Gr–Sil–Qz assemblage. In order to circumvent these problems we assume, largely out of necessity, that $P_{fl} = P_l$ and that all gases in cordierite channels behave like water, with the "dry" cordierite content being equal to that at $P_{H_2O} = P$ condition. These assumptions allow calculation of the P–T dependence of the Fe–Mg mineral compositions via Eqs. (4), (5), and (6).

The calculated isopleths of Mg/(Mg + Fe) of coexisting cordierite and garnet are plotted on a P–T diagram (Fig. 10). To account for Al_2SiO_5 phase transitions the data of Holdaway (1971) have been used. It could be seen from Fig. 10 that certain ranges of compositions of cordierite and garnet lie in certain Al_2SiO_5 polymorph stability fields. The most magnesium assemblages fall within the Ky field and the most ferrous assemblages fall within the And field, while the bulk of the compositions is confined to the Sil field. In contrast to calibrations of Hensen and Green (1973), Thompson (1976), and Lee and Holdaway (1978), the exaggerated stability field of Fe-rich cordierites is lacking.

Several lines of field evidence support the calculated relations. Cordierites of

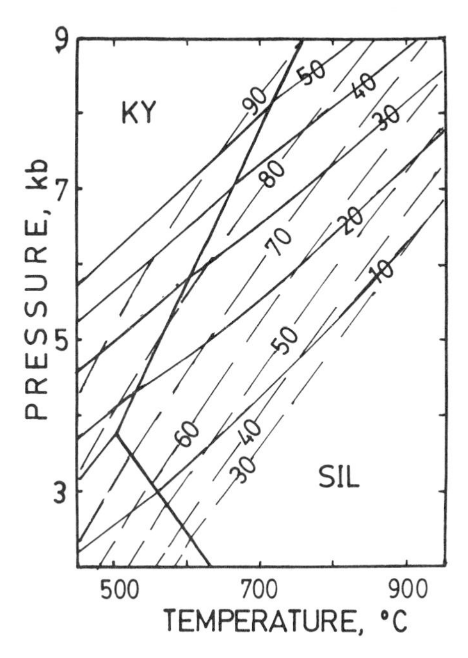

Fig. 10. P–T diagram with N^{Cor}_{Mg} (dashed lines) and N^{Gr}_{Mg} (isopleths for the Cor–Gr–Al_2SiO_5–Qz equilibrium at $P = P_{H_2O}$ condition. Alumina silicates fields from Holdaway (1971).

composition close to Cor_{Fe} were found exclusively in low-pressure rocks (e.g., Lepezin and Melenevsky, 1977), with the most ferrous cordierite ($N_{Mg} = 1$) being described in andalusite-bearing assemblage of a hornfels (Perchuk *et al.*, 1980). In kyanite-bearing rocks only Mg-rich cordierites were reported as stable (Hietanen, 1956; Zotov and Sidorenko, 1968). Grambling (1981) has described the $Cor_{82} + Gr_{25} + Qz$ assemblage in relation to rocks with kyanite, andalusite, and sillimanite in the Truchas Peaks region, New Mexico, and this is in excellent agreement with Fig. 10.

Observations of more than 150 assemblages (Aranovich and Podlesskii, 1982b) have shown that the estimates obtained in terms of Eqs. (4), (5), (8), (12), and (14) are consistent with the Al_2SiO_5 polimorphs stability fields (Holdaway, 1971), with the only exception being the estimates for the compositions from Napier complex, Enderby Land, reported by Ellis *et al.* (1980) (see Fig. 11). It could be seen from Fig. 11 that the estimates for metamorphic complexes or areas fit to narrow trends on a P–T diagram, with each complex being characterized by a peculiar trend. The geological interpretation of the similar trends for the Aldan shield, Chogar, and Khanka complexes have been discussed elsewhere (e.g., Perchuk *et al.*, 1980; Aranovich and Podlesskii, 1982b; Perchuk *et al.*, in press), and here attention will be focussed on consideration of possible errors in estimating the P–T conditions by the Cor–Gr–Sil–Qz thermometer and barometer.

Consistency of our calibration with the thermochemistry, the most reliable

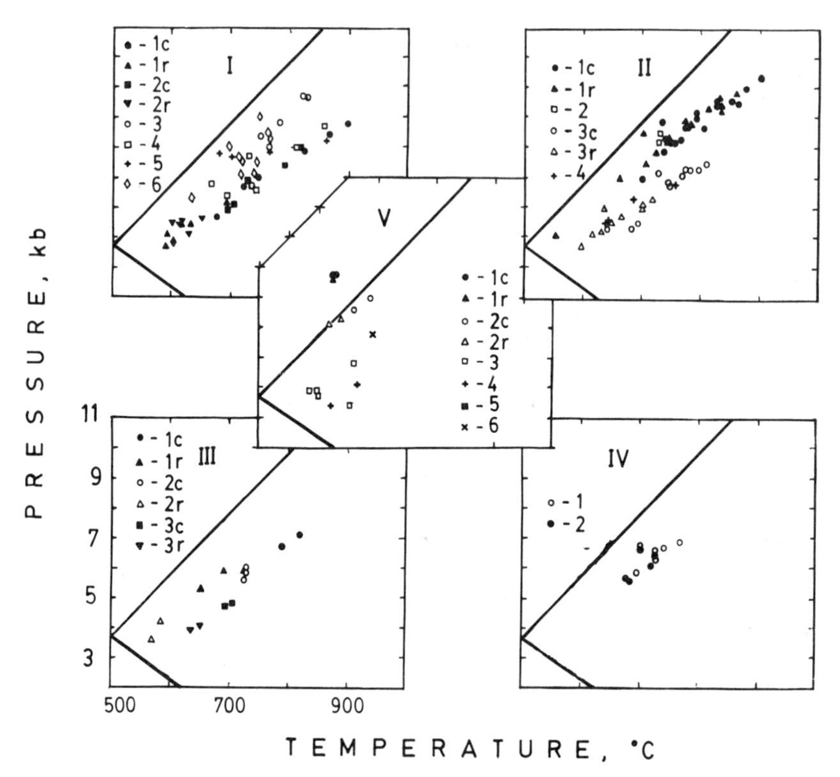

Fig. 11. Estimates of P–T conditions of metamorphism. I. Canada, (1) The Coast Ranges granulites, British Columbia, Cor and Gr compositions from Selverstone and Hollister (1980). (c) Core composition; (r) Rim compositions. (2) The Khtada Lake complex, British Columbia (Hollister, 1977). (3) The Daly Bay complex, Northwest Territory (Hutcheon *et al.*, 1974). (4) Northeastern Saskatchewan, (Kays, 1976). (5) The Opinicon Lake region, Ontario, (Currie, 1971). (6) The Frontenac Axis, Ontarion (Lonker, 1980). II. Finland. (1) The Ivalojoki–Inarijarvi area, Finnish Lapland (Hörmann *et al.*, 1980). (2) Lapland (Korsman, 1977). (3) The Rantasalmi, Sulkewa area, Southeastern Finland (Korsman, 1977). (4) The Attn gneisses, Southwestern Finland (Schellekens, 1980). III. Scottish Caledonides. (1) Glen Skeddle. (2) Huntly Portsoy. (3) Strontian (Ashworth and Chinner, 1978). IV. Norway. (1) The Egersund, Ogua, Southwestern Norway (Hery, 1974). (2) Rogaland (Jacques de Dixmude, 1978). V. Different areas. (1) Enderby Land, Antarctica (Ellis *et al.*, 1980). (2) Southwestern Greenland (Wells, 1979). (3) Sonapahar, Assam, India (Lal *et al.*, 1978). (4) Contact aureole of the Ronda ultramafic intrusions (Loomis, 1979). (5) Contact aureole of the Nain anorthositic complex, Labrador (Berg, 1977). (6) The Rayner complex, East Antarctica (Grew, 1981).

experimental data, and the field evidences implies that qualitative errors are unlikely, while quantitative uncertainty of estimation seems inevitable. Major sources of error are (1) the uncertainty ranges of thermodynamic properties of end member reactions, (2) activity models assumed for the solid solutions involved, and (3) unknown fluid content of cordierite. Errors arising from the uncertainty ranges of parameters of Eqs. (4), (8), and (12) have been determined as $\pm 15^\circ$C and ± 1 kbar for the estimates of temperature and pressure, respectively.

Isomorphism of mineral-forming elements of natural cordierites may be considered, to a high degree of accuracy, as confined to the Cor_{Mg}–Cor_{Fe} solid-solutions series. The mineral chemistry observations (e.g., Lepezin *et al.*, 1975) show that the content of manganese, which is the most significant of the cordierite impurities, is usually negligible. Thus, it seems reasonable to ignore errors arising from the use of the X_{Mg}^{Cor} to estimate the *P–T* conditions. Also, the existing data give no indication of nonideality of mixing of the "dry" and fluid-saturated cordierite end members, and, consequently, the related errors can not be evaluated at present.

Observation of much data on the composition of garnet coexisting with Cor + Sil + Qz (Podlesskii, 1981) shows that the total amount of the Gros and Spes components usually does not exceed 10 mol %, with the more manganous or calcic garnets being very rarely met in this paragenesis. This implies that the effect of the garnet impurities on the *P–T* estimates cannot be pronounced. In any event, despite the fact that reliable data on the mixing properties of the Mn-bearing garnet solid solutions are lacking, an attempt has been made to estimate this effect. It is evident that the maximum *P–T* estimates are obtained with the X_{Mg}^{Gr} substituted into Eq. (8), while the employment of the Mg/(Mg + Fe + Ca + Mn) ratio provides the minimum values. The *P–T* diagram of Fig. 12 shows that the difference between the values thus obtained is negligible for the garnets from Massachusetts (Tracy *et al.*, 1976). Similar patterns have been obtained for most of the other assemblages of the garnets with low Ca and Mn content. The difference has been found considerable ($\backsim$1 kbar) only for the manganous garnets (up to 30 mol % Spes) from hornfelses of the Steinach aureole (Okrusch, 1971) and from the Ryoke gneisses (Ono, 1977). This difference may be not as large as shown in Fig. 13 due to the probable positive mixing energy of pyrope and spessartite.

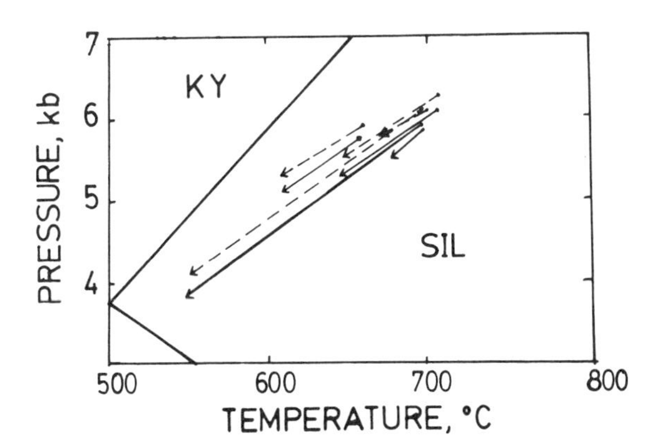

Fig. 12. Comparison of *P–T* estimates obtained for metamorphic rocks of Central Massachusetts (Tracy *et al.*, 1976) by different means. Dashed arrows: Spes and Gros contents of garnets were not taken into account. Solid arrows: additional components taken into account.

Estimation of errors related to the unknown fluid content of cordierite presents some problems. It is easy to calculate that the pressure estimates for "dry" conditions" ($P_{fl} = 0$) are some 3 kbar below those for "wet" conditions ($P_{H_2O} = P$). This is too large a difference to allow any petrogenetic application and, in addition, it is based on the unlikely assumption of the absence of metamorphic fluid. In our opinion, the uncertainty range in this case should be the difference between the estimates based on Eqs. (12) and (13), i.e., for $P_{H_2O} = P$ and $P_{CO_2} = P$ conditions. This assumption is supported by comparison of the pressure estimates based on Eqs. (12) and (13) and that obtained by using the n values determined on the cordierites from the assemblages of Ivalojki–Inarijarvi area (Hörmann *et al.*, 1980) (see Table 3). The analysis shows that the $CO_2/(CO_2 + H_2O)$ ratio in cordierites is high which implies a high CO_2 content in the coexisting fluid. Therefore we should expect that Eq. (13) is more useful in simulating real conditions than Eq. (12). The corresponding data of Table 3 and the estimates obtained from the plagioclase–garnet geobarometer (Aranovich and Podlesskii, 1980) support this assumption. Unfortunately, few data on the fluid content of cordierite coexisting with Gr + Sil + Qz are available to date and it is difficult to make any conclusions concerning the applicability of Eqs. (12) or (13) in describing the natural occurrences. If we consider general increase of $CO_2/(CO_2 + H_2O)$ ratio

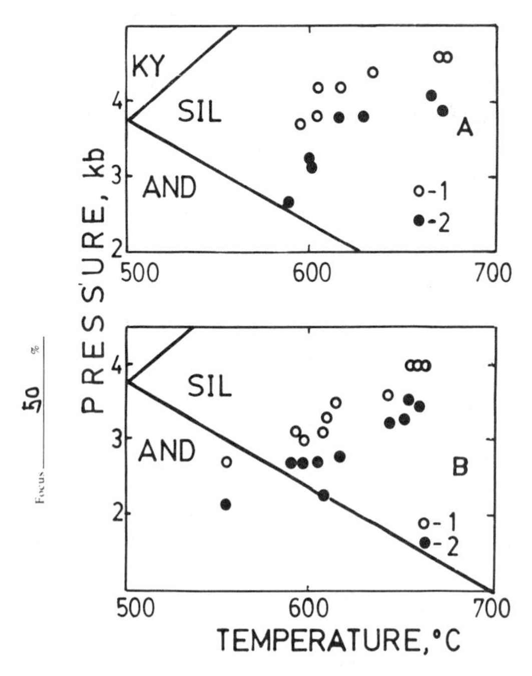

Fig. 13. Comparison of P–T estimates obtained for gneisses of Ryoke (A) (Ono, 1977) and contact hornfelses of Steinach Aureold (B) (Okrush, 1971) by different means. (1) Spes and Gros contents of garnet were not taken into account. (2) Additional components taken into account.

Table 3. Calculated equilibrium pressures for Ivalojoki–Inarijarvi granulites.

N[a]	P_1[b]	P_2	P_3	P_4	n_1[c]	n_2	n_3
89.V^c_r	9.0	7.4	7.2	8.4	0.22	0.49	0.53
	8.8	7.2	7.0	8.0	0.22	0.49	
158.I	7.3	5.8	6.1	—	0.21	0.57	0.43
158.II	7.7	6.1	6.5	7.2	0.23	0.56	0.44
158.IV^c_r	7.2	5.5	5.8	—	0.22	0.59	0.49
	6.8	5.1	5.5	—	0.22	0.61	
161.II^c_r	8.5	6.9	7.0	8.4	0.23	0.52	0.50
	8.2	6.5	6.7	7.4	0.23	0.54	
194.III	7.6	5.9	6.6	6.0	0.23	0.57	0.38
196.III^c_r	7.3	5.7	6.1	7.9	0.23	0.58	0.45
	6.0	4.3	4.9	6.1	0.22	0.68	

[a]Rock sample number according to Hörmann *et al.* (1980).

[b]P_1 calculated at $P = P_{H_2O}$ condition, P_2 calculated at $P = P_{CO_2}$ condition, P_3 calculated using H_2O and CO_2 contents of the cordierites analyzed by Hörmann *et al.* (1980), P_4 using garnet–plagioclase barometer modified by Aranovich and Podlesskii (1980).

[c]n_1 corresponding to P_1, n_2 corresponding to P_2, n_3 derived from the cordierite analyses (Hörmann *et al.*, 1980, and personal communication from M. Raith).

of the metamorphic fluid with depth (Perchuk, 1979), Eq. (12) seems more valid for the assemblages of the medium- (or low-) pressure rocks, than for that of the high-pressure rocks. This is just the opposite of Eq. (13). Further studies are needed to solve these problems.

Acknowledgments

The authors are grateful to L. L. Perchuk for permanent support and fruitful discussions, to M. Raith for providing unpublished data of mineral compositions from the Ivalojoki–Inarijarvi area, and to V. K. Melnikova for typing the manuscript. Authors particularly thank Prof. R. C. Newton for critical review of the manuscript and helpful suggestions.

References

Aranovich, L. Ya. and Podlesskii, K. K. (1980) Garnet–plagioclase geobarometer, *Dokl. Acad. Nauk SSSR* **251**, 1216–1219 (in Russian).

Aranovich, L. Ya. and Podlesskii, K. K. (1982a) Equilibrium: garnet + sillimanite + quartz = cordierite. Experiments and calculations, *Mineral J.* **4**, 20–32 (in Russian).

Aranovich, L. Ya. and Podlesskii, K. K. (1982b) Equilibrium: garnet + quartz = cordierite. Thermobarometry of natural associations, *Miner. J.* **4**, 14–20 (in Russian).

Aranovich, L. Ya., Podlesskii, K. K., and Schepochkina, N. I. (1981) Experimental determination of the CO_2 solubility in cordierite, *Dokl. Akad. Nauk SSSR* **261**, 723–730.

Ashworth, J. R. and Chinner, G. A. (1978) Coexisting garnet and cordierite in migmatites from the Scottish Caledonides. *Contrib. Mineral. Petrol.* **65**, 379–394.

Beltrame, R. J., Norman, D. I., Alexander, E. C., Jr., Sawkins, F. Y. (1976) Volatiles released by step heating a cordierite to 1200°C, *EOS Trans. Amer. Geoph. Union* **57**, 352.

Berg, J. H. (1977) Regional geobarometry in the contact aureoles of the anorthositic complex, Labrador, *J. Petrology* **18**, 399–430.

Burnham, C. W., Holloway, J. R., and Davis, N. E. (1969) Thermodynamic properties of water to 1000°C and 10 000 bars, *Geol. Soc. Amer. Spec. Pap.* **132**, 96.

Charlu, T. V., Newton, R. C., and Kleppa, O. J. (1975) Enthalpies of formation at 970 K of compounds in the system $MgO–Al_2O_3–SiO_2$ from high temperature solution calorimetry, *Geochim. Cosmochim. Acta* **39**, 1487–1497.

Currie, K. L. (1971) The reaction 3 cordierite = 2 garnet + 4 sillimanite + 5 quartz as a geological thermometer in the Opinicon Lake region, Ontario. *Contrib. Mineral. Petrol.* **33**, 215–226.

Currie, K. L. (1974) A note on the calibration of the garnet–cordierite geothermometer and geobarometer, *Contrib. Mineral. Petrol.* **44**, 35–44.

Duncan, I. J. and Greenwood, H. J. (1977) The effect of molecular water on the thermodynamic properties and stability relations of Mg–cordierite, *Geol. Assoc. Canada Annual Meeting (Abstr.)* **2**, 17.

Ellis, D. J., Sheraton, J. W., England, R. N., and Dallwitz, W. B. (1980) Osumilite–sapphirine–quartz granulites from Enderby Land Antarctica, Mineral Assemblages and reactions, *Contrib. Mineral. Petrol* **72**, 123–144.

Engi, M. (1978) Mg–Fe exchange equilibria among Al–Gr spinel, olivine, orthopyroxene and cordierite, *Diss. Doct. Natur. Sci. Swiss. Fed. Inst. Technol. Zurich,* **130**.

Grambling, J. A. (1981) Kyanite, andalusite, sillimanite, and related mineral assemblages in the Truchas Peaks region, New Mexico, *Amer. Mineral.* **66**, 702–722.

Grew, E. S. (1981) Granulite–facies metamorphism at Molodezhnaya Station East Antarctica, *J. Petrology* **22**, 297–336.

Gunter, A. E. (1977) Water in synthetic cordierite and its significance in the experimental reaction: aluminous–biotite + sillimanite + quartz = iron–cordierite + sanidine + water, *Geol. Assoc. Canada Annual Meeting (Abstr.)* **2**, 22.

Henry, J. (1974) Garnet–cordierite–gneisses near the Egersund–Ogna anorthositic intrusion, Southwestern Norway, *Lithos* **7**, 207–216.

Hensen, B. J. (1977) Cordierite–garnet-bearing assemblages as geothermometers and barometers in granulite facies terranes, *Tectonophysics* **43**, 73–88.

Hensen, B. J. and Green, D. H. (1971) Experimental study of the stability of cordierite and garnet in pelitic compositions at high pressures and temperatures. I. Compositions with excess alumino-silicate, *Contrib. Mineral. Petrol.* **33**, 309–330.

Hensen, B. J. and Green, D. H. (1973) Experimental study of the stability of cordierite and garnet in pelitic compositions at high pressures and temperatures. III. Synthesis of experimental data and geological application, *Contrib. Mineral. Petrol.* **38**, 151–166.

Hietanen, A. (1956) Kyanite, andalusite and sillimanite in the shist in Boehls Butte quadrangle, Idaho. *Amer. Mineral.* **41**, 1–27.

Holdaway, M. J. (1971) Stability of andalusite and the aluminum silicate phase diagram, *Amer. J. Sci.* **271**, 97–131.

Holdaway, M. J. and Lee, S. M. (1977) Fe-Mg cordierite stability in high-grade pelitic rocks based on experimental, theoretical, and natural observations, *Contrib. Mineral. Petrol.* **63**, 175–198.

Hollister, L. S. (1977) The reaction forming cordierite from garnet, the Khtada Lake metamorphic complex, British Columbia, *Can. Mineral.* **15**, 217–229.

Hörmann, P. K., Raith, M., Raase, P., Ackermand, D., and Seifert, F. (1980) The granulite complex of Finnish Lapland: Petrology and metamorphic conditions in the Ivalojoki–Inarijarvi area, *Bull. Geol. Surv. Finl.* **308**, 110.

Hutcheon, I., Froese, E., and Gordon, T. M. (1974) The assemblage quartz–sillimanite–garnet–cordierite as an indicator of metamorphic conditions in the Daly Bay complex, N.W.T., *Contrib. Mineral. Petrol.* **44**, 29–34.

Jaques de Dixmude, S. (1978) Geothermométrie comparée de roches du facies granulite du Rogaland (Norvege méridionale), *Bull. Mineral.* **101**, 57–65.

Johannes, W. and Schreyer, W. (1981) Experimental introduction of CO_2 and H_2O into Mg-cordierite, *Amer. J. Sci.* **281**, 299–317.

Ivanov, I. P., Kapustin, N. V., Litvinov, A. V., Mitirev, P. A., Mischenchuk, O. A., Fonarev, V. I., and Chernysheva, C. K. (1977) High-pressure apparatus with externally heated reactors, *Contributions to Physico-Chemical Petrology,* Vol. 6, pp. 79–96, Nauka, Moscow.

Kays, M. A. (1976) Comparative geochemistry of migmatized, interlayered quartz feldspathic and pelitic gneisses: A contribution from rocks of Southern Finland and North-eastern Saskatchewan, *Precambrian Res.* **3**, 433–462.

Korikovskii, S. P. (1979) *Facies of Metamorphizsm of Metapelites*, P. 263, Nauka, Moscow (in Russian).

Korsman, K. (1977) Progressive metamorphism of the metapelites in the Rantasalmi-Sulkewa area, south eastern Finland. *Geol. Surv. Finl. Bull.,* **290**, 82.

Korzhinskii, D. S. (1936) Paragenetic analysis of the quartz-bearing low-calcium schists of Archaean complex of Southern Pribaikalje, *Zapiski Mineral Obsch., Ser. 2,* **65**, 247–280.

Kurepin, V. A. (1976) Thermodynamics of water-bearing cordierite and related mineral equilibria, *Geokhimia* **1**, 49–60 (in Russian).

Lal, R. K., Ackermand, D., Seifert, F., and Haédar, S. K. (1978) Chemographic relations in sapphirine bearing rocks from Sonapahar, Assam, India, *Contrib. Mineral. Petrol.* **67**, 169–187.

Lavrent'eva, I. V. and Perchuk, L. L. (1981) Cordierite–garnet thermometer. *Dokl. Akad. Nauk SSSR* **259**, 697–700 (in Russian).

Lee, S. M. and Holdaway, M. J. (1978) Significance of Fe–Mg cordierite stability relations on temperature, pressure, and water pressure in cordierite granulites in *The Earth's Crust: Its Nature and Physical Properties,* edited by Heacock, J. G., Am. Geoph. Union. Mon. No. 20, pp. 79–94, Washington, D.C.

Lepezin, G. G. (1969) Importance of water in cordierite in natural mineralogenesis, *Dokl. Acad. Nauk SSSR* **186**, 122–125 (in Russian)

Lepezin, G. G., Lavrent'ev, Yu. G., and Pokachalova, O. S. (1975) Peculiarities of cordierite chemistry, in *Minerals and Mineral Parageneses of Metasomatic and Metamorphic Rocks,* pp. 85–90, Nauka, Leningrad (in Russian).

Lepezin, G. G. and Melenevsky, V. N. (1977) On the problem of water diffusion in the cordierites. *Lithos* **10**, 49–57.

Lonker, S. W. (1980) The Conditions of metamorphism in high-grade pelitic gneisses from the Frontenac Axis, Ontario, Canada, *Can. J. Earth Sci.* **17**, 1666–1684.

Lonker, S. W. (1981) The *P-T-X* relations of the cordierite–garnet–sillimanite equilibrium, *Amer. J. Sci.* **281**, 1056–1090.

Loomis, T. P. A. (1972) Contact metamorphism of pelitic rocks by the Ronda ultramafic intrusion, southern Spain, *Bull. Geol. Soc. Amer.* **83**, 2449–74.

Loomis, T. P. A. (1979) A natural example of metastable reactions involving garnet and sillimanite, *J. Petrology* **20**, 271–292.

Marakuschev, A. A. (1965) *Problems of Mineral Facies of Metamorphic and Metasomatic Rocks,* p. 327, Nauka, Moscow (in Russian).

Martignole, J., Sisi, J.-C. (1981) Cordierite–garnet–H_2O equilibrium: A geological thermometer, barometer and water fugacity indicator. *Contrib. Mineral. Petrol.* **77**, 38–46.

Mirwald, P. W., Maresch, W. V., and Schreyer, W. (1979) Der Wassergehalt von Mg-Cordierit zwischen 500° and 800°C sowie 0.5 und 11 kbar, *Fortschr. Mineral.* **57**, 101–102.

Mirwald, P. W. and Schreyer, W. (1977) Die stabile und metastabile Abbaureaktion von Mg-cordierit in Talk, Disthen und Quartz und ihre Abhangigkeit vom Gleichgewichtswassergehalt des Cordierits, *Forschr. Mineral.* **55**, 95–97.

Newton, R. C. (1972) An experimental determination of the high-pressure stability limits of magnesian cordierite under wet and dry conditions, *J. Geol.* **80**, 398–420.

Newton, R. C., Charlu, T. V., and Kleppa, O. J. (1974) A calorimetric investigation of the stability of anhydrous magnesian cordierite with application to granulite facies metamorphism, *Control. Mineral. Petrol.* **44**, 295–311.

Newton, R. C. and Wood, B. J. (1979) Therodynamics of water in cordierite and some petrologic consequences of cordierite as a hydrous phase, *Contrib. Mineral. Petrol.* **68**, 391–405.

Okrusch, M. (1971) Garnet–cordierite–biotite equilibria in the Steinach Aureole, Bavaria, *Contrib. Mineral. Petrol.* **32**, 1–23.

Ono, A. (1977) Temperature and pressure of the Ryoke gneisses estimated by garnet–cordierite geothermometer, *J. Japan. Assoc. Min. Petr. Econ. Geol.* **72**, 114–117.

Perchuk, L. L. (1972) Problems of thermodynamic conditions of mineral equilibria in the deep zones of the earth's crust and the upper mantle, in *Magmatism, Rock Formations, and the Earth's Depths,* pp. 169–176, Nauka, Moscow (in Russian).

Perchuk, L. L. (1973) *Thermodynamic Regime of Depth Petrogenesis,* p. 318, Nauka, Moscow (in Russian).

Perchuk, L. L. (1977) Thermodynamic control of metamorphic processes, in *Energetics of Geological Processes,* edited by Saxena, S. K., pp. 118–141, Springer-Verlag, New York.

Perchuk, L. L., Lavrent'eva, I. V., Podlesskii, K. K., and Aranovich, L. Ya. Biotite–garnet–cordierite equilibria and evolution of metamorphism, Nauka, Moscow (in Russian) (in press).

Perchuk, L. L., Mishkin, M. A., Kotel'nikov, A. R., Lavrent'eva, I. V., Girnis, A. V., Podlesskii, K. K., and Gerasimov, V. Yu. (1980) Thermodynamic conditions of the Khanka Massif metamorphism, in *Contributions to Physico-Chem. Petrology,* Vol. 9, pp. 139–167, Nauka, Moscow (in Russian).

Perchuk, L. L., Podlesskii, K. K., and Aranovich, L. Ya. (1981) Calculation of thermodynamic properties of end-member minerals from natural parageneses, in *Advances in Physical Geochemistry,* pp. 111–129, Springer-Verlag, New York.

Podlesskii, K. K. (1981) Phase correspondence in the cordierite–garnet–sillimanite–quartz system (cordierite-garnet barometer), Unpublished Ph.D. Dissertation, Inst. of Exp. Mineralogy, Chernogolovka, 179 (in Russian).

Popp, R. K. and Frantz, J. D. (1978) Diffusion of hydrogen in gold, *Carnegie Inst. Wash. Year Book* **76**, 662–664.

Richardson, S. W. (1968) Staurolite stability in a part of the system Fe–Al–Si–O–H, *J. Petrology* **9**, 467–488.

Schellekens, J. H. (1980) Application of the garnet-cordierite geothermometer and geobarometer to gneisses of Attu, S.W. Finland an indication of *P* and *T* conditions of the lower granulite facies, *N. Jhb. Miner. Mh.* **h.1**, 11–19.

Schmid, R., Gressey, G., Wood, B. J. (1978) Experimental determination of univariant equilibria using divariant solid solution assemblages, *Nat. Envir. Res. Council,* **Publ. D**, 86–89.

Schreyer, W. and Yoder, H. S. (1964) The system Mg–cordierite–H_2O and related rocks, *N. Jb. Miner. Abh.,* **101**, 271–342.

Selverstone, J. and Hollister, L. S. (1980) Cordierite-bearing granulites from the Coast Ranges, British Columbia: *P–T* conditions of metamorphism, *Can. Mineral.* **18**, 119–129.

Thompson, A. B. (1976) Mineral reactions in pelitic rocks: II. Calculation of some *P–T–X* (Fe–Mg) phase relations, *Amer. J. Sci.* **276**, 425–454.

Tracy, R. J., Robinson, P. R., and Thompson, A. B. (1976) Garnet composition and zoning in the determination of temperature and pressure of metamorphism, Central Massachusetts, *Amer. Mineral.* **61**, 762–775.

Wells, P. R. A. (1979) Chemical and thermal evolution of archaen sialic Crust, Southern West Greenland, *J. Petrology* **20**, 187–226.

Weisbrod, A. (1973a) Cordierite-garnet equilibrium in the system Fe–Mn–Al–Si–O, *Carnegie Inst. Wash. Yearbook* **72**, 515–518.

Weisbrod, A. (1973b) Refinements of the equilibrium conditions of the reaction Fe–cordierite–almandine–quartz–sillimanite (+H_2O), *Carnegie Inst. Wash. Yearbook* **72**, 518–521.

Wesibrod, A. (1973c) The problem of water in cordierite, *Carnegie Inst. Wash. Yearbook* **72**, 521–523.

Zotov, I. A. and Sidorenko, A. G. (1968) On magnesian gedrite from the S. W. Pamirs. Dokl. Akad. Nauk SSSR, **180** (in Russian).

Zimmermann, J. L. (1981) La libération de l'eau, du gaz carbonique et des hydrocarbures des cordierites. Cinétique des mécanismes. Détermination des sites, *Bull. Mineral.* **104**, 325–338.

Chapter 7
Experimental Investigation of Exchange Equilibria in the System Cordierite–Garnet–Biotite

L. L. Perchuk and I. V. Lavrent'eva

Symbols

Thermodynamic Parameters

T: temperature, °K
t: temperature, °C
P: pressure, bar
ΔG_T°: Gibbs free energy of a reaction at given T and 1 bar, cal·mol^{-1}
N_{Mg}^M: molar percentage of Mg component in the solid solution of mineral M: $N_{Mg}^M = 100Mg/(Mg + Fe + Mn)$
X_{Mg}^M: molar fraction of Mg component in the solid solution of mineral M: $X_{Mg}^M = Mg/(Mg + Mn + Fe)$
ΔS_T°: entropy change of a reaction at temperature T and 1 bar, cal·K^{-1}
ΔV: volume change of solids in a reaction, cal·bar^{-1}·mol.$^{-1}$
K_D: distribution coefficient

Statistical Parameters

$\bar{x}_i$: mean arithmetical value
r: the linear regression coefficient
δ: mean square deviation (m.s.d.)
Δ: mean deviation $|(x_i - x_{theor})/n|$; n number of samples
x_i: a given value
x_{theor}: value according to linear equation

Minerals

Alm: almandine
And: andalusite
Ann: annite
Bi: biotite
Cor: cordierite
Est: eastonite

Gr: garnet
Phl: phlogopite
Pyr: pyrope
Qz: quartz
Sil: sillimanite
Sid: siderophyllite
N_{Al}^{Bi} = 100Al/(Al + Mg + Fe + Mn + Si + Ti) in formula of Bi
Hel: Helium gas mixture

Mineral abbreviations with subscript N represent the magnesium composition, e.g., Gr_9 is a garnet with 9 mol % pyrope.

Introduction

Cordierite–garnet and biotite–garnet pairs are the most effective thermometers for estimating the equilibrium temperatures in metapelites. Both of these thermometers are based on the exchange equilibria with relatively large entropy change in the following reactions:

$$Cor_{Fe} + Gr_{Mg} = Gr_{Fe} + Cor_{Mg}, \quad (1)$$
$$Bi_{Fe} + Gr_{Mg} = Gr_{Fe} + Bi_{Mg}. \quad (2)$$

Reactions (1) and (2) were calibrated as mineralogical thermometers by Perchuk (1967, 1969, 1970a,b) and later, in the 1970s, were studied by a number of petrologists (Currie, 1971, 1974; Hensen and Green, 1973; Hensen, 1977; Goldman and Albee, 1977; Holdaway and Lee, 1977; Ferry and Spear, 1978; Perchuk, 1977; Perchuk *et al.* 1981). Generally (Perchuk, 1969, 1977), the temperature increase should transfer Mg from hydrous silicates (Cor and Bi) to anhydrous ones (Gr), with opposite for Fe. The negligible effect of Ca, Mn, and Fe^{3+} contents in garnet in comparison with the temperature effect on reaction (2) was demonstrated by Perchuk (1970). Predominance of the Mg $\rightleftharpoons$ Fe isomorphism in natural minerals lets us define the mole fraction as follows: X_{Mg}^{Gr} = Mg/(Mg + Fe + Mn). Introduction of Ca into this expression (together with Mn) gave worse empirical results. Recently Perchuk (1981) established the correction for the Mn $\rightleftharpoons$ Mg isomorphism in garnet: $T = (3650/\ln K_D^{(2)} + 2.57) + 252.25(X_{Mn}^{Gr} - 0.035)$, where X_{Mn} = Mn/(Mn + Fe + Mg).

Many of these theoretical results are consistent with those cited above. Only Currie (1971, 1974) found the reverse temperature dependence of the distribution coefficient for Fe and Mg in the Cor + Gr pair. The experimental study of the exchange reaction (2) has been carried out by Ferry and Spear (1978) for pure system Alm + Phl = Ann + Pyr. They found a good linear correlation between $1/T$ and $\ln K_D^{(2)}$.

Figure 1 shows existing experimental and theoretical calibration of $\ln K_D$ against $1/T$ for both equilibria under consideration. Despite discrepancies in the

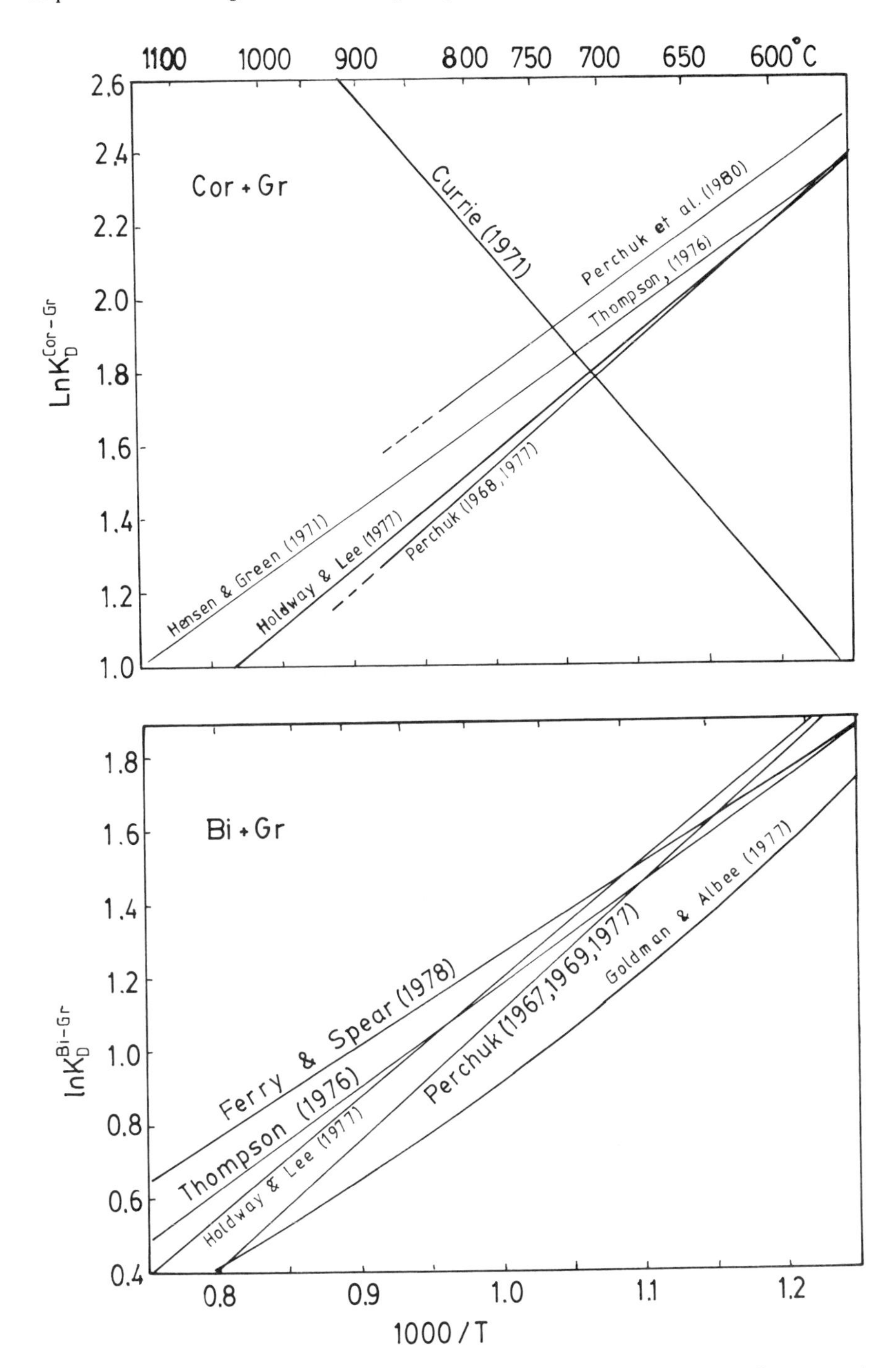

Fig. 1. Plot of ln K_D versus $10^3/T$ for exchange equilibria (1) and (2) according to empirical, theoretical, and experimental data.

correlation shown, all results indicate an ideal of distribution of Fe and Mg in the two systems. A minor deviation from the ideal was predicted (Perchuk, 1970a) for exchange reaction (2) at temperatures below 550°C. This has yet to be established experimentally. The temperature effect on equilibrium $Cor_{Mg} + Bi_{Fe} = Bi_{Mg} + Cor_{Fe}$ is negligible (Perchuk, 1969, 1970b). Lack of reliable experimental data for the system under discussion and complex kinetics of reactions (1) and (2) require that in this chapter we: (1) find a method of equilibration of minerals under *P–T* conditions of the runs; (2) develop an approach for estimating the equilibrium compositions of solid solutions coexisting in a run using the electron-probe analyzer; and finally (3) apply the thermometers to natural metapelites.

Experimental Methods

Procedures and Apparatus

Investigations were conducted in the temperature and pressure ranges of 550–1000°C and 5–7 kbar, respectively. Apparatus was changed according to the temperature of the experiment. Below 800°C Tuttle's vessels were used and above 800°C the gas apparatus was used. The runs were made using Eugster's double capsule technique with the NNO (Ni-NiO) and QFM (quartz–fayolite–magnetite) buffers. The accuracy of temperature was regulated within ±5° and pressure within 50 bar. After each run the material of external gold capsules was studied by optical and X-ray methods to determine the buffer composition. The material from the internal Pt capsules was thoroughly cleaned by a weak HCl solution.

Starting Materials

The helium (hel) preparations as well as mixtures of Fe, Mg, and Al oxalates with amorphous SiO_2 were used as starting materials in experiments at 600 and 650°C. At higher temperatures we used natural minerals mixed with NH_4Cl, Fe, and oxalic acid ($H_2C_2O_4 \cdot 2H_2O$).

The compositions of hel mixtures corresponded to the average chemical compositions of coexisting Bi + Cor + Gr from the medium and high-grade metamorphic rocks (see Table 1).

The oxalates + SiO_2 mixtures corresponded to Gr:Cor = 1:1 at the given X_{Mg} of the system. Reagents were mixed and powdered in the agate mortar and placed in the Pt capsules without any fluid components. In most experiments the natural Cor, Gr, and Bi were used as starting materials. The chemical compositions of some of the minerals and statistical data on homogeneity of the materials ($\bar{\Delta}$) in terms of N_{Mg} are given in Table 2. We used only minerals with a small $\bar{\Delta}$ value. Garnets K-22-5 and Φ-13, cordierite 393/86, biotites K-24, Φ-280, B-74, B-91,

Table 1. Chemical compositions of helium mixtures used in the recrystallization runs with Cor, Gr, and Bi.

	A			*B*		
Oxide	Cor	Gr	Bi	Cor	Gr	Bi
SiO_2	49.47	37.99	35.47	49.92	37.27	35.62
TiO_2	0.01	0.10	2.15	0.12	0.08	3.23
Al_2O_3	33.31	20.74	19.42	32.67	20.74	17.98
Fe_2O_3	0.25	1.13	1.48	0.74	3.11	2.07
FeO	7.57	32.07	19.60	7.07	29.47	17.21
MgO	8.76	5.47	10.77	8.87	6.77	10.29
MnO	0.06	1.17	0.17	0.05	1.24	0.06
CaO	0.14	1.31	0.24	0.08	1.24	0.44
Na_2O	0.34	0.02	0.53	0.37	0.07	0.36
K_2O	0.09	—	8.62	0.10	0.00	8.70
Total	100.00	100.00	97.45	99.99	99.99	95.96

A: the medium-grade mineral assemblages.

B: the high-grade assemblages.

Φ-10, and M 2002/4 were included in this group of starting material. Their chemical analyses will be given in the next part of this chapter.

Table 3 shows the compositions of fluid components used in experiments with Bi + Cor + Gr. In choosing these components we took into account the high solubility of Bi in the chloride fluids, which buffers the chemical potentials of K_2O, H_2O, FeO, and MgO during the run. Only in this way were we able to obtain relatively large grains of Cor and Gr in each run. Thus, Bi was a participant in the exchange reaction, as well as a mineral buffer.

The choice of the compositions of the starting materials depended on the aim of each run at the given temperature. Exchange reactions (1) and (2) do not depend upon activities of any additional components in the fluid. We thought that Cl could penetrate into the structural channels of cordierite with K_2O and H_2O. We could not, however, detect a remarkable concentration of Cl in Cor with a microprobe analyzer.

At temperatures of 600 and 700°C the hel + oxalic acid mixture runs showed a very sluggish crystallization of the minerals. Moreover, the ionic exchange between the minerals did not take place. This prompted the development of a method of recrystallization of natural minerals in the complex fluid mixtures. Our experience showed that recrystallization of Cor and Gr without Bi was very sluggish, with formation of very small grains of these minerals. Addition of a great amount of Bi facilitated the recrystallization of fine-grained starting material to large crystals varying in size from 75 to 300 μ. In many runs we added some amount of pure Fe to starting mixture in order to keep the bulk composition of the charge in relatively Fe-rich field. During the run the Fe dissolved in the fluid completely.

Table 2. Chemical analyses of Gr, Cor, and Bi used in the runs as starting materials.

Oxide	Gr				Cor				Bi	
	F-146	F-7-6	B-640	AO-66-5	Fe-3	64417	1284-1	M5278	F-335	B-91
SiO_2	39.03	37.16	36.88	40.50	45.63	45.93	48.64	49.04	35.00	36.40
TiO_2	0.07	0.390	—	—	—	—	—	—	1.86	1.23
Al_2O_3	20.04	21.97	20.33	22.54	30.52	30.76	34.77	33.45	19.70	22.83
Fe_2O_3	2.33	0.39	2.43	5.07	—	—	—	0.95	1.91	—
FeO	26.13	31.45	30.78	10.45	18.42	18.27	9.81	3.30	16.81	11.55
MgO	8.42	2.18	4.90	16.40	0.57	0.43	7.06	9.67	10.33	13.78
MnO	0.32	2.56	2.79	0.38	1.05	1.28	0.08	—	0.07	0.12
CaO	2.96	4.09	2.00	3.51	—	—	—	0.92	0.52	0.30
Na_2O	—	0.32	—	0.02	1.08	—	0.35	—	0.33	0.43
K_2O	—	0.02	—	0.34	—	—	—	—	8.21	8.58
H_2O	—	—	—	0.03	1.95	—	0.87	0.91	3.54	4.62
Total	99.30	100.53	100.11	99.24	99.24	96.67	101.58	98.24	98.28	99.84
N_{Mg}^{Cha}	34.7	10.9	20.9	66.1	5.2	4.0	56.2	80.6	49.9	68.0
N_{Mg}^{MP}	29	11.3	11.0	66.2	6.6	5.1	58.3	79.6	49.1	62.7
$\overline{\Delta}(\pm)$	5.7	2.07	—	0.97	1.4	1.1	0.58	0.17	2.02	0.3

Cha: Wet chemical analysis.

MP: Electron-probe analysis.

Table 3. The compositions of starting crystallizing mixtures ("fluids") in the runs on recrystallization of coexisting Cor, Gr, and Bi.

	Composition (mg)			
Index	$FeCl_2 \cdot 4H_2O$	NH_4Cl	Fe	$H_2C_2O_4 \cdot 2H_2O$
А	—	5	8	—
Б	—	5	2	—
В	—	5	5	—
Г	—	22	20	—
Д	—	5	5	—
Е	—	5	5	1
Ж	5	—	2	—
З	10	—	—	—
И	—	10	10	1
К	15	—	—	—
Л	—	—	—	7

Microprobe Standards

The results of this chapter are based upon data of analyses of almost 500 runs. These analyses were made using the microprobes Cameca MS-46, Camebax, JXA-5A, and JXA-50A.

In some cases we have applied the Bence–Albee or Afonin (Afonin *et al.*, 1971) programs, which incorporate nonlinearity of the characteristic X-ray radiation as function of the chemical composition of silicates. Afonin's program gives better results, but it is very complicated to use.

To accelerate probe analyses of natural and synthetic Cor, Gr, and Bi, we used these minerals as standards. Using the microprobe we determined inhomogeneity of Gr and chose the best wet chemical analysis of the most homogeneous garnets, as standard. Probe analyses of a dozen randomly chosen garnet grains had entirely the same composition as that obtained by the wet chemistry.

To find the best correlation between the Mg and Fe concentrations with the characteristic X-ray radiation of the elements in garnets, we chose 10 garnets with known chemical analyses recalculated to 100 wt % after excluding Na, K, Cr, etc.[1]

From such data we found the following linear equations:

$$C_{Fe}^{Gr}\ \text{wt}\ \% = 0.529 + 0.04163(I_{Fe}); \quad r = 0.994, \qquad (3)$$
$$\bar{\Delta} = \pm 0.34; \delta = \pm 0.42,$$
$$C_{Mg}^{Gr}\ \text{wt}\ \% = 0.2025 + 0.01621(I_{Mg}); \quad r = 0.9996, \qquad (4)$$
$$\bar{\Delta} = \pm 0.16; \quad \delta = \pm 0.22,$$

where C_i are weight concentrations.

[1]Tables of standards may be obtained from the authors.

All the samples were analyzed for six elements, Fe, Mg, Ca, Mn, Al, and Si, using the probe at 15 keV. The garnet standards have been strictly chosen in order to reflect the wide range of their compositions.

Equations (3) and (4) cannot be used to calculate the Fe and Mg contents in garnets because the equations are given in terms of weight percents (C_{Fe} and C_{Mg} depend upon many factors, the probe analyzer setting in particular). Thus, using Eqs. (3) and (4) we found new formulas in the form of relative intensities i_{Fe} and i_{Mg}:

$$i_{Fe}^{Gr} = I_{Fe}^{sample} / I_{Fe}^{K\text{-}22\text{-}5}, \quad (5)$$

$$i_{Mg}^{Gr} = I_{Mg}^{sample} / I_{Mg}^{A0\text{-}139}, \quad (6)$$

where $I_{Fe}^{K\text{-}22\text{-}5}$ and $I_{Mg}^{A0\text{-}139}$ are parameters for the most Fe-rich (K-22-5) and Mg-rich (AO-139) homogeneous garnets, respectively. Correlations of these parameters can be expressed as follows:

$$C_{Fe}^{Gr}\text{ wt \%} = 0.48 + 36.116(i_{Fe}^{Gr}); \quad r = 0.99931, \quad (7)$$
$$\bar{\Delta} = \pm 0.36; \quad \delta = \pm 0.44 \text{ (wt \%)},$$
$$C_{Mg}^{Gr}\text{ wt \%} = 0.25 + 20.234(i_{Mg}^{Gr}); \quad r = 0.9996, \quad (8)$$
$$\bar{\Delta} = \pm 0.19; \quad \delta = \pm 0.25 \text{ (wt \%)}.$$

The error in N_{Mg} could be ± 0.5 mol %. Besides, the values N_{Mg}^{Gr} calculated using the Bence–Albee program as well as that of Alfonin appear to have a small deviation from the linear regression. The value

$$\Delta N_{Mg} = N_{Mg}^{Bence} - N_{Mg}^{Cha}$$

(Cha denotes wet chemical analysis) changes regularly with N_{Mg}^{Gr} obtained from wet analyses. Fortunately the deviations are negligible and we could use a formula based on (7) and (8) for estimations of N_{Mg}^{Gr} from the microprobe intensities.

All cordierites used as standards are characterized by high degree of homogeneity ($\Delta N_{Mg} = \pm 0.25$ mol %). On the basis of the analytical data and one zero point, the following expressions were deduced:

$$C_{Fe}^{Cor} = 0.18 + 18.28(i_{Fe}^{Cor}); \quad r = 0.999, \quad (9)$$
$$\bar{\Delta} = \pm 0.17; \quad \delta = \pm 0.24,$$
$$C_{Mg}^{Cor} = 0.06 + 9.91(i_{Mg}^{Cor}); \quad r = 0.9995, \quad (10)$$
$$\bar{\Delta} = \pm 0.14, \quad \delta = \pm 0.18,$$

where

$$i_{Fe}^{Cor} = I_{Fe}^{sample} / I_{Fe}^{\text{“Fe-2”}}, \quad i_{Mg}^{Cor} = I_{Mg}^{sample} / I_{Mg}^{\text{“z”}}; \quad (11)$$

N_{Mg}^{Cor} was calculated using formula based on (9) and (10). $\Delta N_{Mg} = N_{Mg}^{Pa} - N_{Mg}^{Cha}$, with $\bar{\Delta} = \pm 0.37$ mol % (where superscript Pa denotes probe analyses) shows that the method used is sufficiently accurate. Thus, (9) and (10) were employed for estimating the composition of synthetic and natural cordierites with two standards.

Homogeneity of biotite standards have been checked with a number of measurements of several grains of Bi in each sample. We tested only concentrations of Fe and Mg in the biotites. The variations of i_{Fe}^{Bi} and i_{Mg}^{Bi} are ± 1.9 and 1%. The values of relative intensities are given by the equations:

$$i_{Fe}^{Bi} = I_{Fe}^{sample}/I_{Fe}^{K\text{-}24} \text{ and } i_{Mg}^{Bi} = I_{Mg}^{sample}/I_{Mg}^{M\text{-}2002/4}, \tag{12}$$

where $I_{Fe}^{(K\text{-}24)}$ and $I_{Mg}^{(M\text{-}2002/4)}$ are absolute intensities of Fe and Mg in two samples of biotites.

The following equations were also derived:

$$\begin{aligned} C_{Fe}^{Bi} &= 0.09 + 19.333(i_{Fe}^{Bi}); \quad r = 0.999, \\ \overline{\Delta} &= \pm 0.4; \quad \delta = \pm 0.5 \text{ wt \%}. \\ C_{Mg}^{Bi} &= 0.29 + 10.775\,(i_{Mg}^{Bi}); \quad r = 0.996, \\ \overline{\Delta} &= \pm 0.52; \quad \delta = \pm 0.8 \text{ wt \%}. \end{aligned} \tag{13}$$

The mole fraction of the Mg component in biotite has been calculated from the equation based on (13). The variations of N_{Mg}^{Bi} within the whole range of compositions correspond to $\overline{\Delta} = 0.43$ mol %. This is a fairly accurate reflection of the high degree of homogeneity of the biotite standards. As mentioned above, during the measurements with a microprobe we did not use all standards. For each mineral we used only two standards and one zero point. Then we corrected the calculated values, taking into account the N_{Mg} deviations of these samples from equations deduced above (see Eqs. (7), (8), (10), (11), (13)).

Method of Estimation of the Equilibrium Mineral Compositions

This problem was complicated by the small diffusion coefficients of Fe and Mg in all the three minerals. We developed a special method for estimating the equilibrium compositions of inhomogeneous minerals after the runs.

As was already mentioned, most runs were conducted with natural starting minerals. The starting compositions were chosen to approach equilibrium from the opposite sides of an isotherm. In some cases we used Fe-rich Gr and the Mg-rich Cor. In particular, both new grains and rims around the starting seeds were grown during the run. In such a case their compositions shifted in opposite directions to reach an identical N_{Mg}. Thus, our purpose was to analyze a great number of grains in the run products and to make several measurements in each of them. As a rule, we have studied 15–25 grains of minerals, determining in total from 50 to 150 compositions. Large grains (80–300 μ) were studied thoroughly in order to obtain their chemical topology map, enabling us to find the most "shifted" compositions (from starting N_{Mg}) of different minerals considered to have reached equilibrium. After several hours the Fe-rich garnet, almost pure Alm, was the first to form in the runs with $N_{Mg}^{Gr} \gg 20$. Then more Mg-rich Gr rims around this almandine core were grown. If the rims of the same compositions overgrew the starting or more Mg-rich garnets, we accepted their N_{Mg}^{Gr} as the equilibrium value. The same approach was adopted for biotite and cordierite.

In the course of recrystallization, Ca and Mn from Gr dissolved into the fluid, leaving the newly formed garnet grains poor in these components. That was a good indicator of a mineral growth.

The ideal distribution of Mg and Fe in each pair of minerals (Perchuk, 1967; 1970a,b, 1977; Thompson, 1976; Holdaway and Lee, 1977, etc.) enabled us to find a series of locally equilibrated compositions by applying the "rim growth" approach to each run product. Using the available probe measurements, we could construct histograms with several compositional spectra. These spectra probably reflect kinetics of the system at constant P–T during its evolution (dissolution of biotite and metallic Fe in the fluid, partial dissolution of Cor and Gr, etc.). Ostvald's step rule allows estimation of compositional ranges of coexisting minerals by the method of projective correspondence (Korjinskii, 1959). Knowing only two equilibrium compositions of Cor and Gr (or Bi and Gr) and finding the center of projection, we defined the limits of equilibrium compositions of two phases. Other parts of compositional spectra were considered as metastable. We used lots of measurements of Fe, Mg and Ca, and Mn contents in minerals (the latter were measured only for garnets) to construct the 4 mol the % step histograms and the projective correspondence diagrams for each run. Examples of these will be considered in next section.

Experimental Results

Cordierite–Garnet Exchange Equilibrium

This equilibrium was studied in the temperature range of 600–1000°C and at pressure of 6 kbar. In several runs, pressures were varied within 5–7 kbar. Because buffers were used, the variation in oxygen fugacity does not affect K_D and the stability fields of minerals involved. To achieve equilibrium starting from Mg-rich cordierite and Fe-rich garnet, five runs were made at 575°C and 5.5 kbar. After 10 days in three runs, we noted no significant changes (composition of Bi was slightly changed, Run 076). Essential changes occurred in Runs 078 and 079. Here Cor disappeared completely. Fe-rich Gr (N_{Mg}^{Gr} = 4.6–11.8 mol %) and small amount of Chl (N_{Mg}^{Chl} = 90) appeared; Bi also changed its composition to N_{Mg}^{Bi} = 33, although in the 078 run products, one grain with N_{Mg}^{Bi} = 22.4 was found. But in the remaining run products Bi predominated. All of these results were obtained with the following "fluid": NH_4Cl—5 mg, Fe—5 mg, oxalic acid—1 mg.

Judging from the results of the runs, it was not possible to obtain reliable data for Cor–Gr exchange equilibrium at 575°C, but it was possible for the Bi–Gr pair. Five runs were conducted at a temperature of 600°C and a pressure of 6 kbar over the wide range of starting mineral compositions (Gr_9–Gr_{35} and Cor_7–

Cor_{91}). In two runs Cor disappeared. The high Fe content in Run 081 led to the decomposition of Cor to Gr + Qz + Sil (?). Differences in N_{Mg} of starting minerals of Run 082 were too large ($N_{Mg}^{Cor} \ll N_{Mg}^{Gr}$) to shift reaction (1) to the right side: it is known that the reaction is always shifted to the left at different T and P. After the run, we estimated the following compositions: $Cor_{30.7}$ + $Gr_{3.3}$. Although the compositions of Gr_{4-5} are the most widespread, the probe did not detect any rims around $Gr_{3.3}$. In Run 083 the starting material had $N_{Mg}^{Cor} > N_{Mg}^{Gr}$. After the run, Cor changed its composition to $N_{Mg} = 85.7$; Gr initially shifted to almandine and then the Fe-rich crystals developed Mg-rich rims. So, in one grain of Gr we found zoning from $Gr_{15.3}$ in the core to Gr_{18-19} in the rims (starting composition: Gr_{28}). Meanwhile, we detected an idiomorphic grain of Gr_{28} in the core and $Gr_{33.9}$ in the rims with intermediate compositions in the middle portion of the grain. Thus, association $Gr_{33.9}$ + $Cor_{85.7}$ was considered to be at equilibrium. In Run 084 the starting compositions of Cor shifted slightly to the Fe-rich end, the average N_{Mg}^{Cor} being about 55. This composition, however, in two cases developed rims with $N_{Mg}^{Gr} = 43$–49. At the same time, starting Gr_{11} recrystallized to Gr_{4-7}; then Gr_{10-11} appeared again in the rims. So, in grain No. 13 from this run product, rims of $Gr_{9.7-11}$ surrounded the core of Gr_6; in grain No. 27 the core was $Gr_{12.3-12.6}$ and rims were in the range $Gr_{10.3-10.8}$. So, we considered Cor_{43} + Gr_6 as well as Cor_{55} + $Gr_{10.5}$ to be in equilibrium, remembering that the Cor compositions were initiated at the Mg-rich end. Obviously, Cor_{48} corresponds to the stability limit of Cor at the T and P.

Summarizing the results of runs at 600°C, we conclude that the equilibrium compositions of recrystallized minerals are estimated with great difficulty. Only for Gr could we obtain the composition by the "opposite side method" mentioned earlier. The composition of Bi seems to influence the Gr composition, because the paragenesis $Gr_{5-5.6}$ + Bi_{30} is the most stable. In general, the compositional spectra of all the three minerals were present in the run products. The intervals $N_{Mg}^{Gr} = 3.3$–33.9 and $N_{Mg}^{Cor} = 30.7$–85.7 are too wide to characterize the distribution of Fe and Mg between Cor and Gr at the chosen P and T. Part of these spectra seems to refer to metastable equilibria. The most realistic estimates of equilibrium compositions were done for Runs 082 and 083. Using these data we constructed Fig. 2. Using results of four successful runs, we calculated the mean value of $\ln K_D^{Cor-Gr} = 2.417 \pm 0.117$ and that of $\ln K_D^{Bi-Gr} = 1.621 \pm 0.02$.

Hel mixtures, oxalates, and the recrystallization method were used to synthesize Cor + Gr at 650°C and 6 kbar. In ten runs, synthesis of the minerals was done either at the NNO or the QFM buffer. Run duration ranged from 24 to 67 days. The results are listed in Table 4. We failed to control compositions of minerals in equilibrium using the mean mole fraction of Mg in the mixture and in Gr as the most stable phase in the system and, therefore, the crystallization was not complete in the runs. However, in runs $I/1$ and $I/3$ we could observe crystals with ~75 μ in diameter. This size is sufficient enough for measuring the compositions in several points.

In Table 4 two pairs of values of compositions of coexisting minerals are listed for each run, for which two peaks (maxima) in histograms appeared.

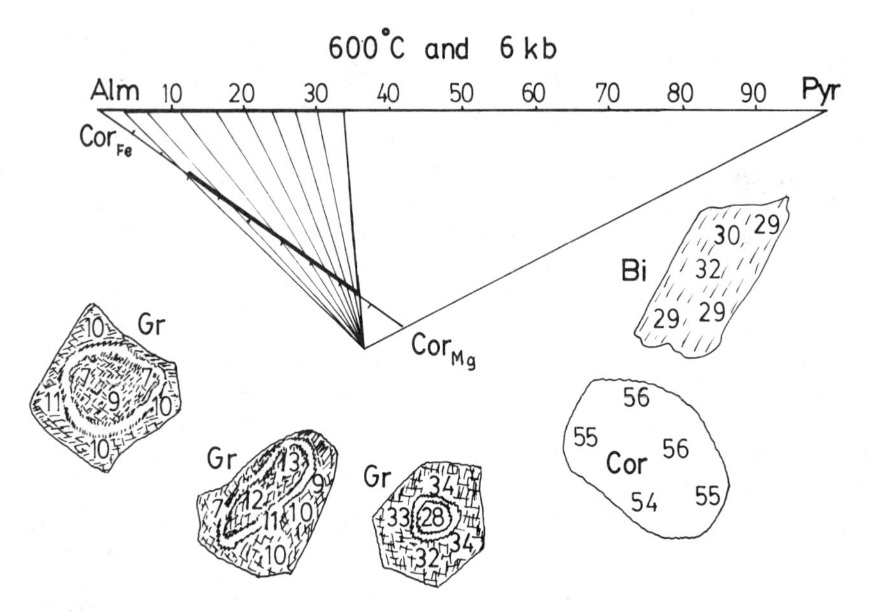

Fig. 2. Zoning developed in some run products at 600 °C and 6 kbars. The diagram illustrates the projective correspondence of Cor and Gr compositions stabilized at given temperature and pressure.

The run duration over the range of 10 to 67 days did not influence compositions of synthesized minerals. ln K_D also does not depend on the run duration. In the composition intervals of N_{Mg}^{Gr} = 4–21 and N_{Mg}^{Cor} = 23–75, the mean ln K_D is 2.25.

In Run 1/1, Cor_{94-87} formed rims with N_{Mg}^{Cor} = 66.3. Rims of Gr_{29-12} developed around starting Gr_{64}; and rims of Gr_{17-19} grew around Gr_{10}. In these cases ln K_D is 2.19, close to the mean value mentioned above.

Table 5 and Fig. 3 show results of the experiments on recrystallization of natural minerals during 7 to 15 days at 650 °C. In Run 032 starting garnet with N_{Mg}^{Gr} = 65 formed the rims of Gr_{29-30} and the latter in its turn was surrounded by Gr_{33}. That composition (Gr_{33}) was common among the run products. Starting Cor changed in composition to Cor_{82-83} and biotite to Bi_{61-55}. Intergrowths of Bi_{61} + Gr_{29} and Bi_{65} + Gr_{33} were found in the run products. In both cases ln K_D^{Gr-Bi} are very similar (i.e., 1.342 and 1.327).

In Run 033 the composition of starting Cor shifted to Cor_{95} and Gr to Gr_{69}, varying within narrow limits (see Table 5). The compositions were determined on adjacent rims of grains: Cor_{94} + Gr_{62}. Biotite retained its primary composition.

For Run 13 P and T were 6 kbar and 650 °C, respectively. The duration was 7 days. The products were analyzed using a microprobe. Part of the run products with the identical fluid composition was again sealed in the capsule and kept under the same conditions for 8 more days. We did not find any changes in the mineral compositions. In the repeated Run 13 (13/1) inhomogeneity in minerals was slightly less and $Gr_{17.5}$ appeared to be in equilibrium with $Bi_{45.5}$ (ln K_D = 1.37).

All runs were conducted with the variable mineral/fluid ratios. But the weight

Table 4. Experimental data on crystallization of Cor and Gr in the helium (Nos. 1–9) and oxide mixtures (Nos. 10–21) in oxalic acid at 650°C and 6 kbar, buffers QFM.

					ΔN_{Mg}[a]		N_{Mg}		
No.	Run	Days	$\overline{N}_{Mg}$	f_{O_2}	Cor	Gr	Cor	Gr	$\ln K_D$
1	16/1	7	56	MW	66–52	17–12	62	14	2.305
2	16/2	7	42	MW	40–69	7–19	58	12	2.315
3	23	7	0.5	MW	60–45	9–14	53	11	2.211
4	GKS	10	0.3	FMI	62–46	13–20.8	52	10	2.277
5	24/2	7	0.8	MI	46–77	8.5–20.6	46	8.5	2.215
6	24/2	7	0.8	MI	46–77	8.5–20.6	61	14	2.262
7	24/1	7	0.5	MI	65–79	6–7	65	17	2.205
8	25/2	4	70	M	71–52	6–10	61	13	2.348
9	26/1	4	47	M	75–60	21–14	68	17	2.339
1	2a	1	47	Mt	40–80	7–13	55	12	2.193
2	2	2.5	47	QFM	47–77	7–23	60	16	2.064
3	2	2.5	47	QFM	47–77	7–23	49	8.5	2.336
4	2d	3	47	Mt	51–83	9–19	61	12.5	2.393
5	3	4	10	QFM	17–23	4–11	23	4	1.970
6	2b	4	47	Mt	41–81	9–19	59	13	2.264
7	2b	4	47	Mt	41–81	9–19	70	19	2.297
8	2b	8	47	Mt	7–23	35–73	71	21	2.220
9	2b	8	47	Mt	7–23	35–73	63	12	2.525
10	28/3	8	80	NNO	75–91	7–15	75	16	2.757
11	1/3	60	30	QFM	5–53	7–17	53	11	2.211
12	1/1	67	78	QFM	63–91	7–21	68	17	2.339

[a]The compositional intervals.

Table 5. The run results on recrystallization of Cor, Gr, and Bi in fluids at 650°C and 6 kbar, buffer NNO.

		N_{Mg} starting		N_{Mg} after run		N_{Mg} equilibrium		
Run	Fluid[a]	Cor	Gr	Cor	Gr	Cor	Gr	$\ln K_D$
028	А	92	28	81–94	17–57.3	92.6	57.3	2.232
032	Д	64	66	78–84	65–19	82	33	2.225
033	Б	64	66	81–95	65–69	93.7	61.6	2.226
034	А	64	66	77–86	66–18	81	31	2.250
13	Г	61	28	65–79	35–15	65	17	2.204
13/1	Г	65	17	63–73	31–15	66	17.5	2.214

[a]See Table 3.

ratios of Bi:Cor:Gr were always 7:5:5. Compositions of the fluids are given in Table 3. Only with these fluids did the essential recrystallization take place and the paragenesis Cor + Gr with similar values of K_D remain stable in spite of the changing bulk composition of system. These values of K_D are close to those obtained in successful runs on synthesis of Cor + Gr from both the hel mixtures and the oxides (see Table 4).

Thus, the K_D values for the Cor + Gr pair at 650°C were obtained by two methods (i.e., synthesis from mixture and recrystallization by the "opposite-side approach") (see Fig. 4, 650°C).

Despite the limited experimental results, the statistics appear to be satisfactory. According to data from Tables 4 and 5, the ln K_D values are statistically meaningful: 2.293 ($\delta = 4.2 \times 10^{-2}$) and 2.225 ($\delta = 2.47 \times 10^{-4}$) for syntheses and recrystallizations, respectively. Uncertainty of the latter value is twice lower than

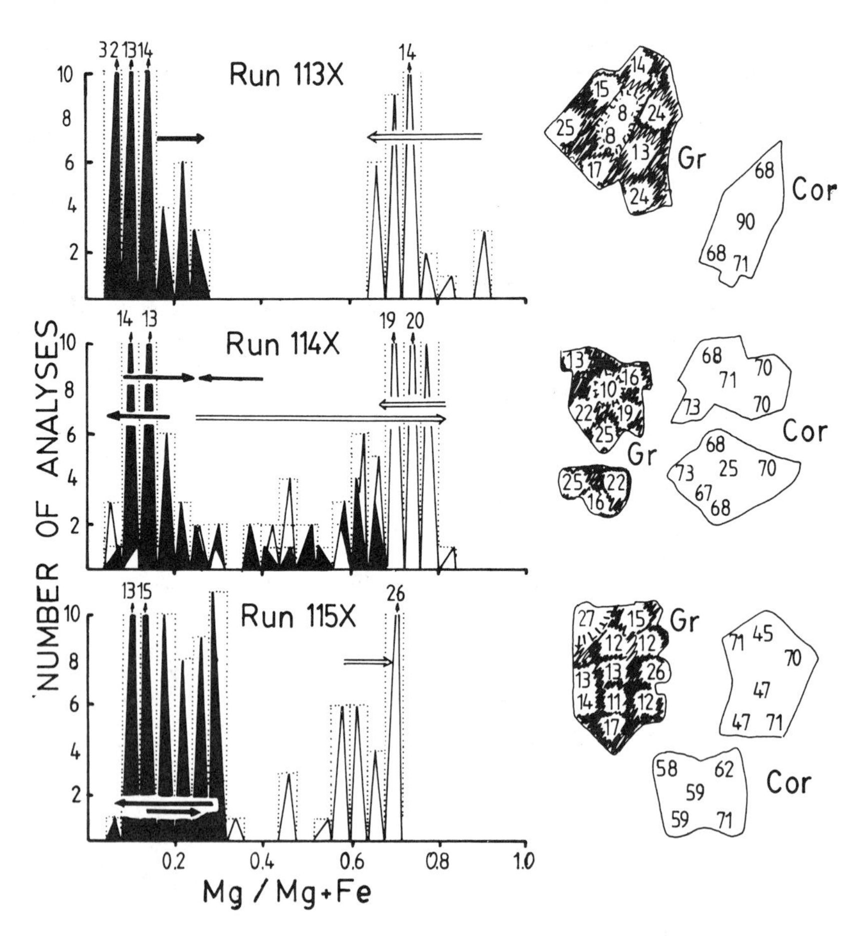

Fig. 3. The X_{Mg} histograms for Cor and Gr from the run products. Arrows show the shift of reaction (1) during the recrystallization at given T and P.

that of the former, suggesting that the recrystallization method is better for the experiments at the chosen P and T.

Table 6 and Fig. 5 show the results of our experiments at 700°C. We present a brief outline of the typical results, dividing them into two groups.

The first group is characterized by sharp differences in N_{Mg} of starting Cor and Gr ($N_{Mg}^{Cor} \gg N_{Mg}^{Gr}$) to bring their composition together during the run. Wide compositional spectra can be observed among run products. Newly formed garnets, however, are always poor in Ca and Mn. This helped to distinguish between the seeds of Gr and recrystallized grains. Run 046 was a typical example of the group: equilibrium compositions (Cor_{68} + Gr_{23}) were identified as those that came the closest in the course of recrystallization starting from Gr_9 and Cor_{92}.

The second group reflects the experimental results with $N_{Mg}^{Gr} \gg N_{Mg}^{Cor}$ in starting materials. This unusual N_{Mg} relation caused sharp shifting of compositions of the minerals. Run 049 is one of the most typical. Large Cor and Gr grains of 300 μ in diameter formed during the run but Bi was completely decomposed. Gr_{63-68} developed very Fe-rich rims (up to Gr_{10}). However, these compositions in turn were surrounded by more Mg-rich garnet rims with maximum $N_{Mg} = 26$. This was accepted as equilibrium composition due to its abundance in the rims of the grains with cores of $N_{Mg} = 63$–68 and also of $26 > N_{Mg} > 10$. Cor_{73} is in equilibrium with Gr_{26} because the Fe-rich cordierite cores formed rims of this composition. In one grain we detected a very complicated compositional growth represented by Cor_{57} in the core and Cor_{62-73} in the rim.

Wide compositional spectra of Cor and Gr at estimated $K_D = 7.7$ in Run 049 allowed us to determine the metastable equilibrium, using a method of projections (see Table 6). By this method we established that the metastable equilibrium compositions were within $N_{Mg}^{Gr} = 15$–35 and $N_{Mg}^{Cor} = 56$–80, but one assemblage Gr_{26} + Cor_{73} represented stable equilibrium.

Thus, at 700°C the equilibrium relations were obtained by the "opposite-side method" in the following compositional ranges: $10 < N_{Mg}^{Gr} < 80$ and $45 < N_{Mg}^{Cor} < 97$ (see Fig. 5, 700°C). The mean ln K_D value is 2.018 with $\delta = 3.57 \times 10^{-3}$ for 43 determinations in 27 runs. A very similar value results if just 27 equilibrium pairs of Cor + Gr in 27 runs are used ($\bar{x} = 2.011$ and $\delta = 4.45 \times 10^{-3}$). Coincidence of these two results shows efficiency of the statistical method used for estimating the N_{Mg} ranges for a pair of coexisting minerals in each run.

At 750°C 15 runs of 10 days duration were carried out. Thirteen of them were successful. The oxygen fugacity was controlled by buffer Ni–NiO at the constant weight ratio Bi:Cor:Gr = 15:15:5 (mg). Products of Run 11X contained many large grains of Cor and Gr. Bi partly dissolved in fluid slightly changing its composition from Bi_{50-52} to $Bi_{48.6}$. Garnet predominated over cordierite with variable compositions from Cor_{56} to Cor_{20}. Equilibrium compositions of these minerals (Cor + Gr) were determined by the "opposite-side method" (see Table 7). Their grains were characterized by well-developed growth zoning. Starting Cor was overgrown by Cor_{80}. But this composition in turn was overgrown by more Fe-rich Cor_{69}. Starting Gr_{66} reached the composition Gr_{8-12} and then changed back to $N_{Mg}^{Gr} = 26$.

In Run 114X the large Cor crystals (up to 300 μ) predominated. There was a

Table 6. The run results on recrystallization of Cor, Gr, and Bi in fluids at 700°C and 6 kbar, buffer NNO.

		N_{Mg} starting		N_{Mg} after run		N_{Mg} equilibrium			N_{Mg}^{Bi}		
Run	Fluid	Cor	Gr	Cor	Gr	Cor	Gr*	ln K_D	Before run	After run	Other minerals present
1	2	3	4	5	6	7	8	9	10	11	12
041	A	92	28	77–91	19–32	77	29	2.103	36	—	Qz
044	B	5	66	83–15	65–23	75.5	29	2.021	63	64	Qz
045	A	5	66	77–71	65–26	77	31	2.008	63	57	Qz
046[a]	B	92	9	92–68	23–9	68	23	1.962	36	33	—
047	B	92	29	93–71	8–35	71	24	2.048	36	33	Qz
048	B	58	9	74–43	7–27	74	27	2.041	36	33	Qz
049	B	58	66	57–79	69–11	73	26	2.041	36	38	Qz
050	B	15	28	7–76	15–31	76	31	1.952	36	—	Qz
050	B	15	28	7–76	15–31	57	15	2.016	36	—	Qz
051	Д	92	9	95–82	7–39	82	39	1.964	63	61	Qz
052	Д	92	28	95–79	21–51	89	51	2.051	63	—	Qz
052	Д	92	28	95–79	21–51	79	35	1.944	63	62	Qz
053	Д	58	9	43–83	9–45	(43)[b]	(9)	2.032	63	60	Qz, Opx
053	Д	58	9	43–83	9–45	56.7	12	2.262	63	60	Qz, Opx
053	Д	58	9	43–83	9–45	85	45	1.935	63	60	Qz, Opx
054	Д	58	66	51.2–85	12–69	52	12	2.072	63	—	Qz, Opx
054	Д	58	66	51.2–85	12–69	80	33	2.094	63	62	Qz, Opx
055	Д	5	28	5–84	17–41	84	41	2.022	63	64	Qz
056	E	30	30	30–86	10.5–33	57	15	2.017	36	34.4	Qz

056	Е	30	30	30–86	10.5–33	79.3	33.1	2.047	36	34.4	Qz
057	Е	80	66	66–84	68–17	74	26.5	2.066	36	36	Qz
057	Е	80	66	66–84	68–17	78	31	2.066	36	36	Qz
058	Е	80	9	65–85	5–33	85	44	1.976	36	—	Qz
059	Е	92	9	94–65	8–68	94	67.9	2.002	36	—	Qz
059	Е	92	9	94–65	8–68	77.2	30	2.066	36	53.7	Qz
060	Е	92	28	89–67	13–51	89	51.5	2.031	36	—	Qz
060	Е	92	28	89–67	13–51	80	34.5	2.027	36	63	Qz
060	Е	92	28	89–67	13–51	66.9	20.9	2.035	36	—	Qz
060	Е	92	28	89–67	13–51	77.2	32.2	2.010	36	—	Qz
060	Е	92	28	89–67	13–51	73.2	27	1.999	36	—	Qz
061	Ж	58	66	57–73	25–69	73.2	25.3	2.087	44	52	Qz
062	Ж	92	7	91–67	56.5–4	90.7	56.5	2.016	44	—	Qz
062	Ж	92	7	91–67	56.5–4	67.1	21.9	1.984	44	48	—
063	Ж	92	66	93–72	66–4	72.1	26	1.995	44	49	Qz
065	Ж	92	66	93–89	69–57	91	56	2.073	53	—	Qz
068	Ж	92	66	95–73	26–66	88(?)	50(?)	1.992	72	74–76	Qz
072	З	92	28	92–69	22–33	69	22	2.066	53	—	—
072	З	92	28	92–69	22–33	77.5	31	2.037	—	58	—
072	З	92	28	92–69	22–33	80	33	2.094	—	58	—
074	К	92	66	95–70	66–70.6	94.5	70.6	1.968	53	—	Qz
075	К	92	28	95–74	33–16	74	33	1.754	53	59	Qz
2/1	Л	92	11	6–16	91–48	48	11	2.011	—	—	—
2/1	Л	92	11	6–16	91–48	58	15.5	2.019	—	—	—

[a]In this and following runs the weight ratios of starting materials are Bi:Gr:Cor = 10:5:5; run duration is 10 days.

[b]The metastable composition at given T and P.

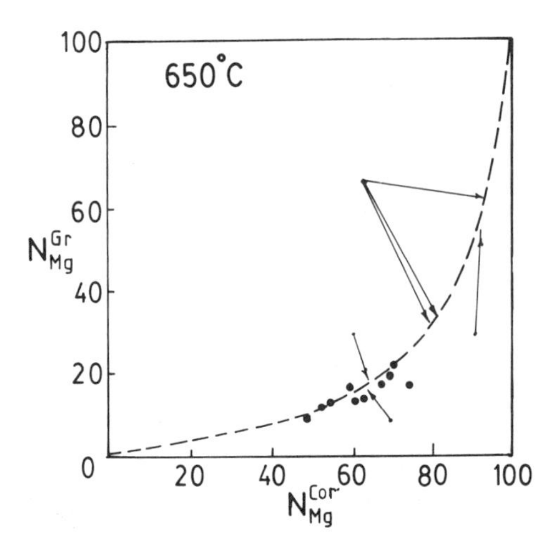

Fig. 4. The distribution of Mg and Fe between Cor and Gr at 650°C and 6 kbars. Arrows show the compositional reversal approach to equilibria by recrystallization of natural minerals in a complex fluid; points reflect the results of recrystallization in helium mixtures.

small amount of garnets with euhedral large grains of 150 μ. Very complicated compositional relations between Cor and Gr resulted from large N_{Mg} differences of starting seeds. Gr reached composition $Gr_{7.4}$ and then turned back to Gr_{25}. Fe-rich Cor converted to Cor_{85-70}, but in the Mg-rich field very complex relationships in compositions were observed. The Mg-rich rims of Cor around ferruginous cores predominated. But the opposite was observed only in two cases: Cor_{71} was overgrown by Cor_{68}. Besides, more Fe-rich cores were also surrounded by similar Cor_{68} rims. Thus, the equilibrium assemblage is $Cor_{68} + Gr_{25}$ with $\ln K_D = 1.852$.

We identified four more maxima in the statistical histogram:

$$Cor_{85.3} + Gr_{47.7}.\ \ln K_D = 1.850; \quad K_D = 6.36;$$
$$Cor_{74.3} + Gr_{30}.\ \ln K_D = 1.893; \quad K_D = 6.63;$$
$$Cor_{63} + Gr_{21}.\ \ln K_D = 1.857; \quad K_D = 6.40;$$
$$Cor_{57.9} + Gr_{18.2}.\ \ln K_D = 1.821; \quad K_D = 6.18.$$

These values characterize dynamics of development of the mineral equilibrium assemblage during Run 114*X* at 750°C according to Ostvald's rule.

Biotites were also analyzed in the run products ranging in composition from Bi_{36} to Bi_{47}.

In many cases estimation of mineral equilibrium at 750°C was made by the "opposite-side method," (i.e., starting from Mg-rich Cor + almandine assemblage) as well as from Fe-rich Cor + Gr_{66} (see Fig. 5, 750°C). In 13 run products more than 1500 analyses of Cor, Gr, and Bi were performed to obtain the mean value of $\ln K_D = 1.832$ ($n = 24, \delta = 1.49 \times 10^{-3}$) found over a wide composition range ($N_{Mg}^{Cor} < 50$).

Exhaustive information on equilibrium of Cor, Gr, and Bi at 800°C and NNO

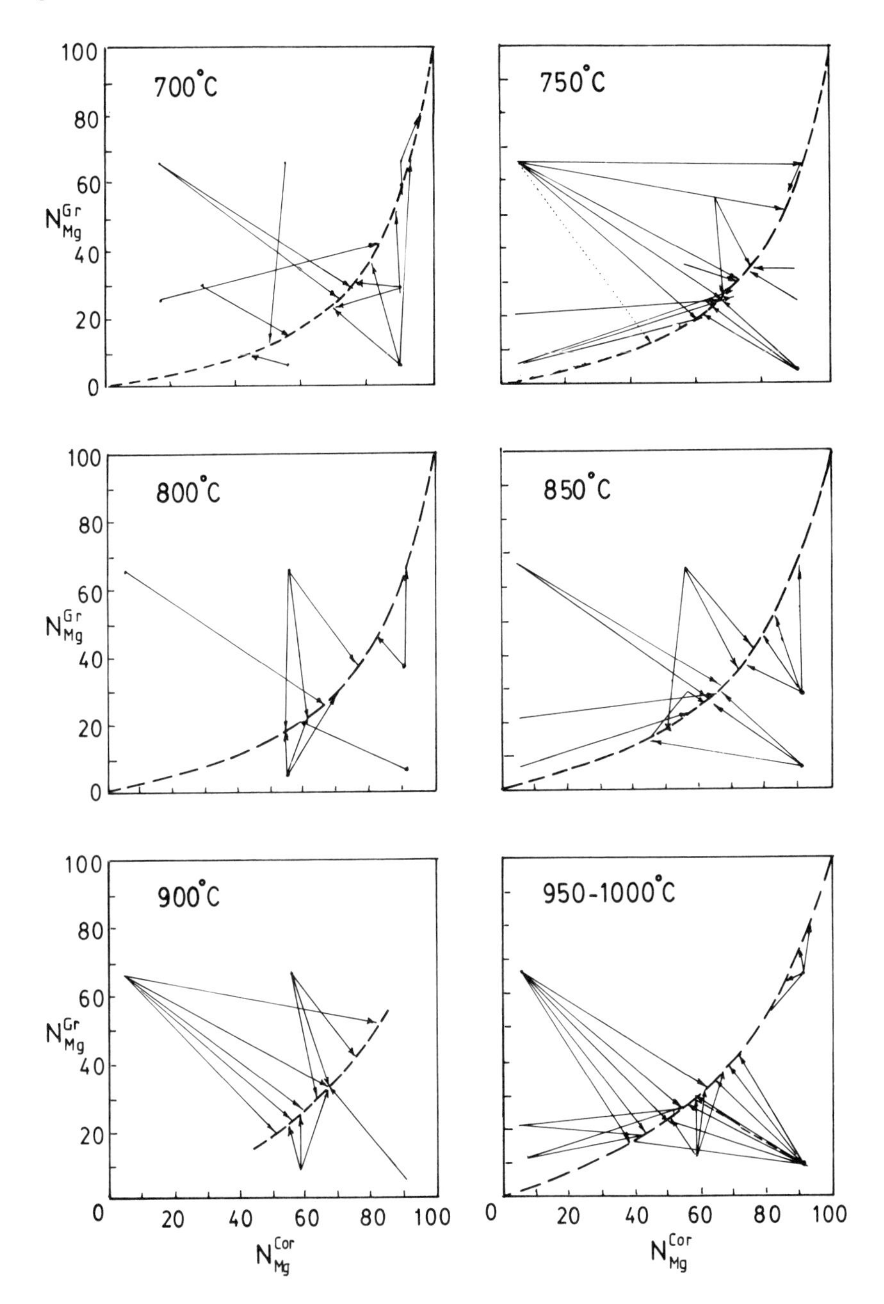

Fig. 5. Isotherms of distribution of Mg and Fe between Cor and Gr obtained by recrystallization of natural minerals in the complex carbonate–chloride fluids at pressure of 6 kbars with NNO buffer.

Table 7. The run results on recrystallization of Cor, Gr, and Bi in fluids at 750°C and 6 kbar, buffer NNO, duration 10 days.

		N_{Mg} starting		N_{Mg} after runs		N_{Mg} equilibrium			N_{Mg}^{Bi}		
Run	Fluid	Cor	Gr	Cor	Gr	Cor	Gr	ln K_D	Before run	After run	Other minerals present
111*X*	E	58	66	56–80	66–8	69	26	1.846	49	48.6	Qz, Opx
112*X*	E	92	29	92–78	15–41	82	41	1.880	94	94	Qz, Opx
113*X*	E	92	9	91–64	6–25	66	24.7	1.778	36	50	Qz
114*X*	E	7	66	5–85	66–7	68	25	1.852	36	42	Qz
115*X*	E	58	29	78–34	6–30	70.8	27	1.880	36	—	Qz
115*X*	E	58	29	78–34	6–30	73.8	30.3	1.868	36	—	Qz
116	E	7	9	5–75	1–30	66	24	1.816	36	46.6	Qz
116	E	7	9	5–75	1–30	74	31.6	1.818	36	—	Qz
117	E	58	21	35–82	11–43	82.4	42.7	1.838	36	66.1	Qz
118	E	92	66	95–82	66–58	90.9	62.4	1.794	94	94	Qz, Opx
119	E	5	21	5–81	6–25	68.6	25.2	1.869	36	48.6	Qz
120	E	92	29	92–65	11–53	70	27	1.842	49	50.6	Opx
120	E	92	29	92–65	11–53	73.2	30.8	1.814	49	54.6	Opx
120	E	92	29	92–65	11–53	78	34	1.929	49	—	Opx
131	E	92 and 7	9	2–79	5–25	68	25.5	1.825	36	—	Qz
131	E	92 and 7	9	2–79	5–25	58.6	18.3	1.843	36	—	Qz
132	E	92 and 7	66	93–64	68–18	61.6	20.5	1.828	49	41	Qz, Opx
132	E	92 and 7	66	93–64	68–18	65	24	1.770	49	—	Qz, Opx
132	E	92 and 7	66	93–64	68–18	74	32	1.799	49	—	Qz, Opx
132	E	92 and 7	66	93–64	68–18	79	38	1.814	49	—	Qz, Opx
132	E	92 and 7	66	93–64	68–18	81	40	1.855	49	—	Qz, Opx
132	E	92 and 7	66	93–64	68–18	88	54	1.832	49	—	Qz, Opx
132	E	92 and 7	66	93–64	68–18	92.4	65.9	1.839	49	—	Qz, Opx
133	И	92 and 7	9	87–67	6–26	66.9	25.8	1.760	36	—	Qz

buffer was obtained from the series of five runs (see Table 8). Equilibration was achieved after 5 days. Large crystals of the minerals (Cor in particular) were grown. Cordierite grains contained many Gr and Bi inclusions. The fluid compositions changed from run to run, but weight ratios of minerals were the same in all the cases, Bi:Cor:Gr = 15:5:5. Cor grains (300–350 μ) in Run 086 were often filled by Gr and Cor inclusions. In the matrix Gr the inclusions were small (10–20 μ) but idiomorphic. Equilibration during the run was clear from the changes in the mineral compositions. For instance, high Mg-rich cordierites (up to Cor_{81}) formed on seeds of Cor_{54-56} and then Cor_{60-61} rims developed around these Mg-rich Cor zones. We identified Cor_{61} as being in equilibrium because of its maximum in the histogram. Gr_6 shifted to Gr_{31}. However, because of lack of large grains of Gr it was impossible, as with Cor, to follow evolution of garnet compositions. Two garnets, $Gr_{22.2}$ and $Gr_{33.5}$, were found among small discrete grains. Equilibrium compositions of Cor and Gr were estimated from Gr intergrowths in large Cor grains. Here Gr_{22} was in contact with $Cor_{60.6}$, but Gr_{33} was included in Cor_{73}. Gr was supposed to change in composition together with Cor, first to Mg-rich field and then again to Fe-rich compositions reaching $N_{Mg}^{Gr} = 22.2$. Thus, Cor_{61} and $Gr_{22.2}$ were in equilibrium with ln K_D of 1.701. Similar value (ln K_D = 1.68) was calculated for pair $Cor_{73} + Gr_{33}$.

As mentioned above, these equilibria (at 800°C) were achieved by the "opposite-side method," illustrated in Fig. 5. According to data from Table 1 the mean value of ln K_D is 1.698 with $\delta = 3.03 \times 10^{-3}$. This accuracy resulted from the good statistics of measurements (study of about 150 grains of coexisting minerals covered by 700 measurements of four elements, Fe, Mg, Ca, and Mn, in Gr and

Table 8. The run results on recrystallization of Cor, Gr, and Bi in fluids at 800°C and 6 kbar, buffer NNO, duration 10 days.

	N_{Mg} starting		N_{Mg} after run		N_{Mg} equilibrium			N_{Mg}^{Bi}		Other minerals present
Run	Cor[a]	Gr[a]	Cor	Gr	Cor	Gr	ln K_D	Before run	After run	
086	58	9	81–51	7–33.5	61	22.2	1.701	36	52	Qz
086	58	9	81–51	7–33.5	73	33.5	1.680	36	—	—
087	7	66	21–66	66–8	66	26	1.709	36	43	Qz
087	7	66	21–66	66–8	62	22.4	1.732	36	40.4	Qz
087	7	66	21–66	66–8	54	18	1.676	36	33.1	Qz
088	92	9	92–58	6–20	58.2	20.3	1.698	36	36	Qz
089	92	28	94–79	26–67	83	47	1.705	94	92	Opx, Qz
089	92	28	94–79	26–67	91.5	66.9	1.673	94	92	Opx
090	58	66	36–77	66–14	76	38	1.698	49	50	Qz
090	58	66	36–77	66–14	61	22	1.712	49	—	Qz
090	58	66	36–77	66–14	55	18.4	1.690	49	—	Qz

[a]Weight ratios of the minerals in starting materials are Cor: Gr: Bi = 5:5:15.

Table 9. The run results on recrystallization of Cor, Gr, and Bi in fluids at 850°C and 6 kbar, buffer NNO, duration 5 days.

Run	Fluid	N_{Mg} starting		N_{Mg} after run		N_{Mg} equilibrium		ln K_D	N_{Mg}^{Bi}		Other minerals present
		Cor[a]	Gr[a]	Cor	Gr*	Cor	Gr*		Before run	After run	
101	E	58	29	54.8–71	13.8–33.2	70	33.2	1.546	36	51	Qz
101	E	58	29	54.8–71	13.8–33.2	58	23	1.531	36	36	
101	E	58	29	54.8–71	13.8–33.2	57	22.2	1.535	36	36	
101	E	58	29	54.8–71	13.8–33.2	54.8	20	1.578	36	—	
101	E	58	29	54.8–71	13.8–33.2	61.1	25.6	1.518	36	—	
102	E	5	66	37.7–67.7	67.3–14.1	62	26	1.536	36	—	Qz
102	E	5	66	37.7–67.7	67.3–14.1	67	30	1.555	36	47.2	
103	E	92	9	91–63.2	16.4–30	65.5	29.6	1.507	36		Qz
103	E	92	9	91–63.2	16.4–30	63.2	25.4	1.631	36	42	
104	E	92	29	90.5–78.8	39.7–66.2	90.5	66.2	1.582	94	—	Qz, Opx
104	E	92	29	90.5–78.8	39.7–66.2	83	51.5	1.525	94	—	
104	E	92	29	90.5–78.8	39.7–66.2	79.8	46.3	1.522	94	—	

105	E	58	66	77.2–66.7	66.8–23.1	77.2	41.9	1.547	49	59.7	Qz
105	E	58	66	77.2–66.7	66.8–23.1	73.1	35.1	1.614	49	—	
106	E	7	9	49.6–66.7	7.2–28.9	66.7	28.9	1.594	36	45.8	Qz
106	E	7	9	49.6–66.7	7.2–28.9	56	21.7	1.524	36	34.6	
107	E	58	11	54.6–72	9.9–33.2	72	33.2	1.643	49	50	Qz
109	E	7	11	17–65.3	9.8–27.5	65	27.5	1.588	36	—	Qz
110	E	92	29	90.7–69.3	15–36.7	73	36.7	1.540	49	—	Qz
110	E	92	29	90.7–69.3	15–36.7	69.6	32.5	1.559	49	—	Qz
111[b]	И	58	66	81.1–49.7	61.5–17.7	49.7	17.7	1.524	49	62.6	Qz, Opx
111[b]	И	58	66	81.1–49.7	61.5–17.7	81.1	46.5	1.596	49	—	Qz
112[b]	И	58	29	44.2–73	37–14.4	44.2	14.4	1.549	36	—	Qz, Sp, Ru
112[b]	И	58	29	44.2–73	37–14.4	52.2	19	1.538	36	—	Qz
112[b]	И	58	29	44.2–73	37–14.4	58.4	23.3	1.531	36	—	Qz
112[b]	И	58	29	44.2–73	37–14.4	65	27	1.613	36	—	Qz
112[b]	И	58	29	44.2–73	37–14.4	69.3	30.6	1.618	36	—	Qz
112[b]	И	58	29	44.2–73	37–14.4	73.0	37	1.527	36	—	Qz

[a]Weight ratios of the minerals in runs 101–110 are as follows: Bi:Gr:Cor = 15:5:5.

[b]In this run Bi:Gr:Cor = 5:2:2.

Fe and Mg in Cor and Bi). These results showed again that the very complicated compositional spectra for Fe–Mg aluminosilicates could be used to estimate their equilibrium compositions. Moreover, this information sheds light on evolution of a system during the run and allows us to understand the reason for the compositional inhomogeneity.

Runs at 850°C were conducted in gas vessels with QFM buffer. Eleven out of twelve runs gave good results summarized in Table 9. The following conclusions can be drawn. The most reliable data were obtained with the "opposite-side" method using high Mg-rich Cor and high Fe-rich Gr as starting materials. The best results were obtained from starting assemblage Gr_{28} + Cor_{56} independently of fluid compositions or weight ratios of starting materials (see Runs 101, 111, and 112). In all of these runs the equilibrium compositions were obtained with good statistics: 28 pairs of coexisting minerals (including metastable equilibria) with the mean value of $K_D = 4.757$ (i.e., $\ln K_D = 1.56$) and $\delta = 1.49 \times 10^{-3}$. The 850°C isotherm was determined by the "opposite-side" approach within a wide range of Mg/Fe ratio (see Fig. 5). About 600 determinations of the mineral compositions were involved in drawing this isotherm.

Five runs were conducted with QFM buffer at 900°C in a gas vessel. Duration of the runs was about 7 days. Mixtures with weight ratios Bi:Cor:Gr = 15:5:5 and fluid composition *E* (see Table 3) were used as starting material. In Run 124 Gr disappeared completely with starting mixture Cor_{91} + Gr_{29} + Bi_{94}: the Mg content in the system was too high for Gr to be stable at 6 kbar pressure and 900°C temperature. Instead of Gr, orthopyroxene, spinel, and high ferruginous Cor_{54} (see Table 10) formed. In Run 121 the phase relationships were very complex despite the high degree of crystallization of the run products. All three minerals formed with crystallographic shapes of grains with sizes up to 150 μ. Complex phase relationships resulted from composition of Cor. The starting N_{Mg}^{Cor} was very similar to the equilibrium value. This Cor recrystallized several times during one run. First, its composition changed to Cor_{45} and then seeds of this mineral were overgrown by more Mg-rich rims, up to Cor_{69}. Simultaneously, many starting Cor_{58} grains developed more Mg-rich rims. Starting Gr_9 changed in composition to a maximum of Gr_{34}. Biotite changed slightly N_{Mg} to become $Bi_{46.8}$ in equilibrium. The rim compositions of the three minerals studied have the following formulas:

$$\text{Cor: } Mg_{1.51}Fe_{0.67}Al_{4.06}Si_{4.75}O_{18},$$
$$\text{Gr: } Mg_{1.07}Fe_{2.08}Mn_{0.03}Ca_{0.01}Al_{1.93}Si_{2.92}O_{12},$$
$$\text{Bi: } K_{0.82}Fe_{1.12}Mg_{0.99}Ti_{0.12}Al_{1.6}Si_{2.92}O_{12}(OH)_2.$$

These equilibrium compositions are in accordance with N_{Mg} maxima for Cor, Gr, and Bi in histograms. As mentioned above, this correlation is a record of the evolution of the system toward equilibrium. We found several metastable Cor + Gr "paragenesis":

$$Cor_{54.3} + Gr_{22.1}\ (\ln K_D = 1.432);\quad Cor_{58} + Gr_{25}\ (\ln K_D = 1.425);$$
$$Cor_{65.4} + Gr_{31}\ (\ln K_D = 1.437);\quad Cor_{69} + Gr_{34}\ (\ln K_D = 1.463);$$
$$Cor_{82.3} + Gr_{52}\ (\ln K_D = 1.535).$$

Table 10. The run results on recrystallization of Cor, Gr, and Bi in fluids at 900°C and 6 kbar, buffer NNO, duration 7 days.

Run	N_{Mg} starting		N_{Mg} after run		N_{Mg} equilibrium		$\ln K_D$	N_{Mg}^{Bi}		Other minerals present
	Cor[a]	Gr[a]	Cor	Gr	Cor	Gr		Before run	After run	
122	7	66	12–70	67–12	61.2	25.8	1.512	49	38	Qz, Opx, Mt
122	7	66	12–70	67–12	51.9	20.2	1.466	49	38	Qz, Opx, Mt
122	7	66	12–70	67–12	56.1	23.9	1.403	49	—	Qz, Opx, Mt
122	7	66	12–70	67–12	67.8	33.5	1.430	49	46	Qz, Opx, Mt
122	7	66	12–70	67–12	82.8	52.0	1.535	49	—	Qz, Opx, Mt
123	92	9	61–92	8–37	68	33.6	1.435	49	—	Qz, Opx, Mt
125	58	66	56–75	66–32	67.2	31.6	1.488	49	47	Qz, Opx, Mt
125	58	66	56–75	66–32	63.4	29.9	1.406	49	47	Qz, Opx, Mt
125	58	66	56–75	66–32	75	41.5	1.441	49		Qz, Opx, Mt
121	58	9	45–69	8–34	69	34	1.463	49	46.8	Qz, Opx, Mt
121	58	9	45–69	8–34	58	25	1.425	—	—	
121	58	9	45–69	8–34	54.3	22.1	1.432			

[a]Weight ratios of the minerals are Cor:Gr:Bi = 5:5:15.

For one of the pairs we obtained crystallochemical formula:

$$Gr_{25}: Mg_{0.76}Fe_{2.29}Mn_{0.16}Ca_{0.16}Al_{2.02}Si_{2.86}O_{12};$$
$$Cor_{58}: Mg_{1.07}Fe_{0.76}Al_{1.14}Si_{5.81}O_{18}.$$

At 900°C and 6 kbar statistics were based on 500 measurements of compositions in Cor and Gr grains from run products: mean value ln K_D was 1.453 with $\delta = 1.6 \times 10^{-3}$. Using the "opposite-side method" of achieving the equilibrium the 900°C isotherm in diagram Fig. 5 was drawn.

Two run series were made at T = 950–1000°C and P = 6–7 kbar in gas vessels with the starting fluid composition: NH_4Cl—5 mg, oxalic acid—1 mg, and Fe—5 mg (see Table 3). The first series of 10 runs was conducted at a pressure of 7 kbar and 950°C for 24 hours and then at 1000°C for one hour. This occurred due to faulty temperature regulation. The second series of five runs was kept under P–T conditions (1000°C and 6 kbar) for 5 days. Under these conditions garnet remained in just two runs. Despite identical weight ratios of starting minerals (Bi:Cor:Gr = 15:5:5), biotite was stable only in three runs (Nos. 091, 092, and 099). Kinetics and phase relationships in the run products of both series are very similar to experimental results at 900°C. Other information is given in Table 11. From this table it is clear that only nine runs were successful enough. A clear zoning and intergrowths developed in Cor and Gr grains. Despite the compositional differences of Cor and Gr in the intergrowths, there were small variations in the distribution coefficients. Good examples are Run 092 (Cor_{62} + Gr_{32} and Cor_{53} + Gr_{25}, with ln K_D = 1.243 and = 1.218, respectively) and Run 098 (Cor_{92} + Gr_{79}, ln K_D = 1.117). In Run 126 $Gr_{38.3}$ is included in Cor_{68}. We consider these minerals in equilibrium. This is supported by rims of Cor_{68} on $Cor_{60\text{–}65}$ cores.

In 12 of the 5-hour runs at 950–1000°C compositional equilibration according to reaction (1) could be achieved within a short time. Even if metastable (because one of the minerals involved could disappear in 5 days), the K_D is almost constant for wide N_{Mg} range of cordierite and of garnet (see Fig. 5 and Table 11).

Mean ln K_D value for seven pairs in the 950°C runs (the first values ln K_D for each run in Table 11) is 1.251 with $\delta = 4.09 \times 10^{-3}$. However, if all 25 pairs of compositions from the table are used, practically the same value of ln K_D (1.257 with $\delta = 2.36 \times 10^{-3}$) is obtained. For runs at 1000°C and 6 kbar (5-day duration), calculated ln K_D is 1.219 with $\delta = 1.6 \times 10^{-4}$.

In summary, our experimental data on equilibria (1) and (2) show the following.

(i) There is no diffusion of Fe and Mg in aluminosilicates under our experimental conditions. There was a wide compositional spectra for each mineral, reflecting slow kinetics of equilibration.

(ii) Equilibrium distribution of Mg and Fe was attained within 2 days and then K_D remained constant, although compositions of coexisting phases changed along the isotherm reflecting the dynamics of recrystallization process (reactions between silicates, iron, and fluid).

(iii) Mineral compositions in equilibrium can be estimated using the "opposite-side" approach.

Table 11. The run results on recrystalliztion of Cor, Gr, and Bi in fluid E at 950–1000°C and 6 kbar, buffer NNO (see Table 3).

Run	t (°C)/ P (kbar)	N_{Mg} starting		N_{Mg} after run		N_{Mg} equilibrium					
		Cor	Gr	Cor	Gr	Cor	Gr	$\ln K_D$	Before run	After run	Other minerals present
091	950/7	58	9	53–65	6–24.5	53.1	24.5	1.247	36	—	—
092	950/7	7	66	32–43	68–19	(43)	19	1.168	36	24	Opx
092	950/7	7	2	2	2	62	32	1.243	36	—	Opx
093	950/7	92	9	93–40	6–42.5	40	15.5	1.290	36	—	—
093	950/7	92	9	93–40	6–42.5	48.9	21.6	1.245	36	—	—
093	950/7	92	9	93–40	6–42.5	53.7	26.1	1.189	36	—	—
093	950/7	92	9	93–40	6–42.5	57.7	29.4	1.187	36	—	—
093	950/7	92	9	93–40	6–42.5	72	42	1.267	36	—	—
096	950/7	7	9 and 66	5–57	6–66	(38.7)	14	1.355	36	—	—
096	950/7	7	9 and 66	5–57	6–66	(37.3)	14	1.295	36	—	—
096	950/7	7	9 and 66	5–57	6–66	(44.5)	18	1.295	36	—	—
096	950/7	7	9 and 66	5–57	6–66	49.9	21.7	1.283	36	—	—
096	950/7	7	9 and 66	5–57	6–66	56	24.6	1.361	36	—	—
097	950/7	58	11	51–66	36–10	66.2	35.7	1.260	49	—	—
097	950/7	58	11	51–66	36–10	65.3	35.3	1.238	49	—	—
097	950/7	58	11	51–66	36–10	59.7	29.9	1.241	49	—	—
097	950/7	58	11	51–66	35–10	54.4	26	1.222	49	—	—
097	950/7	58	11	51–66	36–10	50.7	22.3	1.242	49	—	—
098	950/7	92	66	70–93	10–79	93	79.1	1.256	94	88	Opx
098	950/7	92	66	70–93	10–79	81.7	54.8	1.303	94	—	Opx
098	950/7	92	66	70–93	10–79	85.9	63	1.274	94	—	Opx
098	950/7	92	66	70–93	10–79	90	71.6	1.272	94	—	Opx
098	950/7	92	66	70–93	10–79	92.5	78.5	1.217	94	—	Opx
099	950/7	7	11	7–52	10–25	52	25	1.179	36	—	—
099	950/7	7	11	7–52	10–25	43.9	17.7	1.292	36	—	—
126	1000/6	92	9	73–60	8–38	68	38.3	1.230	36	(85)	Opx, Sp
129	1000/6	92	66	56–64	28–34	56.6	28.1	1.205	94	—	Opx, Sp
129	1000/6	92	66	56–64	28–34	63.6	34	1.221	94	—	Opx

(iv) Synthesized garnets were very close to the binary pyrope–almandine solid solutions with very little Mn and Ca.

Biotite–Garnet Equilibrium

Some information on dynamics of recrystallization of starting material involving Bi + Gr was considered in the preceding section. A number of biotite microprobe analyses in run products made it possible to estimate the compositions of Bi and Gr in equilibrium using the "zoning method" developed for large crystals of the minerals.

As mentioned above, the amount of Bi in all charges was much more than that of Cor + Gr. Being a buffer, the biotite, on solution, regulated the chemical potentials of K_2O and MgO. In any fluid the Bi composition did not shift to the Fe end member showing a high affinity of Mg for dimetasilicates. Addition of pure iron to the system facilitated a shift of Bi composition to Fe-rich field at certain temperatures.

Results of the Bi–Gr equilibrium study were described briefly in the first part of this chapter and the data were presented in Tables 4–11. At the end of the run Bi differed slightly from the starting biotite composition. When a noticeable change in Bi composition occurred during the run, a very clear zoning developed, showing Fe-rich cores and Phl-rich rims. Siderophyllite-rich rims around seeds, as well as reverse zoning, appeared in some runs. For example, Run 062, conducted at 700°C and 6 kbar (see Table 6 and Fig. 6), contains products with wide compositional spectra of Cor, Gr, and Bi. A good reverse zoning developed in garnet: $Gr_7 \rightarrow Gr_4 \rightarrow Gr_{22} \rightarrow Gr_{56.5}$. Bi crystallized as large ($\backsim 200\ \mu$) grains with clear zoning: Mg-rich rims overgrew seeds. In some Bi grains reverse zoning was met.

From Fig. 6 we note that the Bi composition varies in both directions to reach $N_{Mg}^{Bi} \approx 48$ and $N_{Al}^{Bi} = 25.5$ in equilibrium with Gr_{22} and Cor_{67}. The reverse correlation between these two compositional parameters was detected in seven biotite grains. Identical relationships were established in the Bi grains of Run 056 (Table 10), where rims correspond to crystallochemical formula $K_{0.992}Mg_{1.456}Fe_{1.157}Ti_{0.057}Al_{1.723}Si_{2.620}O_{10}(OH)_2$. In Run 061 (Table 6) Bi_{44} converted entirely into $Bi_{70.5}$ with formula $K_{0.96}Mg_{1.97}Fe_{0.83}Ti_{0.05}Al_{1.67}Si_{2.56}O_{10}(OH)_2$. However, this composition was overgrown by more Fe-rich composition with formula $K_{0.94}Mg_{1.29}Fe_{1.2}Ti_{0.06}Al_{1.77}Si_{2.63}O_{20}(OH)_2$ (i.e., $N_{Mg}^{Bi} = 51.8$ and $N_{Al}^{Bi} = 25.5$).

Compositions of Bi and Gr in run products of experiments at 700°C ranged within $15 < N_{Mg}^{Gr} < 56$ and $36 < N_{Mg}^{Bi} < 79$ (see Table 6). The $N_{Al}^{Bi} \approx 25$ value was constant at different Mg/Fe ratios. The mean value of $\ln K_D$ was 1.127 ± 0.035. Equilibrium was achieved by the composition reversal method (see Fig. 5).

About 70 analyses were made for each Bi grain in the run products at 800°C. Several hundreds of compositional determinations of Bi have been obtained at 850°C. The statistics on the homogeneity of Bi compositions in run products at

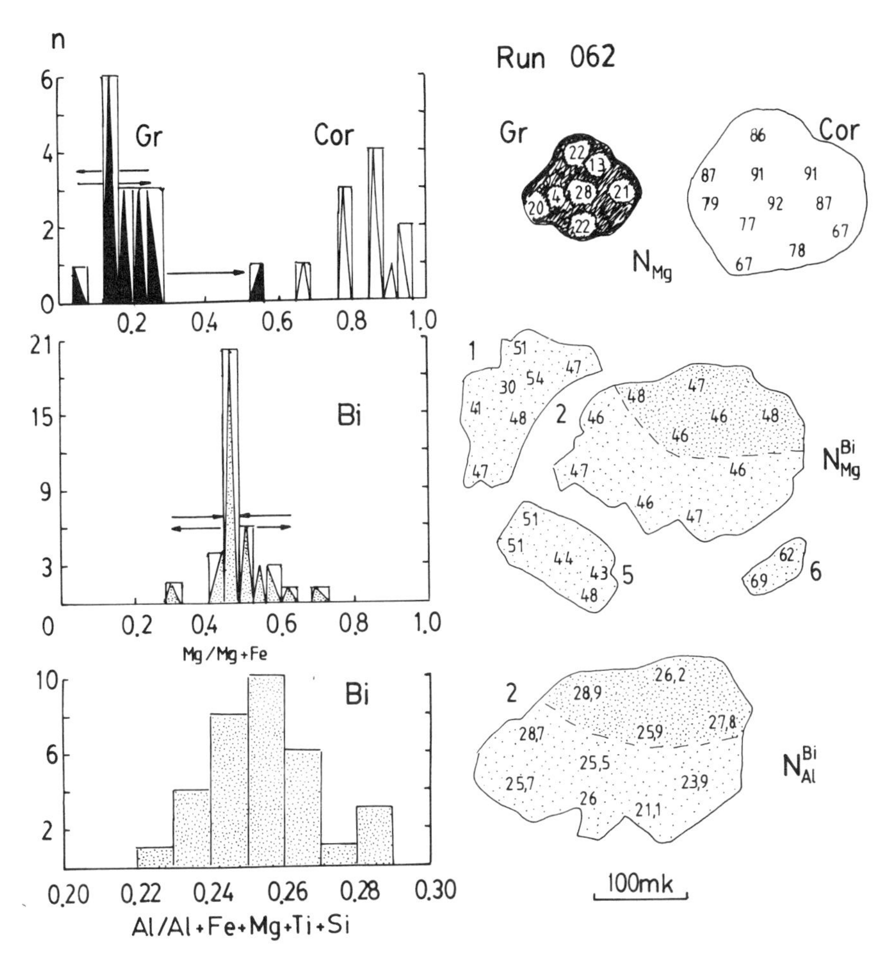

Fig. 6. The mineral compositions of Run 062 at 700°C and 6 kbars. Conventional grain numbers are pointed out on the sketch, and figures inside the grains denote mineral compositions (N_{Mg} or N_{Al}). The lower histogram illustrate reverse relations of N_{Mg} and N_{Al} in two biotite grains. The equilibrium compositions are similar in both grains. mk = micron.

other temperatures were excellent. Everywhere equilibria were achieved by the recrystallization method discussed previously. All of these data are shown in Fig. 7 and Table 13.

Unfortunately, we could not obtain data over a wide range of N_{Mg} in Bi and Gr at all desired temperatures. We consider ideal distribution of Fe and Mg between these minerals at temperatures higher than 600°C (Perchuk, 1969, 1977; Thompson, 1976; Holdaway and Lee, 1977, etc.), and the data presented in Table 13 and Fig. 7 as the experimental basis for calculating the mineralogical thermometer.

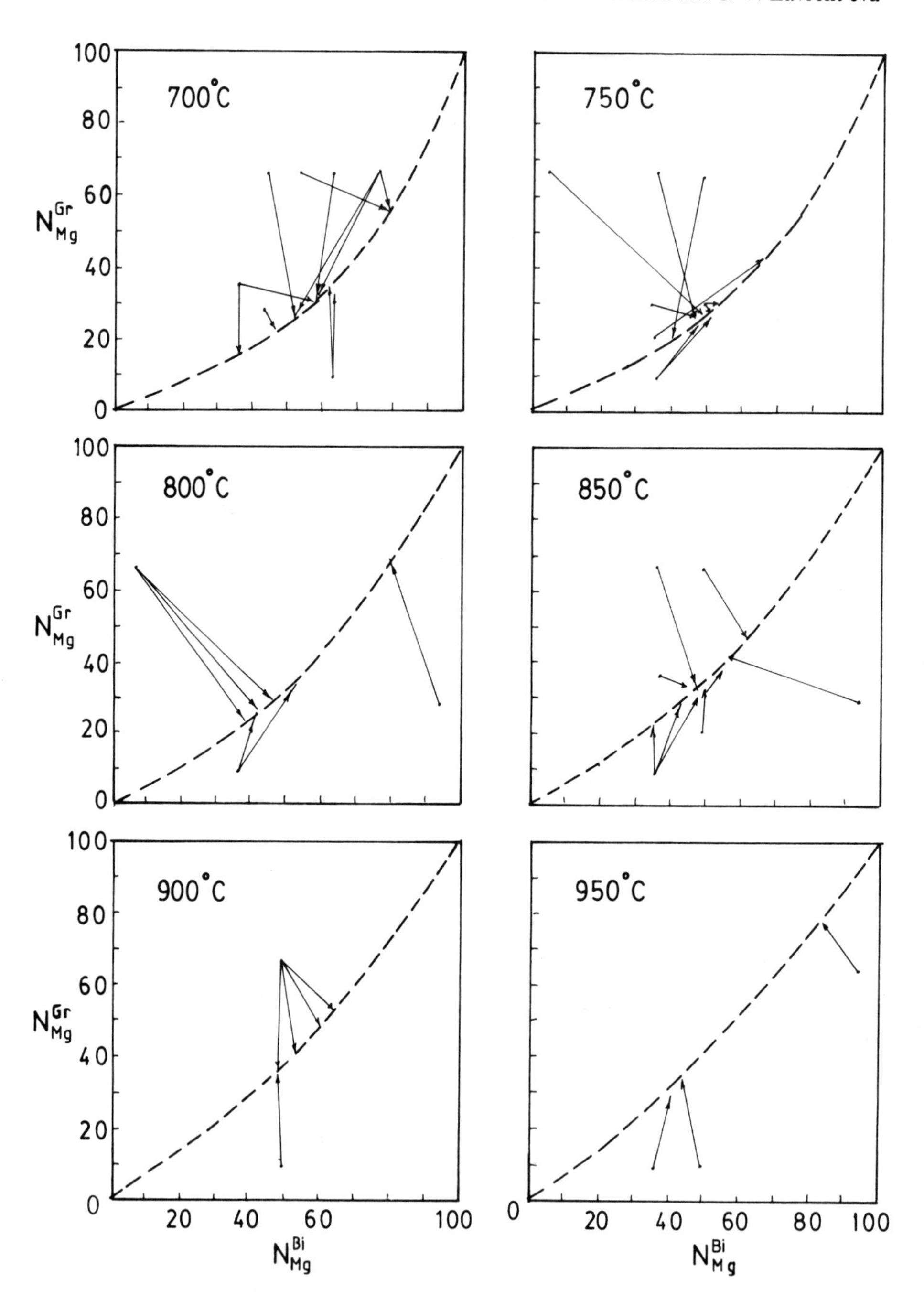

Fig. 7. Isotherms of distribution of Mg and Fe between Bi and Gr obtained using recrystallization of natural minerals in complex fluids at pressure of 6 kbars and NNO buffer.

Temperature Dependence of the Distribution Coefficients for Equilibria (1) and (2)

Cordierite and Garnet

Table 12 summarizes statistical data on the distribution coefficient for equilibrium (1). These data were calculated for two groups of experimental values. The first version covers all the data given in Tables 4–11. Statistical characteristics of the data are indicated in Table 12 by subscript 1 ($\bar{x}_1$, δ_1, $\bar{\Delta}_1$). The second version involves values of $\ln K_D$ calculated for Cor + Gr equilibrium compositions. Metastable compositions were not used. Statistical data for this version indicated by subscript 2 ($\bar{x}_2$, δ_2 and $\bar{\Delta}_2$) are grouped in the right part of Table 12. Both versions show very similar statistical estimates.

Using data from Table 12 the $10^3/T$–$\ln K_D$ correlations were fitted by linear regression to yield the following equations:

(i) First version (n_1 = 156 points):

$$10^3/T = 0.4234 + 0.2387 \ln K_D \tag{14}$$

with the regression coefficient r = 0.9996.

(ii) Second version (n_2 = 78 points):

$$10^3/T = 0.430 + 0.2934 \ln K_D \tag{15}$$

with the regression coefficient r = 0.9988.

These coefficients take into account the uncertainties in the estimates of the mineral composition, but show very strong linear correlation between the two chosen variables. Obviously, both equations are practically identical, but equation (15) is more rigorous. For estimating the MgO and FeO concentrations with microprobe analyzer, only one or two standards were used. This method was discussed in the first part of this chapter. The standards taken, however, have some deviations from linear correlations like (7) and (8). Thus the concentrations of MgO and FeO are uncorrected.

Table 12 is based on such uncorrected values of N_{Mg}. Statistical treatment of the data showed deviations from Eqs. (7) and (8) within $N_{Mg} \approx 1$ mol % both for Cor and for Gr. This is an accuracy of the probe measurements. However, the standards used make these deviations positive for Gr and negative for Cor. These deviations cause an increase of $\ln K_D$ by 0.04 and 0.13 at high and low temperatures, respectively (see Eqs. (14) and (15)). Using these deviations we corrected Eqs. (14) and (15):

(i) First version (156 points, r = 0.999):

$$10^3/T = 0.40495 + 0.32457 \ln K_D. \tag{16}$$

Table 12. Statistics of ln K_D for Cor–Gr exchange equilibrium at 600–1000°C based on uncorrected values of N_{Mg} (see text).

t (°C)	$10^3/T$	n_1	$\bar{x}_1 =$ ln K_D	$\delta_1(^+/_-)$	$\pm \bar{\Delta}_1$	n_2	$x_2 =$ ln K_D	$\delta_2(^+/_-)$	$\pm \bar{\Delta}_2$
600	1.1455	4	2.417	0.013 7	0.053	3	2.454	0.012	0.06
650	1.083	6	2.225	2.47×10^{-4}	0.011	6	2.225	2.47×10^{-4}	0.011
700	1.028	43	2.018	3.57×10^{-3}	0.041	27	2.011	4.37×10^{-3}	0.045
750	0.9775	24	1.832	1.49×10^{-3}	0.029	13	1.834	1.37×10^{-3}	0.023
800	0.932	11	1.698	3.03×10^{-4}	0.013	5	1.702	2.27×10^{-5}	0.004
850	0.890	28	1.560	1.49×10^{-3}	0.051	11	1.560	1.50×10^{-3}	0.031
900	0.8525	12	1.453	1.69×10^{-3}	0.033	4	1.474	1.09×10^{-3}	0.026
975	0.801	25	1.257	2.36×10^{-3}	0.037	7	1.251	4.09×10^{-3}	0.045
1000	0.7855	3	1.219	1.6×10^{-4}	0.009	2	1.217	3.1×10^{-4}	0.013
Total	—	156	—	—	—	78	—	—	—

(ii) Second version (78 points, $r = 0.998$):

$$10^3/T = 0.41155 + 0.31985 \ln K_D, \tag{17}$$

or

$$t^\circ\mathrm{C} = [1000/(0.41155 + 0.31985 \ln K_D^{(1)})] - 273, \tag{18}$$

where

$$K_D = \left(\frac{X_{\mathrm{Mg}}}{1 - X_{\mathrm{Mg}}}\right)^{\mathrm{Cor}} \left(\frac{1 - X_{\mathrm{Mg}}}{X_{\mathrm{Mg}}}\right)^{\mathrm{Gr}} \tag{19}$$

and

$$X_{\mathrm{Mg}} = \mathrm{Mg}/(\mathrm{Mg} + \mathrm{Fe} + \mathrm{Mn}). \tag{20}$$

The pressure effect on $\Delta G_{(1)}$ ($= -RT \ln K_D$) was calculated by applying the thermodynamics to statistical equation (17):

$$\Delta G_T^\circ + P\,\Delta V_{(1)} + RT \ln K_D^{(1)} = 0, \tag{21}$$

where volume change of reaction (1) $\Delta V_{(1)}$ is -0.03535 cal/bar (Perchuk *et al.*, 1981). At $P = 6$ kbar the value $P \cdot \Delta V_{(1)} = -212.1$ cal and

$$\Delta G_T^\circ = 212.1 - RT \ln K_D^{(1)}. \tag{22}$$

From Eq. (17) we obtained:

$$RT \ln K_D = 6212.29 - 2.557T, \text{ cal}. \tag{23}$$

Substituting (23) into (22) we obtained $\Delta\delta G_T^\circ$ at $P = 1$ bar

$$\Delta G_T^\circ = -6000.19 + 2.557T, \text{ cal}. \tag{24}$$

Thus, $\Delta H_{(1)}^\circ = 6000.19$ cal per one exchanging gram-atom, and $\Delta S_{(1)}^\circ = 2.557$ cal $\cdot$ deg.$^{-1}$ $\cdot$ atom^{-1}. These values differ slightly from previous data of Perchuk *et al.* (1981): $\Delta G_T^\circ = 5655.01–2.052T$, but they are almost identical to those of Holdaway and Lee (1977): $\Delta H^\circ = 6150$ cal and $\Delta S^\circ = 2.69$ e.u.

From (24) and (25), we have the Cor–Gr mineralogical thermometer:

$$t^\circ\mathrm{C} = \frac{3020 - 0.018P}{\ln K_D + 1.287} - 273, \tag{25}$$

which can be used to determine temperatures of Cor–Gr equilibria in metapelites. The pressure effect ($0.018P$) is negligible and can be omitted (or estimated approximately) in calculations of temperature.

Biotite and Garnet

The ln K_D at different temperatures along with the statistical characteristics of equilibrium under consideration are listed in Tables 13 and 14). These data yield the following:

$$\ln K_D^{(2)} = (3947.5/T - 2.868) \quad (26)$$

or

$$10^3/T = 0.72657 + 0.2533 \ln K_D \quad (27)$$

with linear regression coefficient $r = 0.9885$.

The data of Table 14 are mostly based on total microprobe analyses of Bi and Gr obtained according to the procedure of Bence and Albee (1968). Treatment and correction of values of N_{Mg}^{Gr} and N_{Mg}^{Bi}, originated from partial analyses, were made using procedure discussed above for Cor–Gr thermometer. The uncertainty of the corrected N_{Mg} value was usually within 1 mol %.

Equations (26) and (27) were obtained for $P = 6$ kbar. Unfortunately, there were no accurate data on ΔV for exchange reaction (2): this value changed with N_{Al}^{Bi}. For Bi the N_{Al}^{Bi} was 0.257. According to the data of Hewitt and Wones (1975) for synthetic series of Est–Sid–Ann–Phl, the volume change for reaction (2) is -0.0246 cal·bar^{-1} (at $N_{Al}^{Bi} = 0.257$). From their data, Perchuk *et al.* (1981) calculated thermodynamic properties of the phases. According to these data $\Delta V_{(2)} = -0.0578$ cal·bar^{-1} per one exchanging gram-atom. Such a great difference in estimates of $\Delta V_{(2)}$ makes it difficult to calculate the standard thermodynamic data (at 1 bar pressure) for exchange equilibria (2) from the equation

$$\Delta G_{r(2)}^{P} = \Delta G_{298}^{\circ(2)} + RT \ln K_D^{(2)} + P\,\Delta V_{(2)}$$
$$= \Delta G_{298}^{\circ(2)} + RT \ln K_D^{(2)} + 6000\,\Delta V_{(2)} = 0. \quad (28)$$

However, we can estimate possible limits of the $\Delta G_T^{\circ(2)}$ variations per one exchanging atom. In accordance with Eq. (26) the following values can be calculated at $P = 6000$ bars:

$$-\Delta G_T = RT \ln K_D = 7843.7 - 5.699T. \quad (29)$$

These data are close to theoretical results (Perchuk, 1977; Perchuk *et al.*, 1981):

$$\Delta G_T = 7642.4 - 6.181T, \text{ cal}. \quad (30)$$

From Eq. (30) we get $\Delta H_{298}^{\circ} = -7843.7 - \Delta V \cdot 6000$ cal, $\Delta S_{298} = -5.699$ cal · deg^{-1} per one exchanging atom. Using these values we obtain the limits of ΔG_{298}°. In the case of $\Delta V_{(2)} = -0.0577$ cal·bar^{-1} we get $-\Delta G_T^{\circ} = 7497 - 5.699T$ cal; $\Delta H_{298}^{\circ} = -7497$ cal while at $\Delta V_{(2)} = -0.0246$ cal·bar^{-1} the enthalpy change will be 200 cal higher:

$$-\Delta G_T^{\circ} = 7696.1 - 5.699T \text{ cal}; \quad \Delta H_{298}^{\circ} = -7696.1 \text{ cal}.$$

Table 13. Equilibrium compositions of biotite and garnet between 575–950 °C and 6 kbar.

t (°C)	Run	N_{Mg}^{Bi}	N_{Mg}^{Gr}	$\ln K_D^{(2)}$
575	076	33	7.4	1.818
	078	22.4	4.9	1.723
	078	33	7.5	1.804
	079	35.7	8.8	1.750
600	081	31	8	1.642
	082	32	8.5	1.622
	083	23.6	6.5	1.751
	084	29	7	1.691
650	032	68	33	1.462
	032	64	29	1.471
	13/7	41	15	1.370
700	045	58	31	1.122
	052	62	35	1.109
	054	61	33	1.155
	056	36	15	1.159
	056	55.7	30	1.076
	059	65	38	1.108
	060	62	34.5	1.130
	061	51.8	25.3	1.154
	062	47	21.9	1.148
	065	79	56	1.083
	067	79	54	1.165
	072	58	31	1.122
	072	61	33	1.155
	060	63	31.2	1.173
	063	51	26	1.085
	068	51	26	1.085
	069	59	32	1.117
	072	48	22	1.185
	075	59	33	1.072
750	IIIX	50	26	0.990
	113X	50	25	1.098
	114X	47	25	0.978
	115X	47	27	0.874
	116X	46.6	24	1.016
	117X	66.1	42.7	0.962
	119X	48.6	25.2	1.031
	120	50.6	27	1.019
	120	54.6	30.8	0.994
	132	41	20.5	0.991

Table 13 (continued).

t (°C)	Run	N_{Mg}^{Bi}	N_{Mg}^{Gr}	$\ln K_D^{(2)}$
800	086	40	22.2	0.848
	086	53	33.5	0.806
	087	40	22.4	0.837
	088	41	25	0.734
	089	83	67	0.877
	090	57	39	0.729
850	101	46	33	0.548
	102	48	31	0.720
	103	47	29.6	0.746
	105	58	41.9	0.650
	106	43	29	0.614
	106	35	21.7	0.664
	107	50	33	0.708
	110	54	36.7	0.705
900	122	46	33.5	0.525
	122	64	52	0.495
	121	48	34	0.583
	125	53	41.5	0.469
	125	60	48	0.485
950	91	41	30	0.483
	97	44	36	0.334
	98	83	76.5	0.290

Table 14. Statistical data on $\ln K_D$ for exchange Bi–Gr equilibrium (2) within 575–950°C and 6 kbar.

t (°C)	$10^3/T$	n	$\bar{x} = \ln K_D^{(2)}$	δ_1 ($\ln K_D$)	$\bar{\Delta}_1$ ($\ln K_D$)	δ_2 (t (°C))	$\bar{\Delta}_2$ (t(°C))	δ_3 (t (°C))	$\bar{\Delta}_3$ (t (°C))
575	1.179	4	1.774	0.045	0.037	8.6	6.7	8.6	6.7
600	1.1455	4	1.676	0.058	0.044	11.1	8.67	12.0	8.67
650	1.083	3	1.434	0.056	0.043	11.2	7.9	11.1	8.07
700	1.028	19	1.127	0.035	0.030	9.83	7.99	19.0	15.8
750	0.9775	10	0.995	0.047	0.031	15.24	9.94	15.2	9.8
800	0.932	6	0.805	0.061	0.049	18.3	14.86	18.5	14.5
850	0.890	8	0.669	0.065	0.050	20.77	15.97	21.9	17.6
900	0.8525	4	0.491	0.058	0.038	19.87	13.6	18.9	13.6
950	0.8177	3	0.369	0.101	0.076	37.73	28.56	37.9	27

Subscripts: 1—deviations from the mean arithmetic value; 2—deviations from linear regression; 3—deviations from the run temperatures.

Thus, general formula for estimating the temperatures of biotite–garnet equilibria in metamorphic rocks is:

$$t^{\circ}\mathrm{C} = \frac{7843.7 + \Delta V(P - 6000)}{1.987 \ln K_D + 5.699} - 273. \tag{31}$$

Using this relation we can estimate uncertainty in $\Delta V_{(2)}$ variations. For example, at $\ln K_D = 1$ and $P = 1$ bar the estimated temperatures are: $t = 792^{\circ}\mathrm{C}$ (for $\Delta V_{(2)} = -0.0577\ \mathrm{cal \cdot bar^{-1}}$) and $t = 766.3$ (for $\Delta V_{(2)} = -0.0246\ \mathrm{cal \cdot bar^{-1}}$). Obviously this difference decreases with pressure to zero at 6000 bar.

Discussion

Equation (31) is based on rather limited N_{Mg} range of the equilibrium compositions covered by experiments. Perhaps, in very Fe-rich and Mg-rich parts of the system, $\ln K_D$ changes with $N_{\mathrm{Al}}^{\mathrm{Bi}}$ and $N_{\mathrm{Mg}}^{\mathrm{Gr}}$, as well as with temperature. We can predict that K_D would decrease with $N_{\mathrm{Al}}^{\mathrm{Bi}}$, because parageneses Phl + Gr from granulites and inclusions from basalts show unrealistic low temperatures. Thus, Bi–Gr thermometer must be studied at different pressures and at various Al contents in Bi.

Cordierite–garnet thermometer was studied over a wide compositional range. Variation of N_{Al} in both of these minerals is small. As Cor has a narrow field of stability in pressure and there is a small volume change in reaction (1), there was no need for further experimental work. Equation (18) can be directly applied for determinations of temperature of mineral equilibria in metapelites. By combination of Eqs. (1) and (2) we obtain

$$\mathrm{Bi_{Mg}} + \mathrm{Cor_{Fe}} = \mathrm{Cor_{Mg}} + \mathrm{Bi_{Fe}}. \tag{32}$$

Using Eqs. (18) and (26) the temperature dependence of $\ln K_D^{(33)}$ is expressed as

$$\ln K_D^{(33)} = 1.587 - 823/T. \tag{33}$$

The value of $\Delta H_{298}^{(33)}$ (823 cal per one exchanging gram-atom) is rather small to expect a noticeable effect of temperature on equilibrium (32). Because $\Delta V_{(33)}$ is unknown (due to questionable $\Delta V_{(2)}$), the pressure effect on this equilibrium remains uncertain.

Let us consider application of Cor + Gr and Bi + Gr thermometers to estimation of temperature of mineral equilibria in metapelites. Several polished thin sections of rocks from the Hankay metamorphic formation (Far East, U.S.S.R.) were studied in detail with a microprobe analyzer. Geology and metamorphism of this formation were described in several publications (Mishkin, 1969; Perchuk *et al.*, 1980).

Figure 8 shows sketch of part of the thin section made from Cor–Gr–Bi–Sil–Qz–Pl gneiss (sample No. 73). A clear zoning in Gr, Cor, and Bi grains is obvious from the sketch. Assuming that the cores were in equilibrium at the peak of pro-

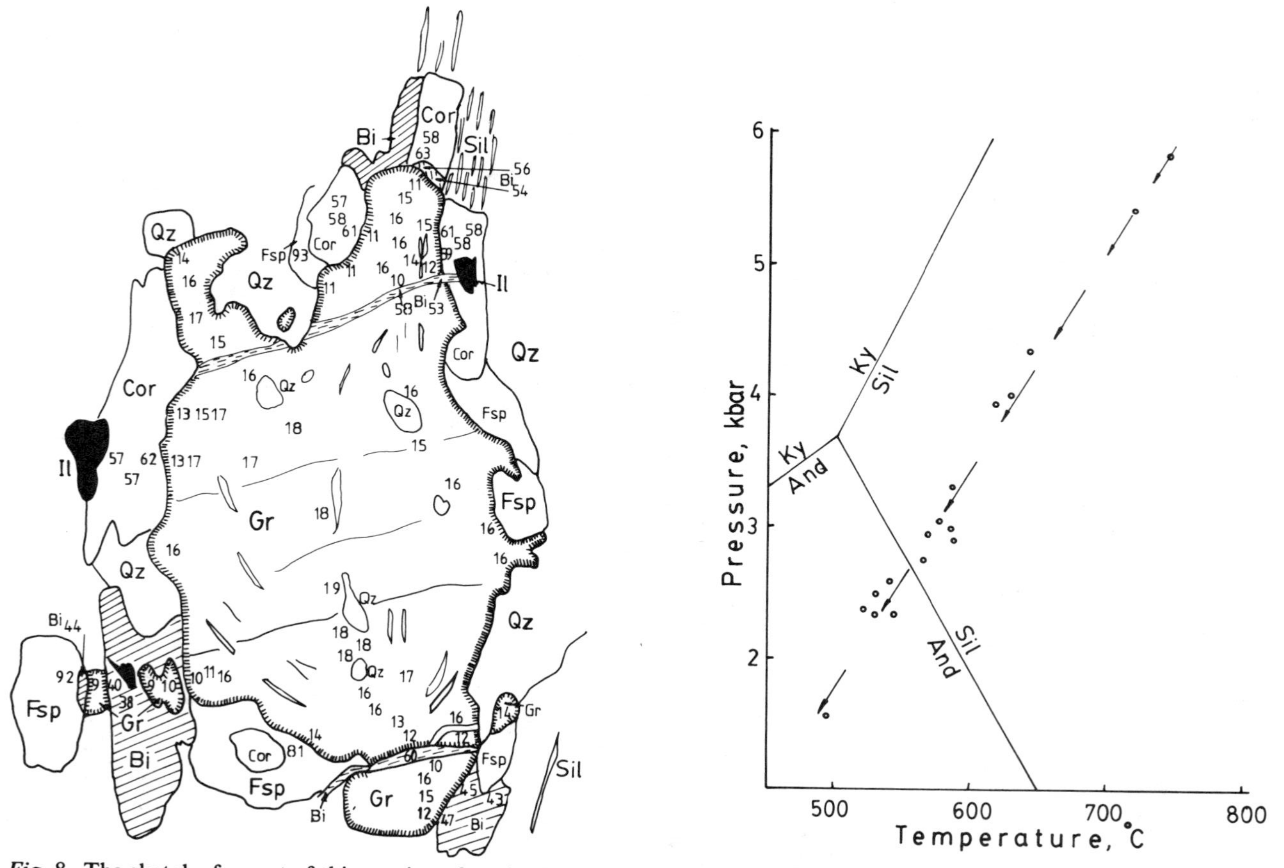

Fig. 8. The sketch of a part of thin-section of gneiss from the Hankay metamorphic formation. Figures inside the grains denote N_{Mg} for Bi, Gr and Cor, and N_{Ab} in Fsp. For other explanations, see text.

grade metamorphism, we can correlate the compositions (N_{Mg}) of adjacent minerals. The rim compositions of the minerals reflect the latest stages of retrograde metamorphism. Using Cor–Gr and Bi–Gr thermometers deduced from experimental data above and the Cor–Gr–Sil–Qz barometer (Aranovich and Podlesskii, 1982), we calculated the *P–T* trend of the retrograde metamorphic process for specimen No. 73. The results are shown in the *P–T* diagram of Fig. 8. Each point of the trend corresponds to *P–T* estimates on two or three compositions of coexisting minerals. Using N_{Mg}^{Bi} and N_{Mg}^{Cor} (see Fig. 8) of grains adjacent to garnet of identical composition (N_{Mg}^{Gr}) we compare the results of temperature estimates by two thermometers:

N_{Mg}^{Gr}	N_{Mg}^{Cor}	N_{Mg}^{Bi}	*t* (°C) (Cor–Gr)	*t* (°C) (Bi–Gr)
13	62	50	577	587
12	59	47	585	593
11	61	54	544	546

This comparison shows quite good agreement between thermometers applied to natural metamorphic rock.

Other points along the *P–T* trend in diagram of Fig. 8 were calculated with the help of the Cor–Gr thermometer and the Cor–Gr–Sil–Qz barometer, as men-

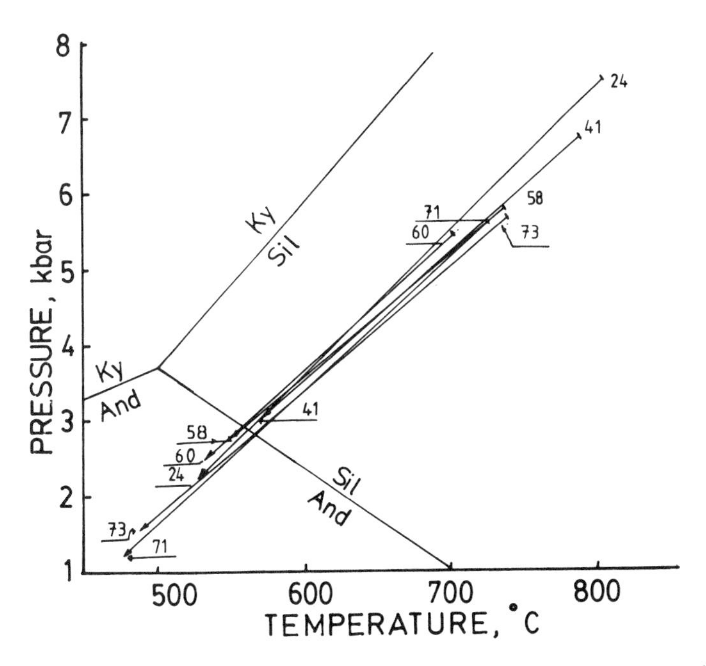

Fig. 9. Retrograde stage of metamorphism of the Hankay formation illustrated by *P–T* trends based on the probe analyses and equilibrium thermometry (Bi + Gr and Cor + Gr) and barometry (Aranovich and Podlesskii, 1983). Numbers on the lines correspond to the specimen numbers.

tioned above. The *P–T* estimates are based on N_{Mg} for cores and rims of adjacent grains, as well as on intermediate compositions, with equilibrium relations of the latter being deduced by applying the method of projections (Fig. 2 has been drawn in a similar way). Despite the provisional character of the method employed, we have obtained meaningful results in Fig. 8. Moreover, the lower part of the trend corresponds to the andalusite field, while Sil is the main rock-forming mineral of the gneiss under consideration. A thorough check of thin sections, however, allowed us to detect the coexistence of Sil and And, as well as pseudomorphs of And formed over Sil.

We have studied several thin sections of metapelites of the Hankay formation in order to see cation redistribution between minerals involved in retrograde metamorphism. Everywhere we observed that with decreasing temperature Mg moves from Gr to Bi and Cor, as predicted on the basis of the general rule (Perchuk, 1969). Thermometers and barometers employed allow us to estimate the extent of retrograde metamorphism in terms of *P–T*. Some of the results are shown in Fig. 9, from which we conclude that each specimen preserves information about its own metamorphic history, as well as about the history of the Hankay formation as a whole.

Acknowledgments

We thank our colleague, Dr. K. K. Podlesskii, for valuable discussions and help in preparation of the manuscript.

References

Afonin, V. P., Perfil'eva, L. A., and Lavrent'ev, Yu. G. (1971) The program for computing the concentrations of elements in materials of different chemical compositions with help of electron microprobe analysis, in *Ezjegodnik—1970*, pp. 398–401, Sibgeochi, Irkutsk (in Russian).

Aranovich, L. Ya and Podlesskii, K. K. (1982) The equilibrium garnet + sillimanite + quartz = cordierite. Experiments and calculations, *Mineral J.* **4**, 20–33 (in Russian).

Currie, K. L. (1971) The reaction 3 Cordierite = 2 Garnet + 4 Sillimanite + 5 Quartz as a geological thermometer in the Opinicon Lake Region, Ontario, *Contrib. Mineral. Petrol.* **33**, 215–226.

Currie, K. L. (1974) A note on the calibration of the garnet-cordierite geothermometer and geobarometer, *Contrib. Mineral. Petrol.* **44**, 35–44.

Ferry, J. M. and Spear, F. S. (1978) Experimental calibration of the partitioning of Fe and Mg between biotite and garnet, *Contrib. Mineral. Petrol.* **66**, 113–117.

Goldman, D. S. and Albee, A. L. (1977) Correlation of Mg/Fe partitioning between garnet and biotite quartz and magnetite, *Amer. J. Sci.* **277**, 750–761.

Hensen, B. J. (1977) Cordierite–garnet bearing assemblage as geothermometers and barometers in granulite facies terranes, *Tectonophysics,* **43**, 73–88.

Hensen, B. J. and Green, D. H. (1973) Experimental study of the stability of cordierite and garnet in pelitic compositions at high pressures and temperatures. III. Synthesis of experimental data and geological applications, *Contrib. Mineral. Petrol.* **38**, 151–166.

Holdaway, M. J. and Lee, S. M. (1977) Fe–Mg cordierite stability in high-grade pelitic rocks based on experimental, theoretical, and natural observations, *Contrib. Mineral. Petrol.* **63**, 175–198.

Korzhinskii, D. S. (1959) *Physicochemical Basis of the Analysis of Paragenesis of Minerals*, Consultants Bureau, New York.

Mishkin, M.A. (1969) *Petrology of the Precambrian Hankay Metamorphic Formation,* Far East. Nauka, Moscow (in Russian).

Perchuk, L. L. (1967) Biotite–garnet geothermometer, *Acad. Nauk USSR, Dokl.* **177**, 411–414 (in Russian).

Perchuk, L. L. (1969) The effect of temperature and pressure on the equilibrium of natural iron-magnesium minerals, *Int. Geol. Rev.* **11**, 875–901.

Perchuk, L. L., (1970a) Equilibrium of biotite a with garnet in metamorphic rocks, *Geochim Int.* **7**, 157–179.

Perchuk, L. L., (1970b) *Equilibria of Rock-Forming Minerals,* Nauka, Moscow (in Russian).

Perchuk, L. L. (1977) Thermodynamical control of metamorphic processes, in *Energetics of Geological Processes,* edited by Saxena, S. K. and Bhattacharji, S., pp. 285–352, Springer-Verlag, New York.

Perchuk, L. L. (1981) Correction of the biotite–garnet thermometer in case of the Mn $\rightleftharpoons$ Mg + Fe isomorphism in garnet, *Dokl. USSR AS* **256**, 38–41.

Perchuk, L. L., Mishkin, M. A., Kotelnikov, A. R., Lavrent'eva, I. V., Girnis, A. V., Podlesskii, K. K. and Gerasimov, V. Yu. (1980) Thermodynamic conditions of metamorphism of rocks of the Hankay formation, Far East, in *Contrib. Physico-Chemical Petrology* Vol. **9**, pp. 134–170 Nauka, Moscow (in Russian).

Perchuk, L. L., Podlesskii, K. K., and Aranovich, L. Ya. (1981) Calculation of thermodynamic properties of end-member minerals from natural paragenesis, in *Advances in Physical Geochemistry* Vol. **1**, pp. 111–130, Springer-Verlag, New York.

Tihomirova, V. I. and Volkova, T. M. (1974) Preparation of the gomogenious multicomponent mixtures closed in composition to biotite and garnet, in *Contributions in Physico-Chemical Petrology Vol. 4,* pp. 186–189, Nauka, Moscow (in Russian).

Thompson, A. B. (1976) Mineral reactions in pelitic rocks: II. Calculation of some *P–T–X* (Fe–Mg) phase relations, *Amer. J. Sci.* **276**, 425–454.

Chapter 8
Thermodynamics of Complex Phases

Roger Powell

Introduction

The gathering of thermodynamic data for crystalline and liquid silicates goes on apace. Yet research into the formulation of activity–composition (a–X) relations does not seem to have the same impetus. This is unfortunate because smoothing and interpolation/extrapolation of thermodynamic data can only be accomplished satisfactorily using thermodynamics.

There are several aspects of the formulation of a–X relationships which warrant a research effort. Notable among these is the problem of accounting for strong short-range order effects, for example, Al-avoidance in silicates (Mazo, 1977; Kerrick and Darken, 1976). The regular and Bragg–Williams models (e.g., Powell, 1977) have simple algebraic forms but they do involve the logical inconsistency of assuming the random distribution of the atoms in the structure while including nonzero interaction energies between these atoms. These models, as approximations to the first order quasi-chemical models (Guggenheim, 1952, 1966), break down long before these energies are large enough to produce unmixing. The standard asymmetric model, the subregular model (e.g., Powell, 1974), obviously involves the same logical inconsistency as the regular model. It also assumes a linear bulk composition dependence for the interaction energy involved.

There is a suspicion that a–X formulations are chosen for mathematical simplicity rather than physical veracity. However, the complexity of the formulation should not be the first consideration, rather, the number of adjustable parameters involved. Obviously this is the position of the purist. From a practical point of view, for example, in the development of geothermometric/barometric methods, it is reasonable to try and include nonideality in the simplest way possible, for example, using the regular and Bragg–Williams models. On the other hand, thermodynamic data when collected should be examined in a way which is as sophisticated as possible given the precision of the data. The production of simplified models should be a secondary aim which is consequent on a theoretical analysis. This is not to say that our best efforts are necessarily close approximations to the real energetics in the phase, but this is no excuse for using models which are even further away from reality.

The purpose of this chapter is to present a simple formalism for developing mixing models and to illustrate this approach with a variety of examples, particularly trying to provide a clear exposition of the thermodynamics of the quasi-chemical model. The approach involves as a first step the deduction of a physical approximation for the system of interest. As a closed formula for the Gibbs energy has to be derived for this physical approximation, some awareness of what is possible is required at this stage. The formulation of the Gibbs energy and thus by differentiation, the chemical potentials, forms the second step. The formalism allows the separation of the first step into two parts. The first part involves identifying which atoms, nearest neighbor pairs, molecules, etc., are involved in the mixing process, and what "sites" they are mixing on. Formulas for the chemical potentials of interest can then be derived *before* formulating the way in which these units mix and interact in and between the sites. The second part involves formulating the Gibbs energy in terms of these mixing/interaction relations. The advantage of this breakdown of the work involved is that the general form of the chemical potentials are known at an early stage in the analysis, and can then be used for all the model variants which may be of interest.

Formalism

Consider an r-component phase, the number of moles of a component, i, being N_i. The mole fraction of a component, i, is $X_i = N_i/\Sigma N_k$. The phase is represented by a set of s structure elements, the set being sufficient to construct the phase. The number of moles of a structure element, j, is n_j, and the mole fraction is $x_j = n_j/\Sigma n_k$. The set of structure elements is chosen so that the formulation of the Gibbs energy can be made as naturally as possible. Sometimes these structure elements can be chosen to be made up of atoms, groups of atoms, vacancies, etc. having such a composition that the fixed relation between the various sites required by the structure of the phase is not changed if these combinations are added to the crystal. If this is not possible then (4) below requires extra terms.

Following the logic of the seminal paper of Kroger *et al.* (1959), virtual chemical potentials, ξ_j, can be written for these structure elements if a virtual Gibbs energy, G^* is constructed which is a first-order homogeneous function in n_j. The virtual Gibbs energy corresponds to the observable Gibbs energy, G, for internal equilibrium in the phase and for the fixed relationship between the number of each type of site in the phase being obeyed. Thus

$$\xi_j = \left(\frac{\partial G^*}{\partial n_j}\right)_{nk(k\neq j)} \quad \text{and} \quad G^* = \sum_{k=1}^{s} n_k \xi_k. \tag{1}$$

One equation which is fundamental to this formalism is an immediate consequence of the above:

$$\mu_i = \left(\frac{\partial G}{\partial N_i}\right)_{Nk(k\neq i)} = \sum_{j=1}^{s} \left(\frac{\partial n_j}{\partial N_i}\right)_{Nk(k\neq i)} \xi_j. \tag{2}$$

Thus chemical potentials are linear combinations of virtual chemical potentials. The chemical potentials are often much harder to formulate directly than via the virtual chemical potentials for an appropriate choice of the set of structure elements.

The formulation, (2), requires the expression of the n_j in terms of the N_i. However, it is a characteristic feature of the thermodynamics of complex phases that not all of the n_j can be expressed in this way. However, the n_j can be expressed in terms of bulk composition and an arbitrarily chosen set of t of the structure elements, the number of moles of such a structure element being ϕ_k. These are internal parameters and, at equilibrium, they can be found from

$$\left(\frac{\partial G^*}{\partial \phi_k}\right)_{N_j,\phi_l(l\neq k)} = 0. \tag{3}$$

Thus (2) becomes

$$\mu_i = \left(\frac{\partial G}{\partial N_i}\right)_{N_{k(k\neq i)}} = \left(\frac{\partial G}{\partial N_i}\right)_{N_{k(k\neq i)},\theta_l} + \sum_{l=1}^{t}\left(\frac{\partial G}{\partial \theta_l}\right)\left(\frac{\partial \theta}{\partial N_i}\right)$$
$$= \sum_{j=1}^{s}\left(\frac{\partial n_j}{\partial N_i}\right)_{N_{k(k\neq i)},\theta_l} \xi_j \tag{2a}$$

so the differentiation can be performed at constant values of the internal parameters. The form of (3) is inconvenient because the virtual Gibbs energy must be expressed in terms of bulk composition and the internal parameters before the differentiation is undertaken. The method of Lagrangian multipliers (e.g., Dahlquist and Bjorck, 1974) allows the minimization of the virtual Gibbs energy with respect to the internal parameters to be undertaken with the virtual Gibbs energy expressed naturally in terms of the n_j. In this situation, the function to be minimized, f, is the virtual Gibbs energy augmented by a series of terms, C_i, representing the bulk composition relations, number of site relations, charge balance relations and so on. Thus the function to be minimized is

$$f = G^* + \Sigma\lambda_i C_i. \tag{4}$$

The differentiation of f with respect to each n_j gives a set of equations:

$$\left(\frac{\partial f}{\partial n_j}\right)_{n_{k(k\neq j)}} = 0 = \xi_j + \Sigma\lambda_i\left(\frac{\partial C_i}{\partial n_j}\right)_{n_{k(k\neq j)}} \qquad (j = 1, 2 \ldots s). \tag{5}$$

These can be solved to give sufficient equations among the virtual chemical potentials, not involving the Lagrangian multipliers, λ_i, from which all the amounts of structure elements, n_j, can be related to the amounts of components, N_i.

At this stage, the required chemical potentials and the constraint equations have been expressed in terms of the virtual chemical potentials for a particular general model for the phase. Further progress requires the formulation of the Gibbs energy in terms of the amounts of the structure elements and various energy parameters. Then differentiation with respect to the amounts of the structure elements leads to the virtual chemical potentials. The above approach should be clarified by its use below.

One-Site Phase

Binary System

Consider a binary phase between end members, say atoms 1 and 2, with each atom having z nearest neighbors. Using the nearest neighbor approximation, that the energetics of a phase are entirely determined by the interactions between pairs of nearest neighbors (e.g., Guggenheim, 1952), the obvious structure elements are 1–1, 1–2, and 2–2 nearest neighbor pairs. The numbers of these can be written as n_{11}, $2n_{12}$ and n_{22}. The number of 1–2 elements is written as $2n_{12}$ to count the (indistinguishable) orientations, 1–2 and 2–1, separately. Noting that there are $z/2$ nearest neighbor pairs involving each atom, the numbers of atoms of 1 and 2, N_1 and N_2, can be expressed in terms of the n_j:

$$\begin{aligned} N_1 &= \frac{2}{z}(n_{11} + n_{12}), \\ N_2 &= \frac{2}{z}(n_{12} + n_{22}). \end{aligned} \tag{6}$$

Mole fractions can be written

$$\begin{aligned} x_{11} &= \frac{n_{11}}{n_{11} + 2n_{12} + n_{22}}, \\ x_{12} &= \frac{n_{12}}{n_{11} + 2n_{12} + n_{22}}, \\ x_{22} &= \frac{n_{22}}{n_{11} + 2n_{12} + n_{22}}, \end{aligned}$$

and

$$\begin{aligned} X_1 &= \frac{N_1}{N_1 + N_2} = x_{11} + x_{12}, \\ X_2 &= \frac{N_2}{N_1 + N_2} = x_{12} + x_{22}. \end{aligned}$$

From (1), G^* can be written in terms of the structure elements:

$$G^* = n_{11}\xi_{11} + n_{12}\xi_{12} + n_{22}\xi_{22}. \tag{7}$$

From (6), in the absence of charge or other constraints, there is one internal parameter. For example n_{11} and n_{22} can be expressed in terms of n_{12} and X_1. Using (4), (5) to obtain the one required constraint equation, the function to be minimized is

$$f = G^* + \lambda\left(\frac{n_{11} + n_{12}}{n_{11} + 2n_{12} + n_{22}} - X_1\right). \tag{8}$$

The term in brackets ensures that the minimization is undertaken at constant bulk composition. Differentiating f with respect to the n_j and setting the differentials to zero gives

$$0 = \xi_{11} + (1 - X_1)\lambda,$$
$$0 = \xi_{12} + (1 - 2X_1)\lambda,$$
$$0 = \xi_{22} + (-X_1)\lambda,$$

where λ in each case includes the common constant term from the differentiation of the bulk composition term in (8). Combining these equations gives the constraint equation:

$$\xi_{12} = \xi_{11} + \xi_{22}. \tag{9}$$

In order to obtain μ_1 and μ_2 we first express n_{11} and n_{22} in terms of n_{12} and N_1, N_2:

$$n_{11} = \frac{z}{2} N_1 - n_{12},$$
$$n_{22} = \frac{z}{2} N_2 - n_{12}.$$

From (2a)

$$\mu_1 = \left(\frac{\partial n_{11}}{\partial N_1}\right)_{N_2,n_{12}} \xi_{11} + \left(\frac{\partial n_{12}}{\partial N_1}\right)_{N_2,n_{12}} \xi_{12} + \left(\frac{\partial n_{22}}{\partial N_1}\right)_{N_2,n_{12}} \xi_{22}.$$

Giving

$$\mu_1 = \frac{z}{2} \xi_{11} \tag{10a}$$

and similarly

$$\mu_2 = \frac{z}{2} \xi_{22}. \tag{10b}$$

Symmetric Case

At this point an explicit formulation of G^* is required, now that the chemical potentials and the constraint equation are expressed in terms of the virtual chemical potentials. The simplest assigns bulk composition-independent energies to nearest neighbor pairs. This turns out to give mixing properties which are symmetric with respect to composition. The Gibbs energy is

$$G^* = \epsilon_{11} n_{11} + 2\epsilon_{12} n_{12} + \epsilon_{22} n_{22} - RT \ln g, \tag{11}$$

where g is the number of ways of arranging the nearest neighbor pairs for fixed values of n_{11}, n_{12}, and n_{22}. Correct formulation of g is impossibly complex so an approximate approach is followed (e.g., Guggenheim, 1952, pp. 43–44; 1966, pp. 84–85). We start by writing g in terms of the number of ways of mixing n_{11}, $2n_{12}$, and n_{22} *independent* pairs:

$$g = h\left(\frac{n!}{n_{11}!(n_{12}!)^2 n_{22}}\right), \tag{12}$$

where the normalization constant, h, takes account of the fact that the bracketed term is for mixing on n pairs, or $2n$ individual sites, whereas we wish the mixing to be on $2n/z$ sites each with z nearest neighbors. The reason why the expression for g is only approximate is because the pairs of nearest neighbors are assumed to be independent, whereas in fact $z/2$ pairs are involved for each atom, so that the pairs can hardly be considered to be independent.

The normalization constant in (12) is constructed by ensuring that the sum of g over all values of n_j at constant N_i is correct. Summing the bracketed term in (12) in this way is accomplished by replacing the sum by the maximum term in the sum (e.g., Hill, 1960, Appendix II). The maximum term can be found using Lagrangian multipliers. The function to be minimized, converting to logarithms and using Stirling's approximation, is

$$f = n \ln n - n_{11} \ln n_{11} - 2n_{12} \ln n_{12} - n_{22} \ln n_{22} + \lambda_1\left(n_{11} + n_{12} - \frac{z}{2}N_1\right) + \lambda_2\left(n_{12} + n_{22} - \frac{z}{2}N_2\right).$$

Differentiating and setting to zero gives

$$0 = -\ln x^*_{11} + \lambda_1,$$
$$0 = -2\ln x^*_{12} + \lambda_1 + \lambda_2,$$
$$0 = -\ln x^*_{22} + \lambda_2,$$

where the * superscript refers to the x_j value in the maximum term. Combining gives

$$0 = \ln \frac{x^{*\,2}_{12}}{x^*_{11} x^*_{22}}.$$

This can be viewed as (11) formulated for zero energy change. Thus the implication is that the maximum term corresponds to random mixing of the $2n$ atoms of 1 and 2, so that the contribution of this term to h can be written

$$\frac{(2nX_1)!(2nX_2)!}{(2n)!}.$$

By the same logic, the remaining part of the normalization constant is given by the random mixing expression for $2n/z$ atoms of 1 and 2:

$$\frac{((2/z)n)!}{((2/z)nX_1)!((2/z)nX_2)!}.$$

Substituting n_j for nX_i in these expressions and substituting into (12) gives

$$g = \frac{((2/z)n)!}{((2/z)(n_{11}+n_{12}))!((2/z)(n_{12}+n_{22}))!} \cdot \frac{(2(n_{11}+n_{12}))!(2(n_{12}+n_{22}))!}{(2n)!} \cdot \frac{n!}{n_{11}!(n_{12}!)^2 n_{22}}.$$

Simplifying by taking logarithms, applying Stirling's approximation, rearranging, then returning to the original form, gives

$$g = \frac{((2/z)(z-1)(n_{11}+n_{12}))!((2/z)(z-1)(n_{12}+n_{22}))!}{((2/z)(z-1)n)!} \cdot \frac{n!}{n_{11}!(n_{12}!)^2 n_{22}!}. \tag{13}$$

Substituting into (11) gives an expression for G^* in terms of the n_j:

$$G^* = \epsilon_{11}n_{11} + 2\epsilon_{12}n_{12} + \epsilon_{22}n_{22} + RT\{n_{11}\ln x_{11} + 2n_{12}\ln x_{12} + n_{22}\ln x_{22} + \frac{2}{z}(z-1)((n_{11}+n_{12})\ln(x_{11}+x_{12}) + (n_{12}+n_{22})\ln(x_{12}+x_{22}))\}. \tag{14}$$

Differentiating with respect to n_{11} at constant n_{12} and n_{22} in the manner of (1), and so on:

$$\begin{aligned} \xi_{11} &= \epsilon_{11} + RT\left(\ln x_{11} - \frac{2}{z}(z-1)\ln X_1\right), \\ \xi_{12} &= 2\epsilon_{12} + RT\left(\ln x_{12}^2 - \frac{2}{z}(z-1)\ln X_1X_2\right), \\ \xi_{22} &= \epsilon_{22} + RT\left(\ln x_{22} - \frac{2}{z}(z-1)\ln X_2\right). \end{aligned} \tag{15}$$

Using these in the constraint equation (9) gives

$$0 = 2\epsilon_{12} - \epsilon_{11} - \epsilon_{22} + RT\ln\frac{x_{12}^2}{x_{11}x_{22}}$$

or:

$$\frac{x_{12}^2}{x_{11}x_{22}} = \exp\left(-\frac{w}{RT}\right) = K,$$

where $w = 2\epsilon_{12} - \epsilon_{11} - \epsilon_{22}$ is known as the interaction energy. Note that the configuration-independent terms in the activities in (15) do not appear in the constraint equation because they cancel. The constraint equation can be viewed as the equilibrium relation for the balanced chemical reaction, the so-called quasi-chemical reaction, written between the nearest neighbor pairs:

$$2(1-2) = (1-1) + (2-2).$$

Constraint equations always have this form because the function, (4), being minimized takes explicit account of bulk composition, charge balance, and number of sites relations so that the resulting constraint equations are balanced with respect to composition, charge and the numbers of sites.

The constraint equation (16), can be solved for x_{12}, and thus for x_{11} and x_{22}, for given w/RT and bulk composition. Substituting $x_{11} = X_1 - x_{12}$ and $x_{22} = X_2 - x_{12}$ into (16) gives a quadratic in x_{12}. Using (16) in this way is of course equivalent to minimizing G^* with respect to x_{12} at constant bulk composition, Fig. 1. The solution to the quadratic in x_{12} is

$$x_{12} = \frac{-K + \sqrt{K^2 + 4K(1 - K)X_1X_2}}{2(1 - K)}.$$

Defining a parameter, δ, with reference to this solution:

$$\delta = \frac{\sqrt{1 + 4X_1X_2(1-K)/K} - 1}{2}.$$

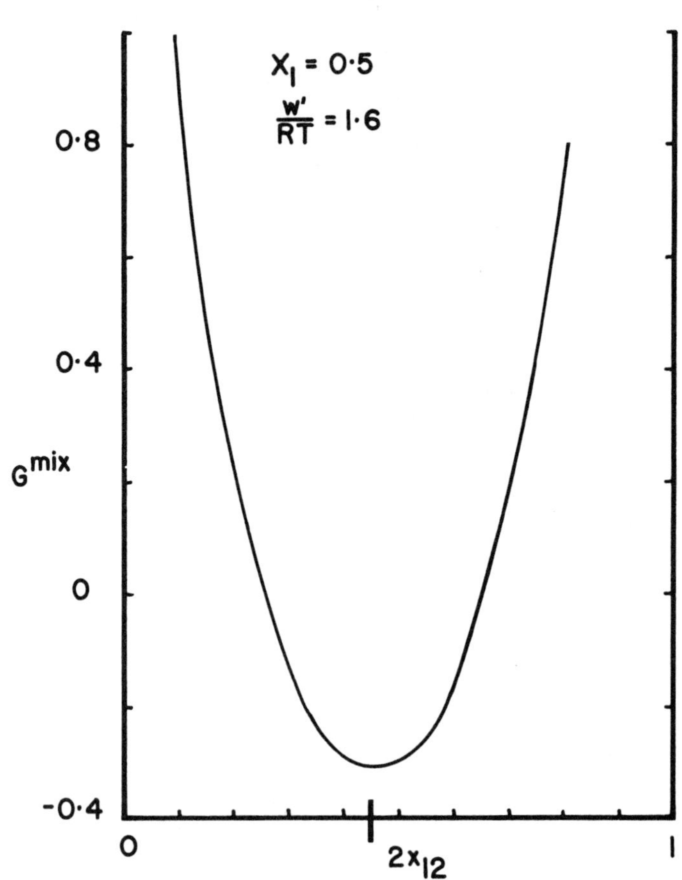

Fig. 1. Graphical minimization of G^{mix} with respect to x_{12} at $X_1 = 0.5$ and $w'/RT = 1.6$. This minimization is normally accomplished using (16).

δ is zero for random mixing ($w = 0, K = 1$), positive for positive deviations from ideality ($w > 0, K < 1$), and negative for negative deviations from ideality ($w < 0, K > 1$). Then, with some manipulation for the second part,

$$x_{12} = \frac{K}{1 - K}\delta = \frac{X_1 X_2}{1 + \delta}. \tag{17a}$$

This shows that x_{12} is smaller than the value predicted for random mixing (X_1X_2) for positive deviations from ideality, and greater than the value predicted for random mixing for negative deviations from ideality. Also

$$x_{11} = X_1 - x_{12} = X_1\left(\frac{X_1 + \delta}{1 + \delta}\right), \tag{17b}$$

$$x_{22} = X_2 - x_{12} = X_2\left(\frac{X_2 + \delta}{1 + \delta}\right). \tag{17c}$$

Using (10, 15 and 17):

$$\mu_1 = \frac{z}{2}\xi_{11} = \epsilon'_{11} + RT\left[\frac{z}{2}\ln X_1\left(\frac{X_1 + \delta}{1 + \delta}\right) + (1 - z)\ln X_1\right],$$

$$\mu_2 = \frac{z}{2}\xi_{22} = \epsilon'_{22} + RT\left[\frac{z}{2}\ln X_2\left(\frac{X_2 + \delta}{1 + \delta}\right) + (1 - z)\ln X_2\right],$$

where terms in square brackets are the natural logarithms of the activities of 1 and 2, respectively. Primed quantities have been multiplied by $z/2$. Further manipulation organizes these terms into a suggestive form:

$$\mu_1 = \epsilon'_{11} + RT\ln X_1 + \frac{z}{2}RT\ln\frac{X_1 + \delta}{X_1(1 + \delta)},$$
$$\mu_2 = \epsilon'_{22} + RT\ln X_2 + \frac{z}{2}RT\ln\frac{X_2 + \delta}{X_2(1 + \delta)}, \tag{18}$$

where the last term in each equation might be identified with the activity coefficient. These equations give the chemical potentials for the symmetric one-site quasi-chemical model. Activity–composition relations for this model are illustrated in Fig. 2. (The development in Guggenheim (1966) leading to his (100–101) (my (18)) appears to contain several errors of sign, though of course the final equations are correct.)

Discussion

The symmetric nature of the mixing properties can be seen from

$$\frac{G^{\text{mix}}}{RT} = (X_1\ln X_1 + X_2\ln X_2) + \frac{z}{2}\left(X_1\ln\frac{X_1 + \delta}{X_1(1 + \delta)} + X_2\ln\frac{X_2 + \delta}{X_2(1 + \delta)}\right) \tag{19}$$

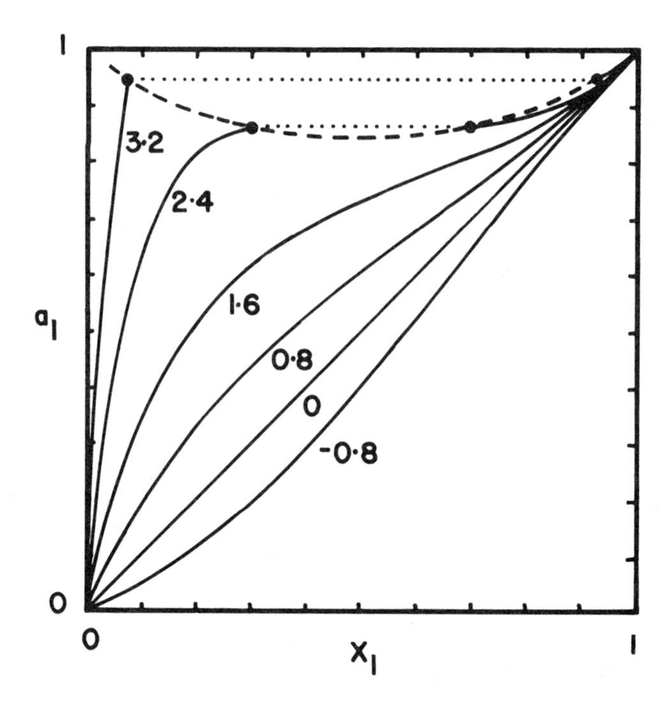

Fig. 2 Activity–composition relations for a series of values of w'/RT. Note the appearance of a solvus at large values of w'/RT. The crest of the solvus is at $X_1 = 0.5$ and $w'/RT = 2.30$.

and from the observation that δ is symmetric with respect to composition (i.e., $\delta(X_1 = y) = \delta(X_1 = 1 - y)$). Also the Henry's Law constants are the same ($X_2 \rightarrow 0$, $\ln \gamma_1 \rightarrow w'/RT$; and $X_1 \rightarrow 0$, $\ln \gamma_2 \rightarrow w'/RT$).

The enthalpy of mixing is

$$H^{\text{mix}} = \left(\frac{\partial(G^{\text{mix}}/RT)}{\partial(1/RT)}\right)_{X_1}$$

$$= \frac{z}{2} X_1 X_2 \frac{1}{(1 + \delta)(1 + \delta(1 + \delta)/X_1X_2)} \frac{1}{K} \left(\frac{\partial(w/RT)}{\partial(1/RT)}\right)$$

and, if $w = w_H - Tw_S$, then

$$H^{\text{mix}} = \frac{X_1X_2}{(1 + \delta)(1 + \delta(1 + \delta)/X_1X_2)} \frac{w'_{\text{H}}}{K}. \tag{20}$$

The form of H^{mix}/X_1X_2 is illustrated in Fig. 3.

The limit, $w = 0$, is of interest. Then $K = 1$ and $\delta = 0$ so that $x_{11} = X_1^2$, $x_{12} = X_1X_2$ and $x_{22} = X_2^2$ which correspond to the expected proportions of the nearest neighbors for random mixing. Thus, $w = 0$ is the random mixing limit. Note that if these proportions are specified initially then the formulation of the thermodynamics is rather different because then the mass relations, (6), are replaced by these relations. This is important because (6) is used in deriving (10). Now, instead of (8)

$$n_{11} = \frac{z}{2}\frac{N_1^2}{N_1 + N_2},$$
$$n_{12} = \frac{z}{2}\frac{N_1 N_2}{N_1 + N_2},$$
$$n_{22} = \frac{z}{2}\frac{N_2^2}{N_1 + N_2}.$$

Instead of (10), we have

$$\begin{aligned}\mu_1 &= \frac{z}{2}((1 - X_2)^2\xi_{11} + 2X_2^2\xi_{12} - X_2^2\xi_{22}) \\ &= \epsilon'_{11} + RT \ln X_1 + X_2^2 w\end{aligned} \tag{21a}$$

and

$$\mu_2 = \epsilon'_{22} + RT \ln X_2 + X_1^2 w, \tag{21b}$$

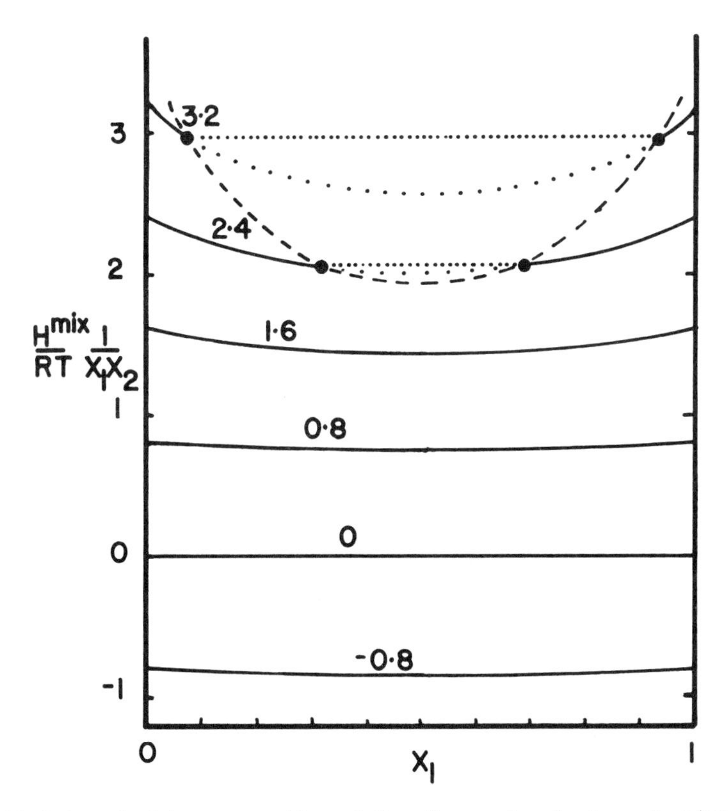

Fig. 3 Enthalpy of mixing–composition relations for a series of values of w'_H/RT. Horizontal lines are predicted for the regular model. The solvus is constructed for $w'/RT = w'_H/RT$.

which are, of course, the regular model equations. These equations are more simply derived for a different choice of structure elements. The relationship of the regular model to the $w = 0$ limit in the quasi-chemical model can be seen by expanding the activity coefficients in (18) in w/RT. To first-order terms this expansion is

$$
\begin{aligned}
K &\rightarrow 1 - \frac{w}{RT}\cdots \quad \text{for } \frac{w}{RT} \ll 1, \\
\delta &\rightarrow X_1X_2\frac{w}{RT}\cdots, \\
\ln \gamma_1 &\rightarrow \frac{z}{2}\ln\left(1 + X_2^2\frac{w}{RT}\cdots\right) \rightarrow X_2^2\frac{w'}{RT}, \\
\ln \gamma_2 &\rightarrow X_1^2\frac{w'}{RT},
\end{aligned}
\tag{22}
$$

so the quasi-chemical model reduces to the regular model for $w/RT \rightarrow 0$. However this restriction does correspond to what would normally be considered as van-

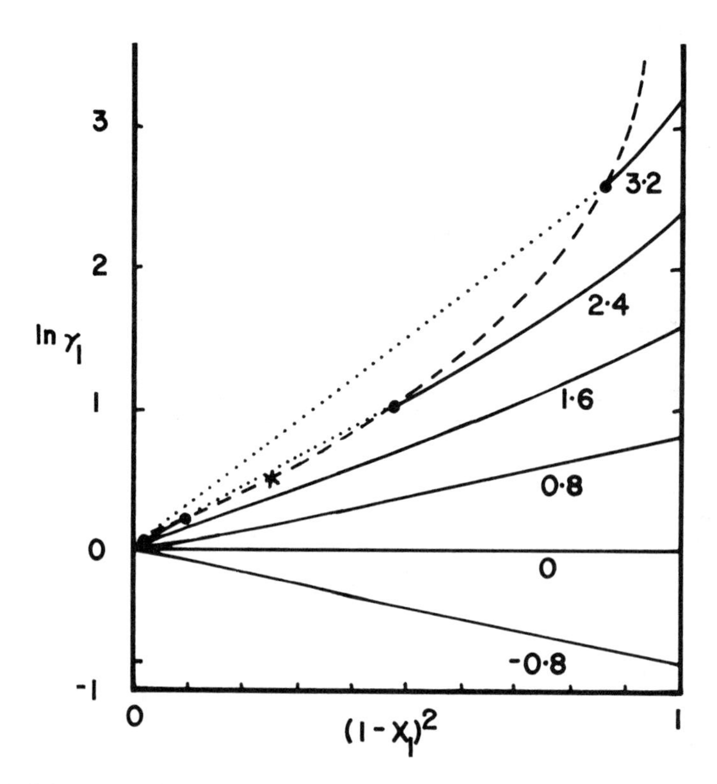

Fig. 4 Activity–composition relationships plotted to illustrate relationships between the quasi-chemical and regular models, see text. The slopes of the curves are the same at $X_1 = 0$ and $X_1 = 1$, thus the curves have considerable curvature near $X_1 = 0$.

ishingly small departures from ideal mixing. This discussion highlights the logical inconsistency of the regular model.

Quasi-chemical model mixing properties can be approximated by the regular model for limited composition ranges. For the regular model

$$\frac{\partial \ln \gamma_1}{\partial (1 - X_1)^2} = \frac{w'}{RT} \quad \text{and} \quad (\ln \gamma_1)_{X_1 \to 0} = \frac{w'}{RT}.$$

For the quasi-chemical model, for small X_1

$$\begin{aligned} \delta &\to X_1 \frac{1-K}{K} - X_1^2 \frac{1-K}{K^2} \dots, \\ \ln \gamma_1 &\to \frac{w'}{RT} - zX_1 \frac{1-K}{K} \dots, \\ \frac{\partial \ln \gamma_1}{\partial (1 - X_1)^2} &\to \frac{z}{2} \frac{1-K}{K} \dots . \end{aligned} \tag{23}$$

The curve of $\ln \gamma_1$ against $(1 - X_1)^2$ is very gently curving from this limit at $X_1 = 0$ sigmoidally to $\ln \gamma_1 = 0$ at $X_1 = 1$ where the slope is the same as at $X_1 = 0$, Fig. 4. Limited composition ranges can be approximated by the regular model. Clearly the amount of nonideality, i.e., size of w/RT, will control the widths of composition ranges which can be considered as approximately linear. If interest is concentrated on such a concentration range, little error is introduced by treating the mixture as regular, but note that the Gibbs energy of the end members of the phase must be corrected to correspond to this reformulation of the mixing properties. For example, if the quasi-chemical mixing properties for X_1 near zero are reformulated with the regular model with interaction energy w^*, then the corresponding ϵ_1^* is just $\epsilon_{11} + w - w^*$.

This discussion has some important consequences. For example, modeling phase equilibria to obtain Gibbs energies and interaction energies for a phase (or phases) when the composition range is limited and does not include the pure end members of interest, will lead to internally consistent thermodynamic data which may bear little resemblance to calorimetrically measured data. This will usually be a problem because the best equations for nonideality we can develop may not be close to the correct ones. The corollary of this is that it will not necessarily be correct to use calorimetrically determined thermodynamic end member data in the extraction of information on *a–X* relations for phases from phase equilibria. Agreement between data sets from the two sources is obviously a necessary but not a sufficient conditions for the applicability of a mixing model.. Another interesting limit is $K \to \infty$ because it should correspond to the case of, say for $X_1 > 0.5$, 2–2 nearest neighbors being forbidden. Mixing properties can be compared to those from the formulation of Iiyama (1974). This is a severe test of the formulation of g in (13) because of the assumption of the independence of pairs being mixed. In the limit $K \to \infty$

$$\begin{aligned} x_{11} &\to X_1 - X_2, \\ x_{12} &\to X_2, \\ x_{22} &\to 0. \end{aligned}$$

Substituting into (14), noting that $x_{22}\ln x_{22} \to 0$, as $x_{22} \to 0$. gives

$$\frac{G}{RT} = \frac{z}{2}((X_1 - X^2)\ln(X_1 - X^2) + 2X_2 \ln X_2) - (1 - z)(X_1 \ln X_1 + X_2 \ln X_2).$$

Iiyama (1974), modelling trace element incorporation in solids, derives an exact formulation for g for $X_2 \to 0$, for a zone of r sites around an atom of 2 from which atoms of 2 are subsequently excluded. In this case

$$g = \frac{N(N - (r + 1))(N - 2(r + 1)) \ldots (N - (N_2 - 1)(r + 1))}{N_2!}.$$

This is only correct for $N_2 \ll N$ because as N_2 gets larger, less than $r + 1$ sites are excluded from those remaining for each atom of 2 placed. Simplifying

$$g = \frac{\left(\dfrac{N}{r + 1}\right)!}{\left(\dfrac{N}{r + 1} - N_2\right)!\left(\dfrac{N_2}{r + 1}\right)!^{1+r}}.$$

In this form, g is seen to be the number of ways of distributing N_2 atoms among $N/(r + 1)$ groups of sites, taking account of the number of ways of placing an atom of 2 on the $r + 1$ sites in each group.

For this formulation

$$\frac{G^{\text{mix}}}{RT} = \left(\frac{X_1 - rX_2}{1 + r}\right)\ln(X_1 - rX_2) + X_2 \ln X_2.$$

The quasi-chemical model, however, gives in the limit $K \to \infty$ and $X_2 \to 0$:

$$\frac{G}{RT} \to \frac{z}{2}(X_1 - X_2)\ln(X_1 - X_2) + X_2 \ln X_2.$$

Exact correspondence occurs for $r = 1$ and $z = 1$. This is hardly surprising because the quasi-chemical formulation of g involves independent nearest neighbor pairs, i.e., $z = 1$, and forbidden 2–2 nearest neighbors, i.e., $r = 1$. In fact, quasi-chemical formulations involving $(r + 1)$-tuplets of atoms, rather than pairs, will produce G^{mix} terms in $(X_1 - rX_2)$ in the appropriate limit.

Another test of the quasi-chemical formulation for $K \to \infty$ is the problem of formulating the thermodynamics of Al-avoidance, for example in feldspars. Si and Al atoms are tetrahedrally coordinated to each other (via oxygens) so $z = 4$. Al avoidance precludes Al–Al coordination. For albite, $X_{\text{Al}} = 0.25$, and for mixing over four sites,

$$\begin{aligned} S^{\text{mix}} &= -4R[2(\tfrac{1}{2}\ln\tfrac{1}{2} + \tfrac{1}{2}\ln\tfrac{1}{2}) - 3(\tfrac{3}{4}\ln\tfrac{3}{4} + \tfrac{1}{4}\ln\tfrac{1}{4})] \\ &= 13.0\ JK^{-1}. \end{aligned}$$

This value should correspond to the difference in entropy between low and high albite if this is due to complete and absent Al-avoidance respectively. The value calculated here is the same as that calculated by Mazo (1977) because his model, although more complex than the one used here, reduces to this one in the limit he used in his calculation. The value compares favourably with the value of -12.2 $J\,K^{-1}$ calculated by Kerrick and Darken (1976, p. 1437) in their novel approach to Al-avoidance.

Asymmetry

Examining the development of the simple symetric quasi-chemical equations above, it is easy to see how asymmetry can be built in to the model. Note that the energies, ϵ_j, in (13), appear explicitly only in the differentiation of (14) to give (15) in the derivation of the constraint equation (16) and the chemical potentials (18). In particular they are not involved in the formulation of g and, consequently, they are not involved in the activity part of the chemical potentials (18). Thus the energies, ϵ_j, in (14) can be made composition dependent providing asymmetry in the mixing properties, and the only change required in the equations for the chemical potentials (18), is the addition of terms resulting from the differentiation of the composition dependent ϵ_j in (14).

The obvious choice of compositional dependence is to make ϵ_j linear in bulk composition:

$$\begin{aligned} \epsilon_{11} &= X_1\epsilon_{11(1)} + X_2\epsilon_{11(2)}, \\ \epsilon_{12} &= X_1\epsilon_{12(1)} + X_2\epsilon_{12(2)}, \\ \epsilon_{22} &= X_1\epsilon_{22(1)} + X_2\epsilon_{22(2)}, \end{aligned} \tag{24}$$

where the subscript, (i), refers to the energy at $X_i = 1$. We maintain the definition, $w = 2\epsilon_{12} - \epsilon_{11} - \epsilon_{22}$, but note that it is now composition dependent:

$$w = w_{(1)}X_1 + w_{(2)}X_2 \qquad \begin{cases} w_{(1)} = 2\epsilon_{12(1)} - \epsilon_{11(1)} - \epsilon_{22(1)}. \\ w_{(2)} = 2\epsilon_{12(2)} - \epsilon_{11(2)} - \epsilon_{22(2)} \end{cases}$$

It is helpful to define the differences which reflect the amount of asymmetry:

$$\begin{aligned} \Delta\epsilon_{11} &= \epsilon_{11(1)} - \epsilon_{11(2)}, \\ \Delta\epsilon_{12} &= \epsilon_{12(1)} - \epsilon_{12(2)}, \\ \Delta\epsilon_{22} &= \epsilon_{22(1)} - \epsilon_{22(2)}, \end{aligned}$$

and logically

$$\Delta w = w_{(1)} - w_{(2)}.$$

Differentiating (14) gives

$$\begin{aligned} \xi_{11} &= \epsilon_{11} + RT\ln x_{11} + \frac{2}{z}(z-1)\ln X_1 + X_2A, \\ \xi_{12} &= 2\epsilon_{12} + RT\ln x_{12}^2 + \frac{2}{z}(z-1)\ln X_1X_2 + (X_2 - X_1)A, \\ \xi_{22} &= \epsilon_{22} + RT\ln x_{22} + \frac{2}{z}(z-1)\ln X_2 - X_1A, \end{aligned} \tag{25}$$

where

$$A = x_{12}\,\Delta w + X_1\,\Delta\epsilon_{11} + X_2\,\Delta\epsilon_{22}.$$

The constraint equation from (9), is identical to (16) because the terms in A cancel. This means that the development from (16)–(18) is applicable here. The only difference is that K is now composition dependent. The chemical potentials become

$$\mu_1 = \epsilon'_{11(1)} + RT\ln X_1 + \frac{z}{2}RT\ln\frac{X_1+\delta}{X_1(1+\delta)} + X_2^2\left(\frac{X_1\,\Delta w'}{1+\delta} + \Delta\epsilon'_{22} - \Delta\epsilon'_{11}\right),$$

$$\mu_2 = \epsilon'_{22(2)} + RT\ln X_2 + \frac{z}{2}RT\ln\frac{X_2+\delta}{X_2(1+\delta)} + X_1^2\left(-\frac{X_2\,\Delta w'}{1+\delta} + \Delta\epsilon'_{22} - \Delta\epsilon'_{11}\right),$$

(26)

where the last term in each equation is the additional term in the activity coefficient. The corresponding additional term in (19) and (20) is simply $X_1X_2(\Delta\epsilon'_{22} - \Delta\epsilon'_{11})$.

The limit $w/RT \to 0$ gives

$$\ln\gamma_1 \to \frac{w'}{RT}X_2^2,$$

$$\ln\gamma_2 \to \frac{w'}{RT}X_1^2,$$

which are the subregular model equations for the activity coefficient. The subregular model is consistent for very small departures from ideality. The infinite dilution limits are

$$\ln\gamma_1 \to \frac{w'_{(1)}}{RT} + (\Delta\epsilon'_{22} - \Delta\epsilon'_{11}),$$

$$\ln\gamma_2 \to \frac{w'_{(2)}}{RT} + (\Delta\epsilon'_{22} - \Delta\epsilon'_{11}),$$

where the second term in each equation might be expected to be much smaller than the first. Thus the Henry's law constants are equal to w/RT for the appropriate end member plus a small constant.

The above approach to asymmetry assumes that the ϵ_j are linear in composition. However it could be argued that this is not the most sensible composition dependence. There are many alternative possibilities, including dependence on the x_j. Distinguishing between different compositional dependences would require comparison with data of a precision not normally available or indeed possible for most phases of geological interest.

Multicomponent System

The extension of the quasi-chemical formulation to multicomponent systems is straightforward. For an n-component system

$$X_i = \sum_{k=1}^{n} x_{ik} \quad (i = 1, 2 \ldots n).$$

The constraint equations, of which there are $\binom{n}{2}$, arise from minimizing

$$f = G^* + \sum_{j=1}^{n-1} \left(\frac{\Sigma n_{jk}}{n} - X_j \right) \lambda_j.$$

The constraint equations have the form

$$2\xi_{ij} = \xi_{ii} + \xi_{jj} \quad (i, j = 1, 2 \ldots n;\ i \neq j).$$

Formulation of g is straightforward leading to

$$G^* = \Sigma\Sigma n_{ij}\epsilon_{ij} + RT\left(\Sigma\Sigma n_{ij} \ln x_{ij} - \frac{2}{z}(z - 1)\Sigma N_i \ln X_i \right).$$

Differentiation gives chemical potential equations of the form

$$\mu i = \frac{z}{2}\xi_{ii} = \epsilon'_{ii} + RT \ln X_i + \frac{z}{2} RT \ln \frac{x_{ii}}{X_i^2}. \tag{27}$$

The constraint equations become

$$0 = w_{ij} + RT \ln \frac{x_{ij}^2}{x_{ii}x_{jj}}, \tag{28}$$

where $w_{ij} = 2\epsilon_{ij} - \epsilon_{ii} - \epsilon_{jj}$. The constraint equations cannot be solved algebraically so they must be solved numerically. The extension of the development to include asymmetry in the mixing properties is simple.

Two-Site Phase

Binary System

Consider a phase involving end members 1 and 2 on two sites A and B. The two sites are points on two intersecting lattices such that an atom on site A is surrounded by z atoms all on site B, and an atom on site B is surrounded by z atoms all on site A. The formula of end member 1 involves two atoms of 1; the formula of end member 2 involves two atoms of 2. The obvious structure elements to use involve nearest neighbor pairs, each pair consisting of an atom on site A and an atom on site B. Thus each structure element obeys the number of sites restriction.

The numbers of these nearest neighbor pairs are n_{11}, n_{12}, n_{21}, and n_{22}, where the first subscript refers to site A, the second to site B. These numbers are related to the numbers of atoms, N_1 and N_2, by

$$
\begin{aligned}
N_1 &= \frac{1}{z}\left(n_{11} + \frac{n_{12}}{2} + \frac{n_{21}}{2}\right), \\
N_2 &= \frac{1}{z}\left(\frac{n_{12}}{2} + \frac{n_{21}}{2} + n_{22}\right).
\end{aligned} \tag{29}
$$

The site occupancies can be written in terms of mole fractions on sites:

$$
\begin{aligned}
X_{1A} &= x_{11} + x_{12}, \quad X_{1B} = x_{11} + x_{21}, \\
X_{2A} &= x_{21} + x_{22}, \quad X_{2B} = x_{12} + x_{22},
\end{aligned}
$$

and

$$
\begin{aligned}
X_1 &= \frac{X_{1A} + X_{1B}}{2} = x_{11} + \frac{x_{12} + x_{21}}{2}, \\
X_2 &= \frac{X_{2A} + X_{2B}}{2} = x_{22} + \frac{x_{12} + x_{21}}{2}.
\end{aligned} \tag{30}
$$

The Gibbs energy is

$$G^* = n_{11}\xi_{11} + n_{12}\xi_{12} + n_{21}\xi_{21} + n_{22}\xi_{22}.$$

The constraint equations, of which there are two because there are four n_j but only two N_i, arise from minimizing:

$$f = G^* + \lambda\left(\frac{n_{11} + (n_{12} + n_{21})/2}{n_{11} + n_{12} + n_{21} + n_{22}} - X_1\right).$$

Differentiating with respect to the n_j, setting to zero and combining, gives the constraint equations

$$
\begin{aligned}
\xi_{12} &= \xi_{21}, \\
\xi_{12} + \xi_{21} &= \xi_{11} + \xi_{22}.
\end{aligned}
$$

The chemical potentials are derived as for (10):

$$
\begin{aligned}
\mu_1 &= z\xi_{11}, \\
\mu_2 &= z\xi_{22}.
\end{aligned}
$$

Symmetric Case

The Gibbs energy can be written

$$G^* = \epsilon_{11}n_{11} + \epsilon_{12}n_{12} + \epsilon_{21}n_{21} + \epsilon_{22}n_{22} - RT \ln g.$$

The main task is formulating g. The development completely mirrors the progress from (13) to (14) in the single-site case. Thus

$$g = \frac{\left(\frac{1}{z}n\right)!}{\left(\frac{1}{z}X_{1A}n\right)!\left(\frac{1}{z}X_{2A}n\right)!} \cdot \frac{\left(\frac{1}{z}n\right)!}{\left(\frac{1}{z}X_{1B}n\right)!\left(\frac{1}{z}X_{2B}n\right)!} \cdot \frac{(nX_{1A})!(nX_{2A})!}{n!} \cdot \frac{(nX_{1B})!(nX_{2B})!}{n!} \cdot \frac{n!}{n_{11}!n_{12}!n_{21}!n_{22}!} \tag{31}$$

where the first and third terms are the contributions to the normalization constant from mixing on the A site, and the second and fourth terms are the contributions from the B site. Substituting x_j for X_i in g and substituting into the Gibbs energy gives

$$\begin{aligned} G^* = {} & \epsilon_{11}n_{11} + \epsilon_{12}n_{12} + \epsilon_{21}n_{21} + \epsilon_{22}n_{22} \\ & + RT\{(n_{11}\ln x_{11} + n_{12}\ln x_{12} + n_{21}\ln x_{21} + n_{22}\ln x_{22}) \\ & - \frac{1}{z}(z-1)((n_{11}+n_{12})\ln(x_{11}+x_{12}) + (n_{21}+n_{22})\ln(x_{21}+x_{22}) \\ & + (n_{11}+n_{21})\ln(x_{11}+x_{21}) + (n_{12}+n_{22})\ln(x_{12}+x_{22}))\}. \end{aligned} \tag{32}$$

Differentiating with respect to the n_j gives

$$\begin{aligned} \xi_{11} &= \epsilon_{11} + RT\left(\ln x_{11} - \frac{1}{z}(1-z)\ln X_{1A}X_{1B}\right), \\ \xi_{12} &= \epsilon_{12} + RT\left(\ln x_{12} - \frac{1}{z}(1-z)\ln X_{1A}X_{2B}\right), \\ \xi_{21} &= \epsilon_{21} + RT\left(\ln x_{21} - \frac{1}{z}(1-z)\ln X_{2A}X_{1B}\right), \\ \xi_{22} &= \epsilon_{22} + RT\left(\ln x_{22} - \frac{1}{z}(z-1)\ln X_{2A}X_{2B}\right). \end{aligned}$$

Substituting into the constraint equations (29) gives

$$0 = \epsilon_{12} - \epsilon_{21} + RT\ln\frac{x_{12}}{x_{21}} - \frac{1}{z}(z-1)RT\ln\frac{X_{1A}X_{2B}}{X_{1B}X_{2A}}, \tag{33}$$

$$0 = \epsilon_{12} + \epsilon_{21} - \epsilon_{11} - \epsilon_{22} + RT\ln\frac{x_{12}x_{21}}{x_{11}x_{22}}. \tag{34}$$

The site preference energy, $s = \epsilon_{12} - \epsilon_{21}$, is a reflection of the differences between the two sites. If $s = 0$ it means that the two sites are identical, any site preference of the atoms of 1 and 2 being caused by the interaction energy, $w = \epsilon_{12} + \epsilon_{21} - \epsilon_{11} - \epsilon_{22}$. Ideal mixing occurs when $w = 0$ and any site preference is due to s, Fig. 5. These constraint equations must be solved numerically for the x_j, e.g., for Fig. 6.

The chemical potential equations are

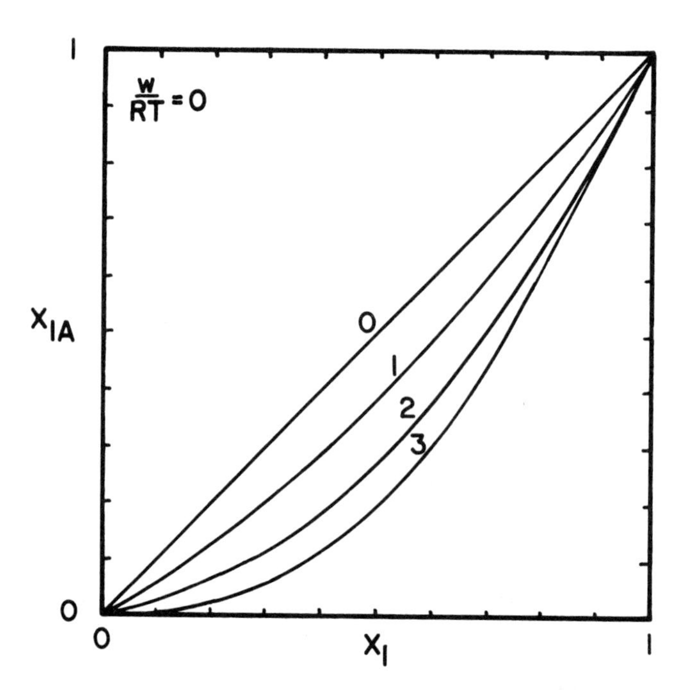

Fig. 5 Site distributions for w'/RT for a series of values of s'/RT.

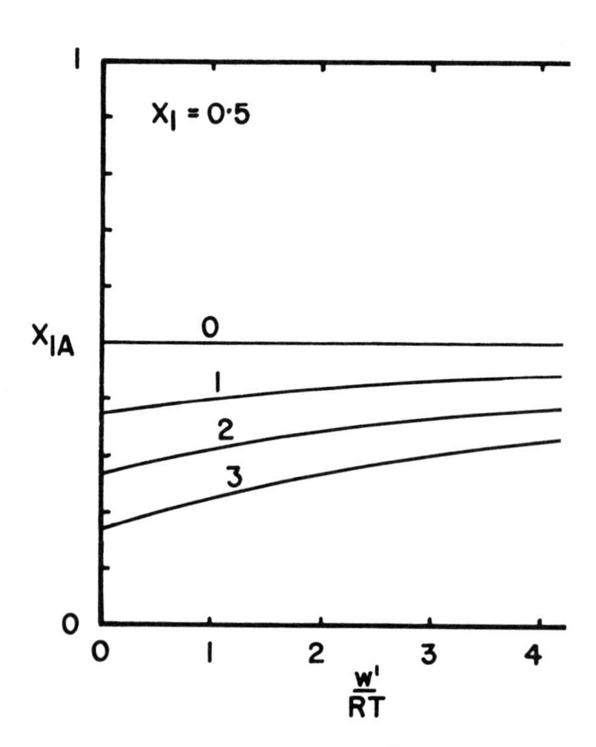

Fig 6. Site distributions for $X_1 = 0.5$ as a function of w'/RT for a series of values of s'/RT. Solvi interrupt this simple pattern for $w'/RT > 4$.

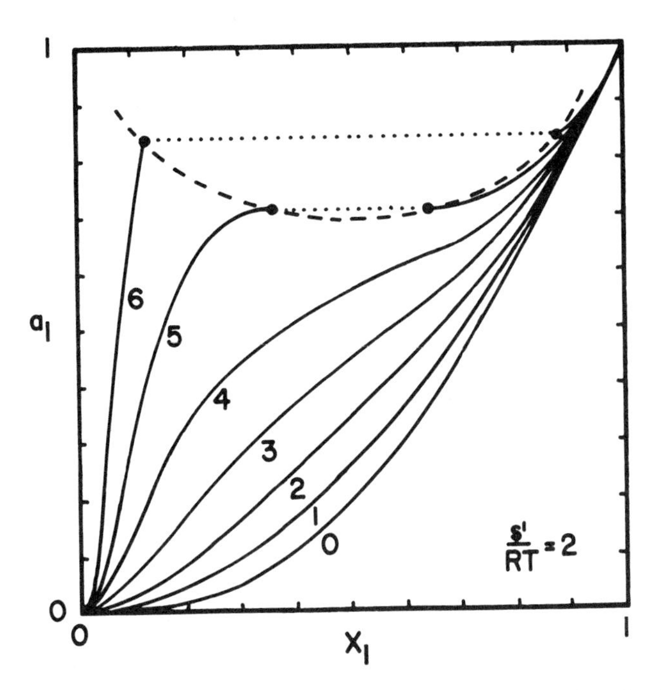

Fig. 7. Activity–composition relations for $s'/RT = 2$ and for a series of values of w'/RT. The zero contour corresponds to ideal mixing. Note the presence of the solvus at large values of w'/RT.

$$\mu_1 = \epsilon'_{11} + RT \ln X_{1A}X_{1B} + zRT \ln \frac{x_{11}}{X_{1A}X_{1B}}, \tag{35}$$

$$\mu_2 = \epsilon'_{22} + RT \ln X_{2A}X_{2B} + zRT \ln \frac{x_{22}}{X_{2A}X_{2B}}.$$

The activity–composition relationships are illustrated in Fig. 7.

Discussion

The behavior of this model in the obvious limits is more difficult to ascertain because of the form of the constraint equations. However, some discussion is possible. For $s = 0$, the two sites are distinguishable only if the occupancies of the sites are different. For positive and small negative w/RT, the site occupancies will be identical, but as w/RT is decreased an order–disorder transition involving the appearance of site preferences will occur. For $w = 0$, (34) becomes

$$\frac{x_{12}x_{21}}{x_{11}x_{22}} = 1.$$

This implies random mixing on each site, $x_{ij} = X_{iA}X_{jB}$. Therefore (33) becomes

$$\frac{X_{1A}X_{2B}}{X_{1B}X_{2A}} = \exp\left[-\frac{s'}{RT}\right],$$

where primed quantities have been multiplied by z. Therefore the site preference energy is solely responsible for site preferences for $w = 0$, and the site occupancies can be calculated from this equation. If, in addition, $s = 0$, then $X_{1A} = X_{1B}$ and $X_{2A} = X_{2B}$ and the site occupancies are identical. For $w = 0$, the chemical potentials are, as expected

$$\mu_1 = \epsilon'_{11} + RT \ln X_{1A}X_{1B},$$
$$\mu_2 = \epsilon'_{22} + RT \ln X_{2A}X_{2B}.$$

In the limit, $w \rightarrow 0$, it would be expected that the model would reduce to the Bragg–Williams model, but this has not been demonstrated analytically because of the difficulties referred to above but it has been demonstrated numerically. Similar problems have been experienced with the limits $|s| \rightarrow \infty$ (very strong site preference, e.g., Si between octahedral and tetrahedral sites), and $w \rightarrow \infty$ (very strong short-range order, e.g., Al–Al short-range order between octahedral and tetrahedral sites). The required numerical experiments will be featured in a forthcoming paper.

No particular formulation problems are involved in having different numbers of each type of site in the structure, in introducing asymmetry into the mixing properties, or in adding further components to the system. Complex phases often have nearest neighbor relationships which involve "one-site" and "two-site" behavior and these features are easily accounted for by combining the appropriate chemical potential equations, although some care is required in deriving the constraint equations.

Other Examples

These examples, all involving simplistic approaches to silicate liquids, are used to illustrate several other types of situation where the formulation of the thermodynamics of a system is assisted by using the formalism introduced earlier. Each development will be relatively brief.

Consider the system $MO–SiO_2$. The classic way of modelling this system is in terms of mixing of the different types of oxygen, O^0, O^- and O^{2-}, each bonded to two, one and no silicons respectively (e.g., Bottinga *et al.*, 1980). Denoting the numbers of these types by n_0, n_1, and n_2, then, by bulk composition and by charge balance,

$$N_O = N_{MO} + 2N_{SiO_2} = n_0 + n_1 + n_2,$$
$$2N_{MO} = 2n_2 + n_1.$$

Combining

$$4N_{SiO_2} = 2n_0 + n_1.$$

Writing

$$G^* = n_0\xi_0 + n_1\xi_1 + n_2\xi_2.$$

The constraint equation is derived from

$$f = G^* + \lambda_1(2n_2 + n_1 - 2N_{MO}) + \lambda_2(2n_0 + n_1 - 4N_{SiO_2}),$$

giving

$$2\xi_1 = \xi_0 + \xi_2,$$

equivalent to

$$2O^- = O^0 + O^{2-}.$$

Writing

$$n_0 = 2N_{SiO_2} - \frac{n_1}{2}.$$
$$n_2 = N_{MO} - \frac{n_1}{2},$$

gives

$$\mu_{SiO_2} = 2\xi_0,$$
$$\mu_{MO} = \xi_2 .$$

For random mixing

$$G^* = \epsilon_0 n_0 + \epsilon_1 n_1 + \epsilon_2 n_2 + RT(n_0 \ln x_0 + n_1 \ln x_1 + n_2 \ln x_2).$$

For ϵ_i independent of composition, the constraint equation becomes

$$0 = 2\epsilon_1 - \epsilon_0 - \epsilon_2 + RT \ln \frac{x_1^2}{x_0 x_2}$$

and the chemical potentials

$$\mu_{SiO_2} = 2\epsilon_0 + RT \ln x_0^2,$$
$$\mu_{MO} = \epsilon_2 + RT \ln x_2.$$

For the ϵ_j dependent on composition, these equations will be more complicated. Note particularly that the constraint equation will then have a composition dependent K.

Using the nearest neighbor approximation, there are six n_j, for example, n_{12} is used for the number of O^-–O^{2-} nearest neighbors, and still two N_i so there are four constraint equations. For some average z

$$2N_{MO} = \frac{2}{z}(n_{01} + 2n_{02} + 2n_{11} + 3n_{12} + 4n_{22}),$$
$$4N_{SiO_2} = \frac{2}{z}(2n_{00} + n_{01} - n_{12} - 2n_{22}).$$

The constraint equations are

$$\begin{aligned} 2\xi_{11} &= \xi_{01} + \xi_{12}, \\ 2\xi_{01} &= \xi_{11} + \xi_{00}, \\ 2\xi_{11} &= \xi_{00} + \xi_{22}, \\ \xi_{11} &= \xi_{02}, \end{aligned}$$

and the chemical potentials are

$$\begin{aligned} \mu_{SiO_2} &= \frac{z}{2} 2\xi_{00}, \\ \mu_{MO} &= \frac{z}{2} \xi_{22}. \end{aligned}$$

The formulation of g is straightforward and gives rise to

$$G^* = \sum_{i=1}^{3} \sum_{j=1}^{3} n_{ij}\epsilon_{ij} + RT\left[\sum_{i=1}^{3} \sum_{j=1}^{3} n_{ij} \ln x_{ij} - \frac{2}{z}(z-1) \sum_{k=1}^{3} N_k \ln X_k\right],$$

where X_k is the mole fraction of O^{k-}. The constraint equations are

$$\begin{aligned} 0 &= 2\epsilon_{11} - \epsilon_{01} - \epsilon_{12} + RT \ln \frac{x_{11}^2}{x_{01}x_{12}}, \\ 0 &= 2\epsilon_{01} - \epsilon_{00} - \epsilon_{11} + RT \ln \frac{x_{01}^2}{x_{00}x_{11}}, \\ 0 &= 2\epsilon_{11} - \epsilon_{00} - \epsilon_{22} + RT \ln \frac{x_{11}^2}{x_{00}x_{22}}, \\ 0 &= \epsilon_{02} - \epsilon_{11} + RT \ln \frac{x_{02}}{x_{11}}, \end{aligned}$$

and the chemical potentials are

$$\begin{aligned} \mu_{SiO_2} &= \epsilon_{00}' + RT \ln X_0^2 + \frac{z}{2} \ln \frac{x_{00}}{X_0^4}, \\ \mu_{MO} &= \epsilon_{22}' + RT \ln X_2 + \frac{z}{2} \ln \frac{x_{22}}{X_2^2}. \end{aligned}$$

Formulation of the thermodynamics of silicate liquids in terms of mixtures of polymeric units (e.g., Gaskell, 1977) is also straightforward. For relatively basic compositions and assuming only linear chains, a polymeric unit can be written as

$$Si_aO_{3a+1}^{(2a+2)-}$$

and the amount of this unit as n_a. Then

$$N_O = 2N_{SiO_2} + N_{MO} = \sum_{a=0}^{\infty} (3a+1)n_a.$$

Counting Si atoms gives

$$N_{SiO_2} = \sum_{a=0}^{\infty} a n_a.$$

Subtracting, or, alternatively, for charge balance

$$N_{MO} = \sum_{a=0}^{\infty} (a + 1) n_a.$$

Writing

$$G^* = \sum_{a=0}^{\infty} n_a \xi_a.$$

Minimization gives equations of the form

$$0 = \xi_a + a\lambda_1 + (a + 1)\lambda_2,$$

which can be combined to give the constraint equations

$$\xi_1 + \xi_a = \xi_{a+1} + \xi_0.$$

equivalent to

$$SiO_4^{4-} + Si_a O_{3a+1}^{(2a+2)-} = Si_{a+1} O_{3a+4}^{(2a+4)-} + O^{2-}.$$

Choosing n_a, $a > 1$, as internal parameters

$$n_0 = N_{MO} - 2N_{SiO_2} - \sum_{a=2}^{\infty} (a - 2) n_a,$$

$$n_1 = N_{SiO_2} - \sum_{a=2}^{\infty} n_a.$$

Then

$$\mu_{MO} = \xi_0,$$
$$\mu_{SiO_2} = \xi_1 - 2\xi_0.$$

For random mixing, the constraint equations are

$$0 = \epsilon_{a+1} + \epsilon_0 - \epsilon_a - \epsilon_1 + RT \ln \frac{x_{a+1} x_0}{x_a x_1},$$

where the energy difference can be referred to as w_a. The number of unknowns involved in the constraint equations can be drastically reduced by making assumptions about the relationship of w_a to $w_a + 1$. Alternatively, the w_a values can be obtained from mass spectrometric measurements of actual polymeric distributions. The chemical potentials are

$$\mu_{MO} = \epsilon_0 + RT \ln x_0$$

$$\mu_{SiO_2} = \epsilon_1 - 2\epsilon_0 + RT \ln \frac{x_1}{x_0^2}.$$

Conclusion

In the development of mixing models, it is always preferable to start with a formulation of the Gibbs energy and to derive the activity–composition relationships and the necessary constraint equations from this formulation. The formalism provides a way of doing this which simplifies the work involved and thus reduces the possibility of confusion and mistakes. The formalism certainly allows a clearer exposition of the thermodynamics of the quasi-chemical model than is usually presented. Much still needs to be done within the quasi-chemical approach let alone in trying to develop better models, particularly in the formulation of the thermodynamics of phases in which there is strong short-range order.

References

Bottinga, Y., Weill, D. F., and Richet, P. (1980) Thermodynamic modelling of silicate melts, in *Thermodynamics of Minerals and Melts* edited by Newton, R. C., Navrotsky, A., and Wood, B. J., Springer-Verlag, New York.

Dahlquist, G. and Bjorck, A. (1974) *Numerical Methods*, Prentice–Hall, Englewood Cliffs, N.J.

Gaskell, D. R. (1977) Activities and free energies of mixing in binary silicate melts, *Met. Trans.* **8B**, 131.

Guggenheim, E. A. (1952) *Mixtures*, Oxford Univ. Press, London/New York

Guggenheim, E. A. (1966) *Statistical Thermodynamics*, Oxford University Press, London/New York.

Hill, T. L. (1960) *An Introduction to Statistical Thermodynamics*, Addison–Wesley, Reading, Mass.

Iiyama, J. T. (1974) Substitution, deformation locale de la maille et equilibre de distribution des elements en traces entre silicates, *Bull. Soc. Franc. Min. Crist.* **97**, 143.

Kerrick, D. M. and Darken, L. S. (1975) Statistical thermodynamic models for ideal oxide and silicate solid solutions, *Geochim. Cosmochim. Acta* **39**, 431.

Kroger, F. A., Stieltjes, F. H., and Vink, H. J. (1959) Thermodynamics and formulation of reactions involving imperfections in solids, *Philips Res. Rep.* **14**, 557.

Mazo, R. M. (1977) Statistical mechanical calculation of Al–Si disorder in albite, *Amer. Mineral.* **62**, 1232.

Navrotsky, A. and Luocks, D. (1977) Calculation of subsolidus phase relations in carbonates and pyroxenes, *Phys. Chem. Minerals* **1**, 109.

Powell, R. (1974) A comparison of some mixing models for crystalline silicate solid solutions, *Contrib. Mineral. Petrol.* **46**, 265.

Powell, R. (1977) Activity–composition relationships for crystalline solutions, in *Thermodynamics in Geology*, edited by Fraser, D. G., Reidel, Dordrecht.

Thompson, J. B. (1970) Chemical reactions in crystals, *Amer. Mineral.* **54**, 341.

Index